Nanocellulose and Its Composites for Water Treatment Application

Nanocellulose and Its Composites for Water Treatment Application

Edited by
DINESH KUMAR

CRC Press
Taylor & Francis Group
Boca Raton London New York

CRC Press is an imprint of the
Taylor & Francis Group, an **informa** business

First edition published 2021
by CRC Press
6000 Broken Sound Parkway NW, Suite 300, Boca Raton, FL 33487-2742

and by CRC Press
2 Park Square, Milton Park, Abingdon, Oxon, OX14 4RN

© 2021 Taylor & Francis Group, LLC

CRC Press is an imprint of Taylor & Francis Group, LLC

Library of Congress Cataloging-in-Publication Data

Names: Kumar, Dinesh (Chemist), editor.
Title: Nanocellulose and its composites for water treatment application /
Dinesh Kumar.
Description: First edition. | Boca Raton : CRC Press, 2021. | Includes
bibliographical references and index.
Identifiers: LCCN 2020058436 | ISBN 9780367487331 (hbk) | ISBN
9781003042556 (ebk)
Subjects: LCSH: Water--Purification--Materials. | Nanostructured
materials--Industrial applications. | Cellulose fibers. | Cellulose
nanocrystals.
Classification: LCC TD433 .N28 2021 | DDC 628.1/64--dc23
LC record available at https://lccn.loc.gov/2020058436

ISBN: 9780367487331 (hbk)
ISBN: 9781032023373 (pbk)
ISBN: 9781003042556 (ebk)

Typeset in Times
by KnowledgeWorks Global Ltd.

Contents

Preface

Biological materials have drawn increasing attention from scientists. Cellulose is abundant, renewable, biodegradable, economical, thermally stable, and lightweight, and has applications in pharmaceuticals, coatings, food, textiles, laminates, sensors, actuators, flexible electronics, and flexible displays. Its nano form has extraordinary surface properties like higher surface area than cellulose; hence, nanocellulose can be one alternative solution to cellulose.

Hence, among many other sustainable, functional nanomaterials, nanocellulose attracts growing interest in environmental remediation technologies because of its many unique properties and functionalities. Academic and industrial research bio-based nanomaterials are flourishing, and the output of research papers and reviews is increasing exponentially. Despite this interest, however, there has been no comprehensive book on nanocellulose at an advanced level incorporating the many new structural and technical developments, particularly the recent advancement on the use of nanocellulose and its nanocomposites for water remediation technologies.

This book provides insight into the application of nanocellulose and its nanocomposites for water purification, leading to advancements in water remediation methods. It also covers different classes of nanocellulose—cellulose nanocrystal (CNC), microfibrillated cellulose (MFC), hairy cellulose nanocrystalloid (HCNC), and bacterial nanocellulose (BNC), for their competency with other renewable and carbonaceous materials like pectin, alginate, carbon nanotubes (CNTs), and their uses in water treatment.

The book also covers the status and future perspectives of nanocellulose and nanocomposites of different biodegradable origins. This book provides a description of fundamentals and an up-to-date overview of development in synthesis, characterization methods, properties—chemical, thermal, optical, structural, surface, mechanical structure-property relationships, and crystallization behavior degradability of biodegradable nanocomposites—and finally applications of nanocellulose and its nanocomposites. This book also discusses the applications of different cellulose nanomaterials: as adsorbent—adsorption of organic and inorganic toxicants in aqueous media, as membrane—for removal of bacteria, virus, and ionic impurities and desalination, as photocatalyst—to improve the degradation rate of organic pollutants, and as sensor—to sense water toxicants.

Contributors

Kaushalya Bhakar
School of Chemical Sciences
Central University of Gujarat
Gandhinagar, India

A.K. Bharimalla
ICAR-Central Institute for Research on Cotton
    Technology
Mumbai, India

A. Geetha Bhavani
Department of Chemistry, School of Sciences
Noida International University
Greater Noida, India

Ankita Dhillon
Department of Chemistry
Banasthali Vidyapith
Banasthali, India

Ismail Muhamad Fareez
Department of Oral Biology and Biomedical
    Sciences, Faculty of Dentistry
MAHSA University, Saujana Putra Campus
Jenjarom, Malaysia

Pallavi Jain
Department of Chemistry
SRM Institute of Science and Technology Delhi-
    NCR Campus
Modinagar, India

Ooi Der Jiun
Department of Oral Biology and Biomedical
    Sciences, Faculty of Dentistry
MAHSA University, Saujana Putra Campus
Jenjarom, Malaysia

Dinesh Kumar
School of Chemical Sciences
Central University of Gujarat
Gandhinagar, India

Subrata Mondal
Department of Mechanical Engineering
National Institute of Technical Teachers' Training
    and Research (NITTTR)
Kolkata, India

Sapna Nehra
Department of Chemistry
Dr. K.N. Modi University
Newai, India

Meena Nemiwal
Department of Chemistry
Malaviya National Institute of Technology
Jaipur, India

Jorge Padrão
Centre for Textile Science and Technology
    (2C2T)
University of Minho
Guimarães, Portugal

P.G. Patil
ICAR-Central Institute for Research on Cotton
    Technology
Mumbai, India

Sapna Raghav
Department of Chemistry
Banasthali Vidyapith
Banasthali, India

Kritika S. Sharma
School of Chemical Sciences
Central University of Gujarat
Gandhinagar, India

Rekha Sharma
Department of Chemistry
Banasthali Vidyapith
Banasthali, India

Wu Yuan Seng
Department of Biochemistry, Faculty of
    Medicine, Nursing and Biosciences
MAHSA University, Saujana Putra Campus
Jenjarom, Malaysia

Theivasanthi Thirugnanasambandan
International Research Centre
Kalasalingam Academy of Research and
    Education (Deemed University)
Krishnankoil, India

Makarand Upadhyaya
Department of Management & Marketing
College of Business Administration
Zallaq, Kingdom of Bahrain

N. Vigneshwaran
ICAR-Central Institute for Research on Cotton
Technology
Mumbai, India

W. M. Dimuthu Nilmini Wijeyaratne
Department of Zoology and Environmental
Management, Faculty of Science
University of Kelaniya
Kelaniya, Sri Lanka

Andrea Zille
Centre for Textile Science and Technology (2C2T)
University of Minho
Guimarães, Portugal

Biography

Dinesh Kumar is currently working as Professor in the School of Chemical Sciences at the Central University of Gujarat, Gandhinagar, India. Prof. Kumar obtained his master's and Ph.D. degrees in Chemistry from the Department of Chemistry, University of Rajasthan, Jaipur, in 2002 and 2006, respectively. Prof. Kumar has received many national and international awards and fellowships. His research interest focuses on the development of capped MNPs, core-shell NPs, and biopolymers-incorporated metal-oxide-based nanoadsorbents and nanosensors for the removal and the sensing of health hazardous inorganic toxicants like fluoride and heavy metal ions from aqueous media. Prof. Kumar developed hybrid nanomaterials from different biopolymers like pectin, chitin, cellulose, and chitosan for water purification. His research interests also focus on the synthesis of supramolecular metal complexes, metal chelates, and their biological effectiveness. He has authored and coauthored over 90 articles in journals of international repute, one book, over seven dozen book chapters, and 70 presentations/talks at national/international conferences.

1

Nanocellulose: Synthesis and Characterization Methods

N. Vigneshwaran, A.K. Bharimalla, and P.G. Patil
ICAR-Central Institute for Research on Cotton Technology, Mumbai, India

CONTENTS

1.1 Introduction

Nanotechnology has revolutionized every field of science and technology during the past three decades. The discovery of nanocellulose accelerated development of new tools and techniques for the production of novel nanomaterials from metals, metal oxides, polymers, and carbon, as well as their properties. However, nanocellulose must have been discovered earlier as the term "microfibrillated cellulose" (MFC) was used in the late 1970s by Turbak, Snyder, and Sandberg at the ITT Rayonier labs in Whippany, New Jersey, USA [1]. Although not using the term nanocellulose, many researchers had also recorded the image of cellulose in its nanodimension, and one such image of microfibrils taken in 1978 from our Institute is given in Fig. 1.1 [2]. The world's first pilot plant to produce nanocellulose was started in 2011 by Innventia (Stockholm). However, CelluForce® was the first commercial producer of nanocellulose, and in January 2012, it started a pilot plant at the site of the Domtar pulp and paper mill in Windsor, QC, with a capacity of producing 1 ton nanocellulose per day. This was followed by various industries like Daicel (Japan), UPM Kymmene and VTT (Finland), Borregaard (Norway), Rettenmaier (Germany), and

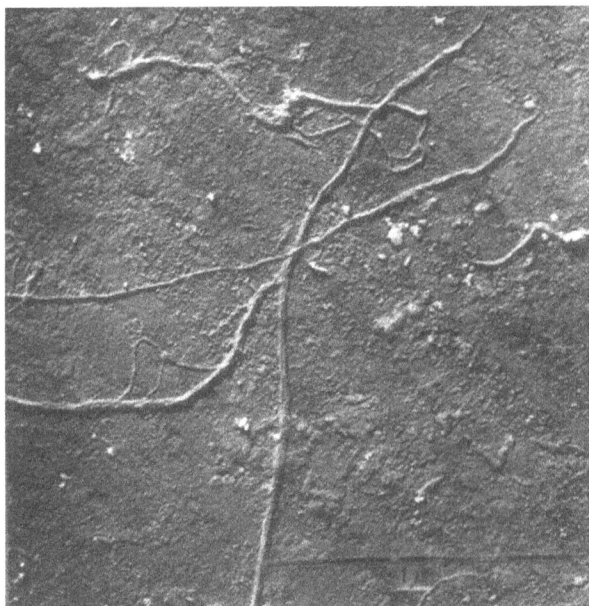

FIGURE 1.1 TEM image of cotton nanocellulose isolated from ICAR-Central Institute for Research on Cotton Technology, Mumbai [2], way back in 1978.

American Process (USA). In India, ICAR-Central Institute for Research on Cotton Technology was the first organization to establish a nanocellulose pilot plant, with a capacity of 10 kg per shift of 8 hours, for the production of nanocellulose from cotton linters and other agroresidues. Nanocellulose is an excellent alternative to various other established nanomaterials like graphene and carbon nanotubes (CNT) because of its organic nature and biological origin. Conventionally, concentrated acid hydrolysis process and mechanical degradation process were used to synthesize nanocellulose resulting in toxic effluent generation and high energy consumption, respectively. The recent developments depend on enzymatic pretreatment of cellulosic biomass, followed by a mechanical process for the sustainable and commercially viable production of nanocellulose. The purest forms of cellulosic raw materials like bacterial cellulose and cotton cellulose are used to produce high-quality nanocellulose. Similarly, the characterization of nanocellulose and the development of suitable standards are very much crucial as the properties change vastly depending on the raw materials and process protocols followed in synthesis. The sustainable development of nanocellulose-based products and their usage will add value to agroresidues and reduce the carbon footprint in the long run.

1.2 Cellulose and Nanocellulose

Cellulose is a naturally available and the most abundant biopolymer on Earth. The chemical formula is $(C_6H_{10}O_5)_n$ consisting of linear chains of β(1→4)-linked D-glucose units. This is a major component of primary cell walls in plants and algae. Some species of bacteria and animals (tunicates) also produce cellulose. The purest forms of cellulose are available from bacteria and cotton fibers. Cellulose is used in paper and paper boards, textile materials, cellophane and rayon, ethanol production, microcrystalline cellulose (MCC), insulators, and building materials. With the recent technological advancements, the production of nanocellulose and its diversified uses are gaining momentum.

The term "nanocellulose" is used to describe the cellulose materials having at least one nanosize dimension. The whiskers and particles of nanocellulose are together called nanocrystalline cellulose (NCC) or cellulose nanocrystals (CNC). The fibrillated forms of nanocellulose are called nanofibrillated cellulose (NFC) or cellulose nanofibrils (CNF). In some literatures, MFC also denotes NFC/CNF. Bacterial nanocellulose (BNC) is a distinct class of nanomaterials and one in the purest form, produced from the cellulose secreted by the bacterium *Acetobacter*.

1.3 Novel Properties of Nanocellulose

Similar to any nanomaterials, nanocellulose also has an extensive surface-area-to-volume ratio (SAR) due to its nanodimension. The nanocellulose with extensive SAR has the following advantages:

a. Enhanced interfacial interaction with the polymer matrix during its use as a filler in composites
b. Uniform distribution as fillers in thin-film composites
c. Potential surface area for immobilization of catalysts or active molecules for any chemical or biochemical process
d. Wider scope for chemical modification or functionalization of the surface for property tailoring
e. Effective material for absorbing and holding the oil in cleaning the oil spill

Another important attribute of nanocellulose is its transparency in the nanodimension. The nanocellulose prepared by sulfuric acid hydrolysis or 2,2,6,6-tetramethylpiperidinyloxy (TEMPO) oxidation is almost transparent. Hence, nanocelluloses can be easily used as fillers in transparent film composites without significantly affecting their light transmission properties.

The enhanced mechanical property of nanocellulose makes it a potential alternative for the existing synthetic high-strength polymers. The strength of nanocellulose is many times more than its bulk counterpart due to the removal of amorphous region during the synthesis process. Hence, the increased crystalline region present in the final product of nanocellulose makes it a high-strength material. The intrinsic configuration of BNC with ultrafine pores makes it a suitable candidate for nanofiltration. The purified BNC can be directly used as a filter without further processing. The complete cellulosic nature of this material makes it inert and an ideal candidate to produce high-performing diaphragm in speakers.

1.4 Raw Materials for Nanocellulose

Fig. 1.2 shows various raw materials like agroresidues, wood, cotton, algae, bacterial cellulose, and tunicates that are being used to produce nanocellulose. Among them, bacterial cellulose and cotton yield the purest forms of cellulose, while other raw materials need to be preprocessed and purified before taken

FIGURE 1.2 Diverse raw materials being used for the extraction of nanocellulose.

up to produce nanocellulose. Depending on raw materials, the cellulose allomorph structure will vary. The cellulose derived from algae and bacteria contains mainly Iα allomorph, while high plants produce mostly Iβ allomorph. Hence, depending on the type of applications and the quality of nanocellulose required, the raw materials need to be chosen.

1.5 Synthesis of Nanocellulose

The synthesis of nanocellulose could be done by two different approaches, as given in Table 1.1.

Both the top-down approach and the bottom-up approach can be used to produce nanocellulose. While the top-down approach starts from the bulk cellulose for nanocellulose production, the bottom-up approach starts with dissolved cellulose or monomers to produce nanocellulose. The mechanical processes are easily scalable for commercial production, but its energy consumption is very high. Many researchers follow the conventional chemical processes like sulfuric acid hydrolysis and TEMPO oxidation to produce nanocellulose. Still, it results in the surface-modified (by sulfate or carboxyl) nanocellulose. The biological process is very much eco-friendly, but difficulties arise with respect to controlling the enzyme's reaction and cost. The electrospinning system of nanocellulose production from dissolved cellulose is very well standardized and produces nanocellulose as a substrate in nanosensors and nanofilters. But the problem is the use of a toxic solvent system for cellulose. Hence, recently, cellulose is being replaced with cellulose acetate to produce nanofibers. BNC is one of the purest forms of nanocellulose, and because of its high cost, it is mostly used for high-end applications. The schematic representation of major processes for nanocellulose synthesis commercially is given in Fig. 1.3.

1.5.1 Chemical Process

Conventionally, nanocellulose is produced by sulfuric acid hydrolysis process (Fig. 1.4). The high concentration of sulfuric acid (64%) hydrolyzes the amorphous cellulose region initially, followed by the crystalline region. Hence, the reaction has to be stopped once the hydrolysis of the amorphous region is completed by adding excess water. This has to be purified before taking for characterization. This process produces a stable form of nanocellulose with sulfate ester groups on its surface. But this is not suitable for biological applications due to sulfate groups. Another important process is TEMPO oxidation, wherein TEMPO-mediated oxidation process is used to produce nanocellulose. In this process, the surface of nanocellulose contains carboxyl groups. Similarly, various acids like hydrochloric acid, phosphoric acid, etc. are also demonstrated to produce nanocellulose. The advantage of the chemical process is the production of stable nanocellulose, while the disadvantage is its unwanted surface chemical modifications.

1.5.2 Mechanical Process

The mechanical synthesis of nanocellulose is achieved using a high-pressure homogenizer, microfluidizer, and mass colloider (friction grinder). Both high-pressure homogenizer and microfluidizer

TABLE 1.1

Different Approaches to the Synthesis of Nanocellulose.

Approaches to Syntheses of Nanocellulose	Processes	Instrument/Protocols
Top-down approach	Mechanical	High-pressure homogenizer, microfluidizer, mass colloider, refiner
	Chemical	Acid hydrolysis, TEMPO oxidation
	Biological	Enzyme hydrolysis, microbial degradation
Bottom-up approach	Physical	Electrospinning system
	Chemical	Cellulose dissolution and regeneration
	Biological	Bacterial production of cellulose

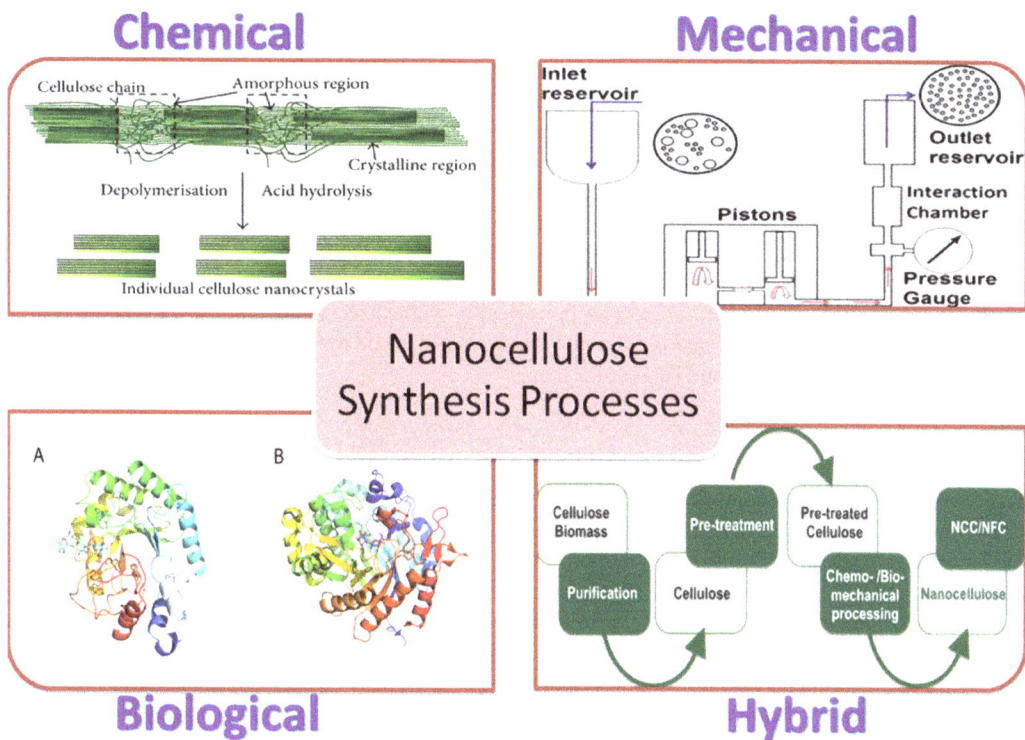

FIGURE 1.3 Schematic representation of four different processes used for the synthesis of nanocellulose.

pressurize cellulose suspension through a nozzle to produce cellulose nanoparticles. In mass colloider, grinding action is used to fibrillate the cellulose fibers to produce CNF. While the energy consumed by mechanical process is very high compared to other processes, productivity is also very high here. To reduce energy consumption, the cellulosic raw materials are pretreated by chemicals or enzymes. At the nanocellulose pilot plant of ICAR-Central Institute for Research on Cotton Technology Mumbai, cotton cellulose is pretreated with chemicals and enzymes before processing in a microfluidizer.

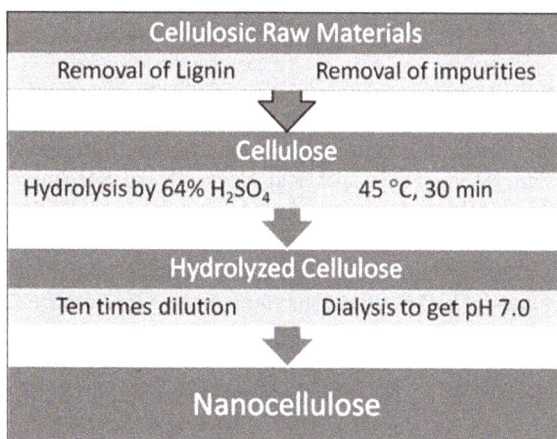

FIGURE 1.4 Schematic representation of sulfuric acid hydrolysis process conventionally used for the synthesis of nanocellulose.

1.5.3 Biological Process

Naturally available bacterial cellulose is the purest form of cellulose, and its conversion to nanodimension results in the formation of BNC. BNC is generally nanofibrillated structure due to the fibrous morphology of its raw material, bacterial cellulose. For other biological processes of nanocellulose, cellulase enzymes are cellulase-secreting microbes used to reduce cellulosic biomass size. The uncontrolled hydrolysis of cellulose by cellulase enzymes results in the production of glucose molecules. Hence, these reactions must be controlled and stopped in between to obtain the nanocellulose. The amorphous forms of cellulose are easily digestible by the cellulase enzyme as compared to crystalline forms. Hence, once it hydrolyzes the amorphous form of cellulose, the cellulase enzyme needs to be inactivated or removed to obtain the crystalline nanocellulose. The biologically synthesized nanocelluloses are eco-friendly, with no chemical surface modification. The only bottleneck shall be the cost involved in using cellulase enzyme and its separation and purification for reuse.

1.5.4 Hybrid Process

Since various synthesis processes have their disadvantages, combinations of different techniques have been used commercially to use their advantages. Chemical or biochemical processes are being used for the pretreatment of cellulosic substrates, followed by a mechanical process for the production of nanocellulose. Here, the pretreatments need to be optimized, depending on the raw materials selected for the production process. Higher amounts of chemicals are required for lignocellulosic biomass, while biochemical methods like the use of enzyme suffice in the case of purified forms of cellulose from cotton and bacterial cellulose. At the nanocellulose pilot plant located in ICAR-Central Institute for Research on Cotton Technology, Mumbai, a novel chemo-mechanical process is employed to produce nanocellulose from cotton fibers and cotton linters. Other agroresidues can also be used in this pilot plant due to the modular design employed in this plant.

1.6 Characterization of Nanocellulose

1.6.1 Size and Size Distribution

For the commercial exploitation of any nanomaterial, its primary properties of size and size distribution need to be understood. To estimate nanoparticles' size distribution in a solution, dynamic light scattering (DLS) technique is widely used. The DLS works on the principle of deriving the hydrodynamic radius of nanoparticles using the Stokes-Einstein equation (Equation 1.1) by analyzing the intensity fluctuation in scattered light by the Brownian motion of the suspended nanoparticles. The dilute nanoparticle suspension in a known solvent is used for analysis. The scattered intensity from a laser light is collected at a 90° angle and the fluctuation in intensity is analyzed in a correlator. The diffusion coefficient obtained is substituted in the Stokes-Einstein equation, and the hydrodynamic radius is then calculated. The minimum size that could be analyzed by this technique is around 0.5 nm. Here, the critical assumption is that all the scattering particles are assumed to be spherical. Hence, the DLS technique can be used to analyze the size and size distribution of spherical nanocellulose. Even if the nanocellulose whiskers are analyzed, it will be assumed to be of spherical particles, and the results will be displayed. Hence, the DLS particle size analysis report needs to be corroborated with the micrographs from scanning electron microscope (SEM), atomic force microscope (AFM), or transmission electron microscope (TEM).

$$R_H = \frac{k_B T}{6\pi \eta D} \tag{1.1}$$

In this Stokes-Einstein equation, R_H represents hydrodynamic radius; k_B, Boltzmann constant; T, absolute temperature; η, solvent viscosity; and D, diffusion coefficient (obtained from DLS technique).

In contrast, the static light scattering (SLS) technique, also called laser diffraction, is used in particle sizing techniques for the particles in the size range of sub-micrometers to several millimeters. This technique measures size distribution by measuring the angular variation in the intensity of scattered light by the sample. Bigger particles scatter light at small angles relative to the source, and vice versa in smaller particles. The size will be calculated using the Mie theory of light scattering, and volume equivalent sphere diameter can be estimated. The minimum size that could be obtained by this technique is around 10 nm. A Fraunhofer approximation is used for more straightforward calculation that provides accurate results for bigger particles (size more than 50 μm). This technique is conventionally used to measure the absolute molecular weight, size, and second virial coefficient of polymers in solution. This technique is used to analyze the equivalent spherical size for large-sized cellulose nanowhiskers and NFC.

The morphology of nanocellulose needs to be studied because the spherical form of nanocellulose is very rare. They will form as whiskers and fibrils due to the crystalline region of molecular arrangement. Besides size, SEM is used to analyze the surface morphology, AFM to get three-dimensional (3D) image and elasticity of nanocellulose, and TEM to understand the crystalline nature of nanocellulose. For analysis by SEM, the sample has to be coated with conducting agents to avoid surface charging by the electron beam. For AFM, nanocellulose can be used as such. For TEM, heavy element staining needs to be done to enhance the contrast for observation. Representative images from literature for SEM, AFM, and TEM of nanocellulose are given in Fig. 1.5 [3, 4, 5].

1.6.2 Zeta Potential

Estimation of the surface charge of nanocellulose is very much important to understand its storage stability. Nanocellulose with higher surface charge (negative or positive) indicates that it is highly stable on storage. The surface charge on each of the particles repels the nearby particles, thus avoiding their aggregation. The surface charge density is defined by the parameter, zeta potential, and reported in the values of mV. The unit can be negative or positive, depending on the surface chemistry of the material. The surface charge of cellulose is generally in the range of −10 to −15 mV, determined by the type of chemical reactions it undergoes during the cellulose extraction and purification process. The process used to convert cellulose into nanocellulose will determine the zeta potential of nanocellulose. The sulfuric acid hydrolysis process will yield the sulfated nanocellulose with negative zeta potential (<−30 mV) imparted by surface-bound sulfate groups (SO_4^{2-}). In the TEMPO-mediated process, the nanocellulose will form with a significant amount of C6 carboxylate groups on the surface, resulting in very high negative zeta potential (<−50 mV). This will result in highly stable nanocellulose by both sulfuric acid and TEMPO processes. The zeta potential of nanocellulose prepared by mechanical processes like homogenization or friction grinding will be modified significantly. Hence, the chances of aggregation will be higher if the nanocellulose is prepared only for a mechanical process.

1.6.3 Chemical Properties

The chemical properties of nanocellulose, like the degree of polymerization (DP), surface modification, and cellulose content, can tell us how it was produced and its purity. This will further help in surface modification for diversified applications and utilization in organic products. The DP is defined as the number of repeating units/monomers in a polymer. But with cellulose, no direct method is available to measure the DP. First, cellulose must be solubilized in cupriethylenediamine solvent, followed by measuring its viscosity. The obtained intrinsic viscosity of cellulose will be converted into DP using the Mark-Houwink-Sakurada equation. The DP of cellulose varies widely, depending on the raw materials and the process used for its separation. The DP of nanocellulose is lower than the cellulose since the molecules get cut during the nanocellulose production process. The surface chemistry of nanocellulose has to be analyzed for sulfate or carboxyl group presence induced by sulfuric acid hydrolysis or TEMPO oxidation process, respectively. The process of conductometric titration can estimate the charged sulfate ester groups on the nanocellulose [6]. The presence of carboxyl groups in cellulose can be estimated by their ion exchange capacity and using a complexometric titration [7].

FIGURE 1.5 (a) SEM image of bacterial nanocellulose (Reproduced with permission from Ref. [4]). (b) AFM image of cotton nanocellulose (Reproduced with permission from Ref. [5]). (c) TEM image of cotton nanocellulose (Reproduced with permission from Ref. [3]. Copyright © Elsevier 2013).

1.6.4 Rheological Properties

The nanocellulose exhibits complex rheological properties and, in general, it shows shear-thinning property wherein the viscosity reduces under shear strain. Also, the rheological property can be modified by their dimensions and surface chemistry [8]. This non-Newtonian behavior of nanocellulose finds extensive applications in paints and whipped creams, while it affects negatively during their use in coating, extrusion, molding, and 3D printing. Hence, a better understanding of the rheological property of nanocellulose suspension is required for their successful utilization. One such example is the use of BNC in gluten-free muffin batters[9]. Here, the appropriate BNC level added to the gluten-free batters results in viscous systems that entrap more air, resulting in a higher volume of the resultant baked products.

1.6.5 Crystal Properties

The crystal structure of nanocellulose depends on the raw material selected for its production. In most cases, the initial crystalline structure (cellulose Iα, cellulose Iβ, cellulose II, cellulose III, or cellulose IV) remains unaltered after the production of nanocellulose. But in cases like the use of harsh chemicals, there are chances of chemical structure changes. Also, high-energy mechanical processing will reduce the crystalline region and increase the amorphous content of nanocellulose. The crystallinity could be analyzed using the diffraction peaks obtained from nanocellulose powder in the X-ray diffractometer.

Further details could be revealed using X-ray crystallography or electron diffraction analysis. The crystallinity affects the transmission of light in the final product. Also, increased crystallinity shows higher mechanical strength of nanocellulose and its products. While the CNC is completely crystalline, the CNF contains both crystalline and amorphous regions.

1.6.6 Mechanical Properties

The novel mechanical attribute of nanocellulose is its increased elastic modulus because of the higher crystalline region. This helps the nanocellulose for potential application in high-performing nanocomposites. It is easy to analyze the mechanical property of the final product from nanocellulose-reinforced composites. But to understand the elastic modulus of a single nanocellulose particle or fiber, we need to depend on the strength of AFM. After imaging the nanocellulose using AFM, in the same position, the cantilever probe needs to be indented into the nanocellulose sample and the simultaneous force exerted has to be analyzed to derive the elastic modulus of nanocellulose. Also, by carefully linking the nanocellulose fibrils between the cantilever probe and the sample stage, the tensile strength can be analyzed. But because of a lot of variations in the sample and noise involved during analysis, there will be a lot of variation in the final values analyzed by AFM for mechanical properties. Also, three-point bending and surface roughness parameters can be analyzed by the AFM technique.

1.7 Standardization in Nanocellulose Characterization

In the evolving literature related to nanocellulose, various names are being used to describe the nanocellulose: cellulose whiskers, cellulose nanowhiskers, CNCs, CNFs, nanocellulose, cellulose microfibrils, microfibrillated cellulose, and nanofibrillated cellulose. Hence, it is confusing to understand and compare the type and nature of nanocellulose described in the literature. The use of different methods of analyses results in different values for a single parameter of a nanocellulose sample. Hence, to avoid these types of differences, standards for nanocellulose started evolving from the year 2009 onward with the initiation from the Technical Association of the Pulp and Paper Industry (TAPPI) organization. The development of standards will support uniform nomenclature, grade development, material specifications and quality control, and toxicity analysis. The International Organization for Standardization (ISO), TAPPI, and Canadian Standards Association (CSA) are the forerunners in developing standards relevant to nanocellulose. The standard ISO/TS 20477:2017 describes the standard terms and their definition for cellulose nanomaterial, while ISO/TR 19716:2016 describes CNC's characterization. The chemical characterization of cellulose nanomaterial is under the purview of the ISO subcommittee ISO/TC 6/SC 2, which works on test methods and quality specifications for paper and board. Usually, a combination of techniques is required to understand the characteristics of nanocellulose. While international standards are already available to characterize the nanocellulose, regulations regarding its usage are under development.

1.8 Conclusion

Nanocellulose, the novel nanomaterial derived from the abundant biopolymer cellulose, has a lot of potential for diversified applications apart from adding a high value to its raw materials. Before considering nanocellulose for commercial use, a lot of bottlenecks need to be overcome like the high cost of production, less stability in biologically synthesized nanocellulose, surface chemical modification in chemically produced nanocellulose, non-uniform size distribution, and contamination because of lignin and hemicelluloses that come from the raw materials. The scope is its advantage in replacing the synthetic materials without affecting the performance parameters. Already many commercial-scale plants have been established all over the world to produce nanocellulose, and pulp and paper production plants are the leaders in their production and utilization. Other industries are yet to understand the impact of nanocellulose usage in their products.

REFERENCES

[1] Turbak, A.F., Snyder, F.W., and Sandberg, K.R. "Microfibrillated cellulose, a new cellulose product: properties, uses, and commercial potential," *Journal of Applied Polymer Science: Applied Polymer Symposia* 37 (1983): 815–827. https://www.osti.gov/biblio/5062478.

[2] Paralikar, K.M. "Morphology and structural factors of cellulosic substrates that influence action of cellulase enzyme," Cotton Technological Research Laboratory Thesis (1978).

[3] Maiti, S., Jayaramudu, J., Das, K., Reddy, S.M., Sadiku, R., Ray, S.S., and Liu, D. "Preparation and characterization of nano-cellulose with new shape from different precursor," *Carbohydrate Polymers* 98 (2013): 562–567. doi: https://doi.org/10.1016/j.carbpol.2013.06.029.

[4] Molina-Ramirez, C., Canas-Gutierrez, A., Castro, C., Zuluaga, R., and Ganan, P. "Effect of production process scale-up on the characteristics and properties of bacterial nanocellulose obtained from overripe Banana culture medium," *Carbohydrate Polymers* 240 (2020): 116341. doi: https://doi.org/10.1016/j.carbpol.2020.116341.

[5] Satyamurthy, P., and Vigneshwaran, N. "A novel process for synthesis of spherical nanocellulose by controlled hydrolysis of microcrystalline cellulose using anaerobic microbial consortium," *Enzyme and Microbial Technology* 52 (2013): 20–25. doi: 10.1016/j.enzmictec.2012.09.002.

[6] Abitbol, T., Kloser, E., and Gray, D.G. "Estimation of the surface sulfur content of cellulose nanocrystals prepared by sulfuric acid hydrolysis," *Cellulose* 20 (2013): 785–794. doi: 10.1007/s10570-013-9871-0.

[7] Fras, L., Stana-Kleinschek, K., Ribitsch, V., Sfiligoj-Smole, M., and Kreze, T. "Quantitative determination of carboxyl groups in cellulose polymers utilizing their ion exchange capacity and using a complexometric titration," *Materials Research Innovation* 8 (2004): 145–146. doi: 10.1080/14328917.2004.11784850.

[8] Moberg, T., Sahlin, K., Geng, S., Westman, G., Zhou, Q., Oksman, K., and Rigdahl, M. "Rheological properties of nanocellulose suspensions: effects of fibril/particle dimensions and surface characteristics," *Cellulose* 24 (2017): 2499–2510. doi: 10.1007/s10570-017-1283-0.

[9] Marchetti, L., Andrés, S.C., Cerruti, P., and Califano, A.N. "Effect of bacterial nanocellulose addition on the rheological properties of gluten-free muffin batters," *Food Hydrocolloids* 98 (2020): 105315. doi: https://doi.org/10.1016/j.foodhyd.2019.105315.

2

Nanocellulose: Chemical, Thermal, and Optical Properties

Ismail Muhamad Fareez, Ooi Der Jiun, and Wu Yuan Seng
MAHSA University, Saujana Putra Campus, Jenjarom, Malaysia

CONTENTS

2.1 Introduction

Nanotechnology has been heralded as a key technology in the drive toward eco-efficient and sustainable development of materials in many industries. The uprising environmental concerns toward using petroleum-based products have led to the search and development of green polymers in the production of bio-renewable nanomaterials. In recent years, there has been surging interest in using cellulose, mainly because of its biodegradability [1]. Nanocellulose (NC), a derivative of cellulose, holds a great potential to surpass the performance of existing fossil-based materials as a source of energy [2], building material [3], packaging [3], and clothing. It is owing to its interesting optical and rheological properties, high specific surface area, tailorable surface chemistry, and undeniably high mechanical strength. It has been exploited to house a range of guest materials to form a nanocomposite. With its highly tunable properties, NC opens a broad range of new functionalities as well as smart applications such as sensors technology [4, 5], enzyme immobilization [6], automotive [7], and electronic devices [8, 9].

The usage of agricultural biomass products as the primary source in NC production is expected to be a valuable addition in nanocomposites' industrial manufacturing. Lignocellulosic agricultural fibers, including cotton stalk, corn husk, fruit wastes, rice straw, marine biomass, and wood residues, have received considerable attention in the isolation and characterization of NC [10]. They are inexpensively available with an inexhaustible supply. Through different processes, the cellulosic fibers can be sorted into nanoscale forms that exhibit exceptional mechanical properties (tensile strength of at least 1.5 GPa and Young's modulus of up to 160 GPa) [11, 12]. The NC-reinforced composite had demonstrated an increase in strength-to-weight ratio. It is eight times better in comparison to stainless steel [13, 14]. Incorporation of NC into various composites, theoretically and experimentally, showed a decrease in the glass transition temperature and exhibited high light transmittance [15]. Further characterization of NC from different sources using various extraction methods could better understand their potential use as a reinforcement material.

Fibers extracted from wood are short in length, in the range of millimeters, and either untwisted or linear, especially when dignified. On the contrary, fibers from bast plants and cotton are long (few centimeters) and twisted [16, 17]. Mechanical shearing converts cellulose fibers into nanoscale units called cellulose nanofiber (CNF). Upon controlled acid hydrolysis or oxidation, the amorphous region of CNF gets cleaved and disintegrated into highly crystalline, nanometer-long needle-shaped substructural units called cellulose nanocrystals (CNCs) [18]. Bacterial NC (BNC) or microbial NC (MBC) is another type of crystallite-structured cellulose produced by bottom-up biosynthesis by the cellulose-producing bacteria, such as *Acetobacteria* species [19]. Besides, the amorphous NC (ANC) and cellulose nanoyarn (CNY) are some recently emerging categories of NC [20].

NC, of different forms, carries unique features, which define their applicability and functionality. The distinct properties of NC depend on the flexible chemical and material processing and potential compatibility with other guest materials, thus allowing tailorable surface properties. The above-mentioned NC features offer an unprecedented level of control during the NC processing with flexibility in final product morphology and function. This chapter discusses NC's chemical properties, including the analysis of its crystallinity and hydrogen bonding of different NCs. We highlight NC's thermal properties, including its specific heat and thermal conductivity, and the influence of its inclusion into composite and other modifications. The regular light transmittance analysis of different NC products is also reviewed.

2.2 Chemical Properties of Nanocellulose

Cellulose $(C_6H_{10}O_5)_n$, a polysaccharide, is built of a linear series of glucose molecules with a flat ribbon-like conformation (Fig. 2.1) [21]. The repeated unit composed of two anhydroglucose rings coupled through oxygen covalently bonded via a β-1,4 glycosidic linkage [22]. Meanwhile, the hydroxyl groups, readily formed hydrogen bonds with oxygen atoms at the adjoining ring molecules, contribute to more solid linkage lying on a linear conformation of the cellulose chain [22, 23]. The degree of polymerization (n) of cellulose is approximately 10,000 (wood-derived cellulose) to 15,000 (cotton-derived cellulose), varying considerably based on source [24].

Cellulose is the main structural component in the rigid cell walls in plants. These cellulosic fibers vary in shapes and dimensions, depending on the plant's parts and types [17, 24]. Essentially, various approaches have been developed to produce NCs by taking into consideration specific physicochemical and mechanical properties, as well as crystallinities and surface chemistries of different sources. There are great active sites available for NC functionalization with an abundance of surface hydroxyl groups. Due to its hydrophilic nature, hydrophilization is performed to improve compatibility and allow uniform dispersion of CNCs in non-polar solvents and the anticipated polymeric components. Polymer grafting is one of the most commonly used methods to modify the CNCs, and it is accomplished either through "grafting-from" or through "grafting-onto" technique. The polymer matrix is readily homogeneous with the modified CNCs that improve nanocomposite materials' strength and exert other unique properties,

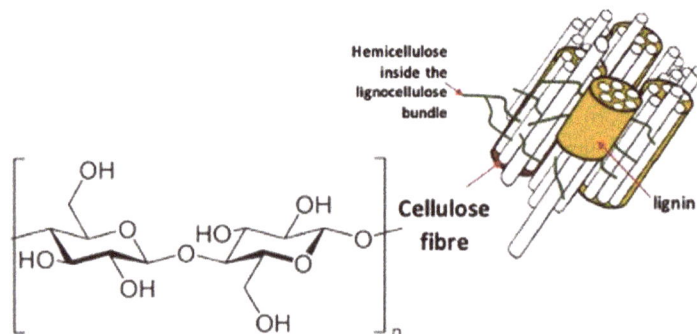

FIGURE 2.1 Schematic of a cellulose microfibril and the chemical structure of cellulose. n is the number of repeating units. Schematic of a cellulose microfibril structure showing both amorphous and crystalline regions [24].

depending on the surface chemistry of grafted CNC chains. This chapter covers NC's chemical properties based on their crystallinity properties and hydrogen bonding interactions in the molecules.

2.2.1 Crystallinity of Nanocellulose

NCs are a high-impact biobased material exhibiting exceptional strength characteristics, low thermal expansion, tailorable surface chemistry, high dielectric properties, and interesting optical properties. Such excellent properties of NCs are rendered by the high crystallinity explained by the packing configuration of cellulose and the uniaxial/nematic molecular alignment in inflexible chain system [25, 26]. The crystallinity index (CI) is a quantitative indicator of crystallinity. It is defined by the mass ratio of crystalline material in a given sample based on the degree of order within the crystals. Since there is no sharp distinction between crystalline–amorphous phase boundaries in their cellulosic structures, it is quite challenging to estimate their CI values [26]. Various techniques such as X-ray diffraction (XRD), wide-angle X-ray scattering (WAXS), small-angle X-ray scattering (SAXS), Fourier transform infrared spectroscopy (FTIR), Raman spectroscopy, and solid-state CP/MAS 13C NMR spectroscopy have been used to define CI of NC [24]. Nevertheless, varying CI values of cellulosic materials are reported, depending on the purity and morphology of NCs and measurement technique and analytical processing.

Celluloses are represented by four major types of polymorphs (I–IV). Their crystallinity attributes depend predominantly on the molecular chain structure. They undergo the irreversible transition from one polymorph to another following treatments, both chemically and thermally. Cellulose I is considered native cellulose, which can be further classified into two structurally different allomorphs, cellulose Iα (unilateral hydrogen bonding sheets) and Iβ (alternating hydrogen bonding sheets) [27]. Cellulose II, or regenerated cellulose, is the most thermodynamically stable and widespread crystalline form and often emerges after mercerization or recrystallization [28, 29]. The arrangement of their atoms distinguishes these two cellulose polymorphs. In cellulose I, the chains are aligned in a parallel direction with no inter-sheet hydrogen bonding. In contrast, in cellulose II, the chains are arranged in an antiparallel direction, with some inter-sheet hydrogen bonding contributing to its uniformly distributed symmetrical structure. On the other hand, cellulose III is generated from cellulose I and cellulose II upon treatment with ammonia, while heat at 260°C induces conversion of cellulose III to cellulose IV [30, 31].

The study of crystallinity of NC is crucial to comprehend the precise influence of the extraction methods and their sources on the nanoparticle's crystal structure. The varying crystallinity or polymorph structures lead to the difference in chain packing configurations [25]. NC's crystalline structure, which holds intricate intra- and interchain molecular hydrogen bonding, provides this nanomaterial with high mechanical strength and resistance to chemical or enzymatic attack [17].

Since cellulose I has significant portions of the highly crystalline and non-crystalline, disordered amorphous region, the variable ratio of these two components and theoretical crystallite size can be analyzed to determine the CI. CNC or BNC possesses relatively higher CrI ranging from 65% to 95% compared to CNF [24, 32]. The controlled acid treatment removes the disordered, amorphous region. This causes the individual cellulose chains to assemble, forming an intrinsically well-ordered crystal graphitic structure. The structure is interconnected by a network of intra- and intermolecular hydrogen bonds, predominantly across the glucose unit, reflecting a higher crystallinity ratio of CNC [17]. In fact, the difference in the crystallite ratio of CNF and CNC can be directly interpreted through their XRD diffraction pattern, as shown in Fig. 2.2. CNF exhibited a lowly visible main peak at $2\theta = 22.5°$ (200) and broad shoulder in the $2\theta = 14.9°$ (110) region, attributable to native cellulose (typical cellulose I) diffractogram [21]. This explicates the lower crystalline ratio of CNF, associated with a very limited crystalline packing process due to the higher amount of less ordered amorphous portion bound to the NC's crystallite domain.

Alkaline treatment, such as bleaching or mercerization, has traditionally been commonly used in natural fiber treatment. Sodium hydroxide (NaOH) is commonly used to weaken and disrupt hydrogen bonds between cellulose microfibrils and depolymerize the cellulose structures. The treatment removes the lignin, wax, and oils that shielded the fiber exteriors, reduces the fiber surface coarseness, and exposes the tiny crystallites [33]. Fareez et al. [21] have previously demonstrated the effect of alkaline treatment

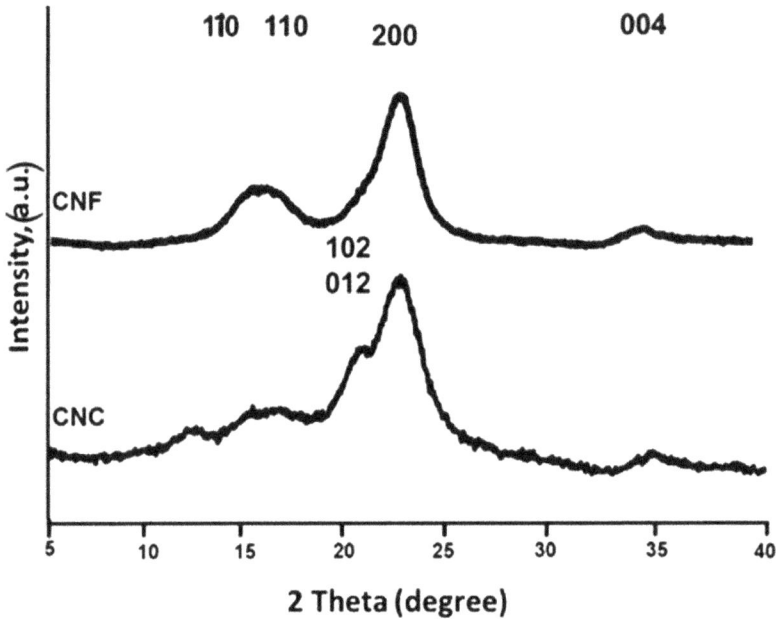

FIGURE 2.2   X-ray diffraction pattern of (a) CNF and (b) CNC.

by hour on the fibers extracted from pineapple leaf fiber (PALF), which mostly contains cellulose I as monoclinic Iβ. A 4-hour treatment with 4.0% sodium hypochlorite and 0.2% NaOH resulted in higher CI (85.41%) over non-treated cellulose (62.8%). It is interesting to note that both approaches keep the original structure of cellulose in the PALF. From XRD analysis, the recorded crystal sizes for bleached PALF samples were in the range of 3.564–3.807 nm. Using transmission electron microscopy (TEM), most CNFs reported in the literature produced nanofiber of 50–100 nm. Further removal of these residues by acid treatment enhances the crystallinity of CNF samples by forming CNC [34].

2.2.2 Hydrogen Network and Bonding

FTIR could be a useful tool to investigate CNC's structural difference, CNF, and BNC via attenuated total reflection measurement that produces the infrared spectra [24]. Briefly, FITR spectra of native cellulose displayed characteristic peaks typical of cellulose moiety at 900–895 cm^{-1} (Fig. 2.3). This is attributed to C—H deformation vibration representing α-glycosidic bonding between the glucose monomer units in cellulose. The FTIR spectra of the CNC display functional group are relatively similar to CNF. The only difference lies in cellulose peak in the CNCs spectra with stronger absorption intensity, signifying higher cellulose content in CNC. This explains the disordered amorphous region's presence that consists of low amounts of hydroxyl groups in CNF. Zheng et al. [35] demonstrated structural changes of raw cellulose isolated from walnut shell (WS) after various treatments. The FTIR spectra obtained showed that cellulose I existed as the predominant form of the raw WS. The stretching vibration peak of the C—H groups at 2897 cm^{-1} belonged to the cellulose molecule. Wide vibration stretch observed at 3350 cm^{-1} of C-3 implies intramolecular hydrogen bonds formed with oxygen (O-5). Meanwhile, in the bleached samples, an absence of the peak at 1738 cm^{-1} in the FTIR spectrum that attributed to the ironic esters and acetyl groups of the hemicellulose shows that the hemicellulose in raw WS was vanished by alkali treatment. While there is no significant difference between the CNC and CNF spectra from raw WS, a slight increase in peak intensity at 1034 cm^{-1} is observed in the CNC.

Due to the large surface distribution of hydroxyl functional groups, cellulose has a strong affinity to itself and materials containing hydroxyl group, especially water. Therefore, cellulose fiber is highly hygroscopic [36, 37]. The interaction of water with cellulose occurs on the surface and in the bulk amorphous core [38]. The water absorbed corresponds to the surrounding relative humidity until equilibrium

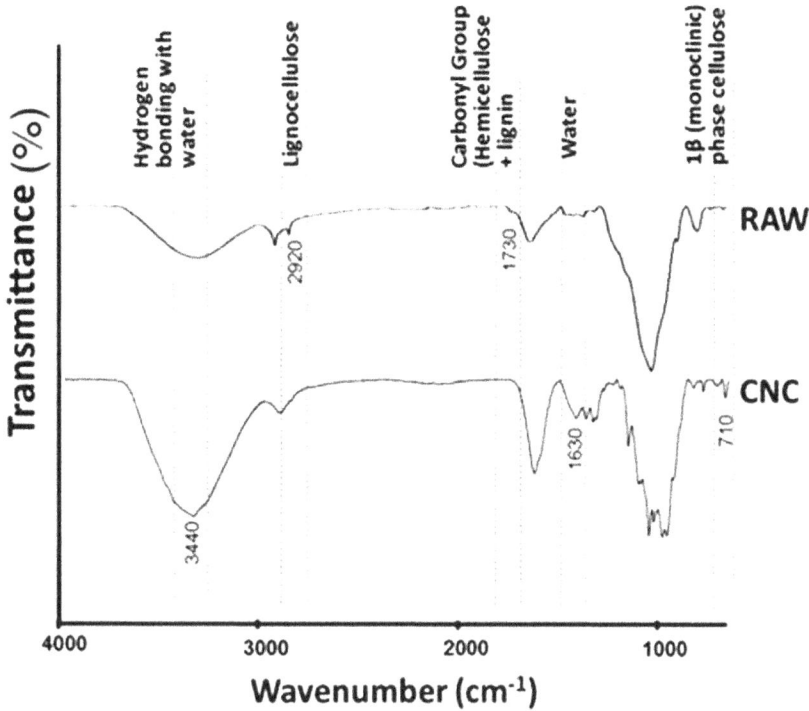

FIGURE 2.3 FTIR Spectra of (a) CNF and (b) CNC.

is reached, as reflected by its purity and degree of crystallinity [39]. This parameter can be analyzed by measuring the moisture sorption isotherm. There is a strong water affinity toward the hydroxyl groups in the amorphous phase.

In contrast, only a small amount of water interaction with the surface hydroxyl groups in crystalline phase is observed. Surface modifications can be done by either grafting certain functional groups or modulating the chemistry of NC itself. The functional groups are introduced onto NC surfaces or used as precursors for further modification either through covalent (i.e. polymer grafting, oxidation, silylation, esterification, and etherification) or non-covalent modifications [40].

Aromatic and aliphatic mono and diisocyanates i.e., phenyl isocyanate (PI) and *n*-octadecyl isocyanate (OI), diphenylmethane diisocyanate (MDI), toluene diisocyanate (TDI), and hexamethylene diisocyanate (HMDI)]—are among the common group of chemicals used to minimize and eliminate the hydrophilicity of cellulose and NC [41]. While this functionalization improves the dispersion of modified CNCs in organic solvents, it facilitates the stress transfer between CNCs and the matrix, leading to enhanced mechanical properties.

2.3 Thermal Properties of Nanocellulose

NC has received significant attention as a new functional material for the development of nanocomposite in many industrial applications. The CNCs and CNFs, in particular, as nanofillers in thermoplastic polymeric matrices, are preferably used when designing cheaper and stronger bio nanocomposite materials. Thermal stability is a key parameter in affecting its performance. Various factors such as temperature, crystallinity, morphology of the cellulose sample, presence of impurities (i.e., metals), and type of atmosphere affect the course of cellulose pyrolysis [42, 43]. The thermal decomposition process of cellulose generates flammable volatiles released by various processes like oxidation, dehydration, hydrolysis, decarboxylation, and transglycosylation. Therefore, the atmospheric conditions during the thermal processing of NC, especially during freezing and air-drying, could interfere with the reactions

TABLE 2.1

Thermal Properties of NC and NC-Based Nanocomposites.

Cellulose Type	Source	Isolation Method	T_{onset} °C	T_{maz} °C	References
CNC	Wood pulp	Acid hydrolysis	300	321	[67]
CNF	Wood pulp	Acid hydrolysis	319	358	[67]
CNC	Corn husk	Acid hydrolysis	257	351	[68]
CNF	Corn husk	TEMPO-mediated oxidation	229	279	[68]
CNF	Corn husk	High-intensity ultrasonication	278	348	[68]
CNC	Corncob	Sulfuric acid hydrolysis	278	313	[69]
CNC	Corncob	Formic acid hydrolysis	334	360	[69]
CNF	Corncob	TEMPO-mediated oxidation	237	305	[69]
CNF	Corncob	FPI refining	287	336	[69]

and influence the product performances [44]. The presence of nitrogen in the surrounding may determine the steps required during NC processing. In addition, time, heating temperature, presence of impurities (i.e., sulfate ester on the cellulose surface), and degree of crystallinity of NC are other crucial factors that affect cellulose pyrolysis [45, 46]. Although the abundance of sulfate esters on the cellulose surface allows better dispersion of individual CNC components in aqueous media, it alleviates NC's thermal stability [43]. The analysis of thermal properties of newly designed NC-reinforced nanocomposites relies on different key parameters such as thermal stability that attributes to its melting point (T_m), glass transition temperature (T_g), thermal decomposition and impurities of NC using thermogravimetric analysis (TGA), differential scanning calorimetry (DSC), dynamic mechanical analysis/spectroscopy (DMA), and differential thermogravimetry (DTG) [25, 47].

Thermal characteristics of NC also determine the right processing temperature of NC-based composites. NC shows significant thermal decomposition at a temperature of around 200–300 °C (Table 2.1). To avoid degradation of NC, the processing temperature should not exceed 200 °C [48]. Nepomuceno et al. demonstrated the importance of thermal stability parameters during the processing of products [49]. The study revealed that the long-term exposure to concentrated sulfuric acid or phosphoric acid during NC processing reduces the thermal stability and the degree of crystallinity of CNC. These may vary because of the large distribution of charged sulfate or phosphate groups on its surface. Nevertheless, to improve CNC's thermal stability, the impact can be reversed by chemical modification of the sulfated NC via desulfonation, oversulfation, and acylation of these ester groups [50]. Apart from that, the source of cellulose and extraction methods may affect the thermal stability as well.

Thermal stability studies and the dynamic mechanical characteristics of a few plants-based NC composites have been studied. Briefly, NC-reinforced composites conferred exceptional mechanical properties, even at a low concentration. The recorded modulus of elasticity could attain up to 150 GPa with a low thermal expansion coefficient (0.01 ppm/K). These signify their distortion values concerning temperature change [50]. The bond strength between the atoms that make up the materials may also affect these. Dominic et al. [51] investigated the effect of NC's inclusion, derived from rice husk, as a nanofiller in natural rubber compounding to replace carbon. The dynamic mechanical analysis (DMA) findings highlighted the composites containing 5 wt% of rice husk—NC experts lower loss tangent (tan δ) value at 60 °C when compared 25 wt% carbon black. This implies that the former leads to low rolling resistance, one of the most important features for manufacturing eco-friendly tire that could act as a viable alternative to petroleum-based products.

2.4 Optical Properties of Nanocellulose

NC is inexpensive, abundant, and biodegradable. These properties render it an ideal sustainable reinforcement material for various emerging transparent composites for photonic applications. Liquid crystalline behavior of CNFs and CNCs contribute to the development of optically transparent stand-alone films and iridescent films by evaporation technique. These materials' unique features depend on the nature and surface chemistry of cellulose, which can be tailored to increase homogeneousness between hydrophilic NC nanofiller and hydrophobic polymer matrix to be used as optically active nanoparticles to be incorporated into optically relevant molecules by covalent modifications [52].

The transparency of NC-based film can be accessed via standard light transmittance by using a UV-visible spectrometer across the 200–1000 nm wavelength range [53]. With a standard optical transmittance at 600 nm wavelength, lower transmittance is expected, depending on NC's concentration, the difference of refractive indices, the size of the particle, and film thickness. The transmittance will further reduce with a decrease in wavelength [54].

CNF nanocomposites, especially in films, are optically transparent and have minimal light scattering when NC is densely packed with tiny interstices between the nanosized cellulosic fibers [55]. The spaces created ensued in light scattering during the loading and unloading of the fiber. Due to the high surface light scattering, the films are often more densely packed and translucent when it is made by slow filtration, drying, and compression [53]. High light scattering diminishes cellulose microfiber (CMF) film transparency and is observed when the material is prepared by mechanical compression. Filtration technique produces NC film of high transparency after being polished with emery paper. In Ref. [56], TEMPO-oxidized MFC sheet of 55-μm thickness is presented with 71.6% of transparency at a wavelength of 600 nm. The transparency of the films derived from hardwood and softwood was approximately 78% and 90%, respectively, when measured at standard transmittance of 600 nm. The slightly lower transparency of hardwood cellulose is attributable to high xylan content, partly affecting uniform dispersion of the fiber in water. With a low inorganic content, the films made up of NC are transparent. However, it turns translucent as the concentration increases, implying NC's light-scattering properties increasingly important [57]. The CNC filler concentration, CNC sources, and its isolation method affect the optical properties of transmittance, haze, and transparency of solutions and composites. In another study, CNC solutions are prepared in both water and dimethylformamide (DMF). In contrast, CNC composites are prepared by utilizing poly(methyl methacrylate) (PMMA) as the matrix with surfactants added to facilitate the complete dispersion of the nanoparticle [58].

Higher amounts of cellulose in solution and solid film have led to a decline in light transmittance, showing an increase in light scattering as measured by clarity and haze. For the CNC solution samples, the same optical patterns have been found as the path's length (sample thickness) increased. The CNCs in DMF presented a similar transmittance to those in water, but significantly reduced light scattering. Similar to CNC-based systems, hydrophobic grafting and surface modification often result in improved compatibility of CNF films. However, because of its high viscosity, even at low concentrations, CNFs are not extensively used in some high-speed coating applications compared to CNCs. CNCs suspension of needle-like morphology also exhibits a specific optical property, owing to their anisotropic diffusion and phase behavior. The nanoparticles in CNCs undergo a phase transition from being isotropic to anisotropic by utilizing polarized light microscopy and viscosity measurements. This is attributed to lyotropic liquid crystalline behavior as the concentration changed [59]. A chiral nematic phase could develop over specific attention. The suspension may gradually undergo evaporation under some particular conditions in producing semitranslucent films in chiral nematic suspensions of cellulose crystallites. Such films display iridescence that reflects the light across a narrow spectrum of wavelengths, as defined by the film's refractive index and chiral nematic pitch [53]. Liquid crystals from derivatives of cellulose appear with either left-handed or right-handed chirality, whereby the chiral nematic order of CNCs is always left-handed, mainly due to the NC right-handed chiral conformation. The left-handed chiral nematic order of CNC films exhibits colors when the helicoidal pitch is approximately the value of the visible light wavelength. Further, since P is very sensitive to surroundings, it is quite manageable in determining film color. Such a phenomenon can also be seen in brilliant naturally occurring photonic colors observed in the *Margaritaria nobilis* fruit and butterfly wings' seed hull. Such optical properties are likely to produce new CNC film applications. The exposure to ultrasound may enhance the chiral nematic pitch, as in both CNC suspension and films, and reflect iridescence with colors ranging from blue-violet to red as the energy raised [60]. The reflective properties can be easily adjusted via the sonication process of different energy inputs and electrolytes' inclusion [61].

Besides its cholesteric structure, CNCs show various other optical functionalities such as UV-blocking [62], fluorescence [63], low refractive index [64], and surface plasmon [65] raising the possibility of many applications, including electronic sensing, particle tracking, anti-counterfeiting technologies, and greenhouse plastics. The optically tunable CNC-based coatings for greenhouse covers were developed by coating polyethylene with a mixture of CNCs, ZnO, and SiO_2 nanoparticles [66]. While enhancing the mechanical strength of the plastic, the addition of ZnO and SiO_2 nanoparticles evidenced to absorb/block harmful UV rays and the infrared, respectively, which both contribute to energy conservation at night.

2.5 Conclusion

The usage of biomass in green resources and products has a huge potential to replace petroleum-based products. The eco-friendly approach would greatly interest the world. Being one of the most accessible natural polymers on earth, it is only recently that cellulose has received considerable attention as a nano-structured fiber as NC. Materials made from NC are carbon-neutral, non-toxic, renewable, recyclable, and highly tunable for specific uses. Following various rigorous research efforts, the commercialization and production of NC on an industrial scale are warranted.

REFERENCES

[1] Klaine Stephen J, Koelmans Albert A, Horne Nina, Carley Stephen, Handy Richard D, Kapustka Larry, Nowack Bernd, and von der Kammer Frank. "Paradigms to assess the environmental impact of manufactured nanomaterials." *Environmental Toxicology and Chemistry* 31 (2012): 3–14. doi: 10.1002/etc.733.

[2] Lasrado Dylan, Ahankari Sandeep, and Kar Kamal. "Nanocellulose-based polymer composites for energy applications: a review." *Journal of Applied Polymer Science* 137 (2020): 48959. doi: 10.1002/app.48959.

[3] Akhlaghi Mohammad Amir, Bagherpour Raheb, and Kalhori Hamid. "Application of bacterial nano-cellulose fibers as reinforcement in cement composites." *Construction and Building Materials* 241 (2020): 118061. doi: 10.1016/j.conbuildmat.2020.118061.

[4] Dai Lei, Wang Yan, Zou Xuejun, Chen Zhirong, Liu Hong, and Ni Yonghao. "Ultrasensitive physical, bio, and chemical sensors derived from 1-, 2-, and 3-D nanocellulosic materials." *Small* 16 (2020): 1906567. doi: 10.1002/smll.201906567.

[5] Naghdi Tina, Golmohammadi Hamed, Vosough Maryam, Atashi Mojgan, Saeedi Iman, and Maghsoudi Mohammad Taghi. "Lab-on-nanopaper: an optical sensing bioplatform based on curcumin embedded in bacterial nanocellulose as an albumin assay kit." *Analytica Chimica Acta* 1070 (2019): 104–111. doi: 10.1016/j.aca.2019.04.037.

[6] Hussin Fathin Najihah Nor Mohd, Attan Nursyafreena, and Wahab Roswanira Abdul. "Taguchi design-assisted immobilization of *Candida rugosa* lipase onto a ternary alginate/nanocellulose/montmorillonite composite: physicochemical characterization, thermal stability and reusability studies." *Enzyme and Microbial Technology* 136 (2020): 109506. doi: 10.1016/j.enzmictec.2019.109506.

[7] Naiman I, Ramasamy D, and Kadirgama K. "Effect of nanocellulose with ethylene glycol for automotive radiator application." *IOP Conference Series: Materials Science and Engineering* 788 (2020). doi:10.1088/1757-899X/788/1/012096.

[8] Lang Augustus W, Osterholm Anna M, and Reynolds John R. "Paper-based electrochromic devices enabled by nanocellulose-coated substrates." *Advanced Functional Materials* 29 (2019): 1903487. doi: 10.1002/adfm.201903487.

[9] Han Lian, Cui Songbo, Yu Hou-Yong, Song Meili, Zhang Haoyu, Grishkewich Nathan, Huang Congguo, Kim Daesung, and Tam Kam Michael Chiu. "Self-healable conductive nanocellulose nanocomposites for biocompatible electronic skin sensor systems." *ACS Applied Materials & Interfaces* 11 (2019): 44642–44651. doi: 10.1021/acsami.9b17030.

[10] Rajinipriya Malladi, Nagalakshmaiah Malladi, Robert Mathieu, and Elkoun Saïd. "Importance of agricultural and industrial waste in the field of nanocellulose and recent industrial developments of wood

based nanocellulose: a review." *ACS Sustainable Chemistry & Engineering* 6 (2018): 2807–2828. doi: 10.1021/acssuschemeng.7b03437.

[11] Kim Joo-Hyung, Shim Bong Sup, Kim Heung Soo, Lee Young-Jun, Min Seung-Ki, Jang Daseul, Abas Zafar, and Kim Jaehwan. "Review of nanocellulose for sustainable future materials." *International Journal of Precision Engineering and Manufacturing: Green Technology* 2 (2015): 197–213. doi: 10.1007/s40684-015-0024-9.

[12] Ismail Muhamad Fareez, Jasni Ainil Hawa, and Jiun Ooi Der. "Fabrications of cellulose nanocomposite for tailor-made applications." *Polymers and Polymer Composites* (2020): 0967391120932055. doi: 10.1177/0967391120932055.

[13] Azizi Samir MAS, Laurent Chazeau, Fannie Alloin, Cavaille J-Y, Dufresne Alain, and Sanchez J-Y. "POE-based nanocomposite polymer electrolytes reinforced with cellulose whiskers." *Electrochimica Acta* 50 (2005): 3897–3903. doi: 10.1016/j.electacta.2005.02.065.

[14] Aulin Christian, Gällstedt Mikael, and Lindström Tom. "Oxygen and oil barrier properties of microfibrillated cellulose films and coatings." *Cellulose* 17 (2010): 559–574. doi: 10.1007/s10570-009-9393-y.

[15] Nogi Masaya, Iwamoto Shinichiro, Nakagaito Antonio Norio, and Yano Hiroyuki. "Optically transparent nanofiber paper." *Advanced Materials* 21 (2009): 1595–1598. doi: 10.1002/adma.200803174.

[16] Klemm Dieter, Kramer Friederike, Moritz Sebastian, Lindström Tom, Ankerfors Mikael, Gray Derek, and Dorris Annie. "Nanocelluloses: a new family of nature-based materials." *Angewandte Chemie International Edition* 50 (2011): 5438–5466. doi: 10.1002/anie.201001273.

[17] Lee HV, Hamid SBA, and Zain SK. "Conversion of lignocellulosic biomass to nanocellulose: structure and chemical process." *The Scientific World Journal* 2014 (2014): 1–20. doi: 10.1155/2014/631013.

[18] Jiang Feng, and Hsieh You-Lo. "Chemically and mechanically isolated nanocellulose and their self-assembled structures." *Carbohydrate Polymers* 95 (2013): 32–40. doi: 10.1016/j.carbpol.2013.02.022.

[19] Ma Lina, Bi Zhijie, Xue Yun, Zhang Wei, Huang Qiying, Zhang Lixue, and Huang Yudong. "Bacterial cellulose: an encouraging eco-friendly nano-candidate for energy storage and energy conversion." *Journal of Materials Chemistry A* 8 (2020): 5812–5842. doi: 10.1039/C9TA12536A.

[20] Hanieh Kargarzadeh, Michael Ioelovich, Ishak Ahmad, Sabu Thomas, and Alain Dufresne. "Methods for extraction of nanocellulose from various sources." *Handbook of Nanocellulose and Cellulose Nanocomposites* 1 (2017): 1–51. doi: 10.1002/9783527689972.ch1.

[21] Fareez Ismail Muhamad, Ibrahim Nur Ain, Yaacob Wan Mohd Hanif Wan, Razali Nur Amira Mamat, Jasni Ainil Hawa, and Aziz Fauziah Abdul. "Characteristics of cellulose extracted from Josapine pineapple leaf fibre after alkali treatment followed by extensive bleaching." *Cellulose* 25 (2018): 4407–4421. doi: 10.1007/s10570-018-1878-0.

[22] Zugenmaier Peter. "Conformation and packing of various crystalline cellulose fibers." *Progress in Polymer Science* 26 (2001): 1341–1417. doi: 10.1016/S0079-6700(01)00019-3.

[23] Conley Kevin, Godbout Louis, Whitehead MA Tony, and van de Ven Theo GM. "Origin of the twist of cellulosic materials." *Carbohydrate Polymers* 135 (2016): 285–299. doi: 10.1016/j.carbpol.2015.08.029.

[24] Fareez Ismail M, Haque Nazmul, Der Juin Ooi, Jasni Ainil H, and Fauziah A Aziz. "Physicochemical properties of nanocellulose extracted from pineapple leaf fibres and its composites." In *Pineapple leaf fibers: processing, properties and applications*, eds. M. Jawaid, M. Asim, P. Md. Tahir, and M. Nasir 167–183. Singapore: Springer. 2020.

[25] Santmarti Alba, and Lee Koon-Yang. "Crystallinity and thermal stability of nanocellulose." In *Nanocellulose and sustainability: production, properties, applications, and case studies*, ed. K.Y. Lee, Boca Raton, FL: CRC Press. 2018.

[26] Daicho Kazuho, Saito Tsuguyuki, Fujisawa Shuji, and Isogai Akira. "The crystallinity of nanocellulose: dispersion-induced disordering of the grain boundary in biologically structured cellulose." *ACS Applied Nano Materials* 1 (2018): 5774–5785. doi: 10.1021/acsanm.8b01438.

[27] Lennholm Helena, Larsson Tomas, and Iversen Tommy. "Determination of cellulose Iα and Iβ in lignocellulosic materials." *Carbohydrate Research* 261 (1994): 119–131. doi: 10.1016/0008-6215(94)80011-1.

[28] Matthews James F, Himmel Michael E, and Crowley Michael F. "Conversion of cellulose Iα to Iβ via a high temperature intermediate (I-HT) and other cellulose phase transformations." *Cellulose* 19 (2012): 297–306. doi: 10.1007/s10570-011-9608-x.

[29] Nishimura Hisao, and Sarko Anatole. "Mercerization of cellulose. III. Changes in crystallite sizes." *Journal of Applied Polymer Science* 33 (1987): 855–866. doi: 10.1002/app.1987.070330314.

[30] Hori Ritsuko, and Wada Masahisa. "The thermal expansion of cellulose II and III$_{II}$ crystals." *Cellulose* 13 (2006): 281–290. doi: 10.1007/s10570-005-9038-8.

[31] Wada Masahisa, Heux Laurent, and Sugiyama Junji. "Polymorphism of cellulose I family: reinvestigation of cellulose IV$_I$." *Biomacromolecules* 5 (2004): 1385–1391. doi: 10.1021/bm0345357.

[32] Blanco Angeles, Monte M Concepcion, Campano Cristina, Balea Ana, Merayo Noemi, and Negro Carlos. "Nanocellulose for industrial use: cellulose nanofibers (CNF), cellulose nanocrystals (CNC), and bacterial cellulose (BC)." In *Handbook of nanomaterials for industrial applications*, ed. C. M. Hussain 74–126. Oxfordshire, UK: Elsevier. 2018.

[33] Mtibe A, Linganiso Linda Z, Mathew Aji P, Oksman Kristiina, John Maya J, and Anandjiwala Rajesh D. "A comparative study on properties of micro and nanopapers produced from cellulose and cellulose nanofibres." *Carbohydrate Polymers* 118 (2015): 1–8. doi: 10.1016/j.carbpol.2014.10.007.

[34] Yu Hou-Yong, Qin Zong-Yi, Liu Lin, Yang Xiao-Gang, Zhou Ying, and Yao Ju-Ming. "Comparison of the reinforcing effects for cellulose nanocrystals obtained by sulfuric and hydrochloric acid hydrolysis on the mechanical and thermal properties of bacterial polyester." *Composites Science and Technology* 87 (2013): 22–28. doi: 10.1016/j.compscitech.2013.07.024.

[35] Zheng Dingyuan, Zhang Yangyang, Guo Yunfeng, and Yue Jinquan. "Isolation and characterization of nanocellulose with a novel shape from walnut (*Juglans regia* L.) shell agricultural waste." *Polymers* 11 (2019): 1130. doi: 10.3390/polym11071130.

[36] Célino Amandine, Fréour Sylvain, Jacquemin Frédéric, and Casari Pascal. "The hygroscopic behavior of plant fibers: a review." *Frontiers in Chemistry* 1 (2014): 43. doi: 10.3389/fchem.2013.00043.

[37] Lovikka Ville A, Rautkari Lauri, and Maloney Thaddeus C. "Changes in the hygroscopic behavior of cellulose due to variations in relative humidity." *Cellulose* 25 (2018): 87–104. doi: 10.1007/s10570-017-1570-9.

[38] Makarem Mohamadamin, Lee Christopher M, Sawada Daisuke, O'Neill Hugh M, and Kim Seong H. "Distinguishing surface versus bulk hydroxyl groups of cellulose nanocrystals using vibrational sum frequency generation spectroscopy." *The Journal of Physical Chemistry Letters* 9 (2018): 70–75. doi: 10.1021/acs.jpclett.7b02729.

[39] Gauthier Helene, Coupas Anne-Cecile, Villemagne Pascal, and Gauthier Robert. "Physicochemical modifications of partially esterified cellulose evidenced by inverse gas chromatography." *Journal of Applied Polymer Science* 69 (1998): 2195–2203. doi: 10.1002/(SICI)1097-4628(19980912)69:11<2195::AID-APP11>3.0.CO;2-Z.

[40] Habibi Youssef. "Key advances in the chemical modification of nanocelluloses." *Chemical Society Reviews* 43 (2014): 1519–1542. doi: 10.1039/C3CS60204D.

[41] Abushammala Hatem, and Mao Jia. "A review of the surface modification of cellulose and nanocellulose using aliphatic and aromatic mono-and diisocyanates." *Molecules* 24 (2019): 2782. doi: 10.3390/molecules24152782.

[42] Kim Ung-Jin, Eom Seok Hyun, and Wada Masahisa. "Thermal decomposition of native cellulose: influence on crystallite size." *Polymer Degradation and Stability* 95 (2010): 778–781. doi: 10.1016/j.polymdegradstab.2010.02.009.

[43] Shafizadeh Fred. "Chemistry of pyrolysis and combustion of wood." *Progress in Biomass Conversion* 3 (1982): 51–76. doi: 10.1016/B978-0-12-535903-0.50006-4.

[44] Peng Yucheng, Gardner Douglas J, Han Yousoo, Kiziltas Alper, Cai Zhiyong, and Tshabalala Mandla A. "Influence of drying method on the material properties of nanocellulose I: thermostability and crystallinity." *Cellulose* 20 (2013): 2379–2392. doi: 10.1007/s10570-013-0019-z.

[45] Wang Neng, Ding Enyong, and Cheng Rongshi. "Thermal degradation behaviors of spherical cellulose nanocrystals with sulfate groups." *Polymer* 48 (2007): 3486–3493. doi: 10.1016/j.polymer.2007.03.062.

[46] Zhang Jing, Nolte Michael W, and Shanks Brent H. "Investigation of primary reactions and secondary effects from the pyrolysis of different celluloses." *ACS Sustainable Chemistry & Engineering* 2 (2014): 2820–2830. doi: 10.1021/sc500592v.

[47] Jordan Jacobs H, Easson Michael W, Dien Bruce, Thompson Stephanie, and Condon Brian D. "Extraction and characterization of nanocellulose crystals from cotton gin motes and cotton gin waste." *Cellulose* 26 (2019): 5959–5979. doi: 10.1007/s10570-019-02533-7.

[48] Trache Djalal, Tarchoun Ahmed Fouzi, Derradji Mehdi, Hamidon Tuan Sherwyn, Masruchin Nanang, Brosse Nicolas, and Hussin M Hazwan. "Nanocellulose: from fundamentals to advanced applications." *Frontiers in Chemistry* 8 (2020): 1–33. doi: 10.3389/fchem.2020.00392/full.

[49] Nepomuceno Neymara C, Santos Amelia SF, Oliveira Juliano E, Glenn Gregory M, and Medeiros Eliton S. "Extraction and characterization of cellulose nanowhiskers from *Mandacaru* (*Cereus jamacaru* DC.) spines." *Cellulose* 24 (2017): 119–129. doi: 10.1007/s10570-016-1109-5.

[50] Gan PG, Sam ST, Abdullah Muhammad Faiq bin, and Omar Mohd Firdaus. "Thermal properties of nanocellulose-reinforced composites: a review." *Journal of Applied Polymer Science* 137 (2020): 48544. doi: 10.1002/app.48544.

[51] Dominic Midhun, Joseph Rani, Begum PM Sabura, Kanoth Bipinbal Parambath, Chandra Julie, and Thomas Sanmariya. "Green tire technology: effect of rice husk derived nanocellulose (RHNC) in replacing carbon black (CB) in natural rubber (NR) compounding." *Carbohydrate Polymers* 230 (2020): 115620. doi: 10.1016/j.carbpol.2019.115620.

[52] Wilts BD, Dumanli AG, Middleton Roxanne, Vukusic P, and Vignolini Silvia. "Invited Article: chiral optics of helicoidal cellulose nanocrystal films." *APL Photonics* 2 (2017): 040801. doi: 10.1063/1.4978387.

[53] Dufresne Alain. "Nanocellulose: a new ageless bionanomaterial." *Materials Today* 16 (2013): 220–227. doi: 10.1016/j.mattod.2013.06.004.

[54] Schütz Christina, Sort Jordi, Bacsik Zoltán, Oliynyk Vitaliy, Pellicer Eva, Fall Andreas, Wågberg Lars, Berglund Lars, Bergström Lennart, and Salazar-Alvarez German. "Hard and transparent films formed by nanocellulose–TiO$_2$ nanoparticle hybrids." *PLOS ONE* 7 (2012): e45828. doi: 10.1371/journal.pone.0045828.

[55] Abral Hairul, Ariksa Jeri, Mahardika Melbi, Handayani Dian, Aminah Ibtisamatul, Sandrawati Neny, Pratama Angga Bahri, Fajri Nural, Sapuan SM, and Ilyas RA. "Transparent and antimicrobial cellulose film from ginger nanofiber." *Food Hydrocolloids* 98 (2020): 105266. doi: 10.1016/j.foodhyd.2019.105266.

[56] Fukuzumi Hayaka, Saito Tsuguyuki, Iwata Tadahisa, Kumamoto Yoshiaki, and Isogai Akira. "Transparent and high gas barrier films of cellulose nanofibers prepared by TEMPO-mediated oxidation." *Biomacromolecules* 10 (2009): 162–165. doi: 10.1021/bm801065u.

[57] Hu Liangbing, Zheng Guangyuan, Yao Jie, Liu Nian, Weil Ben, Eskilsson Martin, Karabulut Erdem, Ruan Zhichao, Fan Shanhui, and Bloking Jason T. "Transparent and conductive paper from nanocellulose fibers." *Energy & Environmental Science* 6 (2013): 513–518. doi: 10.1039/C2EE23635D.

[58] Dong Hong, Strawhecker Kenneth E, Snyder James F, Orlicki Joshua A, Reiner Richard S, and Rudie Alan W. "Cellulose nanocrystals as a reinforcing material for electrospun poly(methyl methacrylate) fibers: formation, properties and nanomechanical characterization." *Carbohydrate Polymers* 87 (2012): 2488–2495. doi: 10.1016/j.carbpol.2011.11.015.

[59] Van Rie Jonas, Schütz Christina, Gençer Alican, Lombardo Salvatore, Gasser Urs, Kumar Sugam, Salazar-Alvarez Germán, Kang Kyongok, and Thielemans Wim. "Anisotropic diffusion and phase behavior of cellulose nanocrystal suspensions." *Langmuir* 35 (2019): 2289–2302. doi: 10.1021/acs.langmuir.8b03792.

[60] Chen Qi, Liu Ping, Nan Fuchun, Zhou Lijuan, and Zhang Jianming. "Tuning the iridescence of chiral nematic cellulose nanocrystal films with a vacuum-assisted self-assembly technique." *Biomacromolecules* 15 (2014): 4343–4350. doi: 10.1021/bm501355x.

[61] Beck Stephanie, Bouchard Jean, and Berry Richard. "Controlling the reflection wavelength of iridescent solid films of nanocrystalline cellulose." *Biomacromolecules* 12 (2011): 167–172. doi: 10.1021/bm1010905.

[62] Feng Xin, Zhao Yafei, Jiang Yaoquan, Miao, Cao Shaomei, and Fang Jianhui. "Use of carbon dots to enhance UV-blocking of transparent nanocellulose films." *Carbohydrate Polymers* 161 (2017): 253–260. doi: 10.1016/j.carbpol.2017.01.030.

[63] Wu Bolang, Zhu Ge, Dufresne Alain, and Lin Ning. "Fluorescent aerogels based on chemical crosslinking between nanocellulose and carbon dots for optical sensor." *ACS Applied Materials & Interfaces* 11 (2019): 16048–16058. doi: 10.1021/acsami.9b02754.

[64] Niskanen Ilpo, Suopajarvi Terhi, Liimatainen Henrikki, Fabritius Tapio, Heikkila Rauno, and Thungström Göran. "Determining the complex refractive index of cellulose nanocrystals by combination of Beer-Lambert and immersion matching methods." *Journal of Quantitative Spectroscopy and Radiative Transfer* 235 (2019): 1–6. doi: 10.1016/j.jqsrt.2019.06.023.

[65] Zhang Shuaidi, Xiong Rui, Mahmoud Mahmoud A, Quigley Elizabeth N, Chang Huibin, El-Sayed Mostafa, and Tsukruk Vladimir V. "Dual-excitation nanocellulose plasmonic membranes for molecular and cellular SERS detection." *ACS Applied Materials & Interfaces* 10 (2018): 18380–18389. doi: 10.1021/acsami.8b04817.

[66] Nevo Yuval, Peer N, Yochelis S, Igbaria M, Meirovitch S, Shoseyov O, and Paltiel Y. "Nano bio optically tunable composite nanocrystalline cellulose films." *RSC Advances* 5 (2015): 7713–7719. doi: 10.1039/C4RA11840E.

[67] Schnell Carla N, Galván María V, Peresin María S, Inalbon María C, Vartiainen Jari, Zanuttini Miguel A, and Mocchiutti Paulina. "Films from xylan/chitosan complexes: preparation and characterization." *Cellulose* 24 (2017): 4393–4403. doi: 10.1007/s10570-017-1411-x.

[68] Yang Xue, Han Fuyi, Xu Chunxia, Jiang Shuai, Huang Liqian, Liu Lifang, and Xia Zhaopeng. "Effects of preparation methods on the morphology and properties of nanocellulose (NC) extracted from corn husk." *Industrial Crops and Products* 109 (2017): 241–247. doi: 10.1016/j.indcrop.2017.08.032.

[69] Liu Chao, Li Bin, Du Haishun, Lv Dong, Zhang Yuedong, Yu Guang, Mu Xindong, and Peng Hui. "Properties of nanocellulose isolated from corncob residue using sulfuric acid, formic acid, oxidative and mechanical methods." *Carbohydrate Polymers* 151 (2016): 716–724. doi: 10.1016/j.carbpol.2016.06.025.

3

Nanocellulose: Structural Properties and Mechanical Structure-Property Relationships

Meena Nemiwal
Malaviya National Institute of Technology, Jaipur, India

Kaushalya Bhakar
Central University of Gujarat, Gandhinagar, India

CONTENTS

3.1 Introduction

Cellulose polymer is mainly obtained from wood and cotton. Cotton holds about 90% of cellulose, while wood contains 30–40% of cellulose. Many other species such as tunicates, algae, bacteria, and some marine animals also contain cellulose. The basic repeating unit of cellulose is cellobiose and is linked by a β-1,4-glycosidic bond between two D-anhydro glucopyranose units at 180° and is equatorial-equatorial glycosidic linkage. Because of the cellulose chain position, the hydrogen bond is formed with the adjacent cellulose chain between the hydrogen of one cellulose chain and the oxygen of the C3 hydroxyl group of another chain. The ribbon-like character in cellulosic macromolecules prevents adjacent cellulose chains' free movement and fits together in an ordered crystalline manner, therefore stiffing the chain. The mechanical properties in the cellulose fibers are due to these crystalline regions. These polymers have both types of hydrogen bonds: intermolecular (between two cellulose polymers) and intramolecular (within the cellulose polymer unit). The intramolecular bond handles stiffness to the polymer chain, while sheet structure is formed due to intermolecular bonds. Cellulose fibers are water-soluble and insoluble in non-polar solvents because of many hydrogen bonds and high crystallinity.

Although biosynthetic mechanism in cellulose synthesis is same in all organisms, their typical crystallinity and dimension depend on their origins [1, 2]. Cellulose constitutes crystalline and amorphous regions [3], and its crystal structure is a mixture of both triclinic and monoclinic structures in nature. Different treatments can transform many other thermodynamically favorable crystalline forms in terms of chains and geometry orientation. The crystalline region exhibits the elastic modulus property, making it a suitable material for the reinforcement of biocomposite formation [4].

The structure of cellulose may be discussed in various aspects. Morphologically, it constitutes the cellulose fiber and its cell wall. At the supramolecular level, ordered crystalline and amorphous regions are involved and at the molecular level, the basic chain structure and its hydrogen bonding with adjacent chains are discussed. The cell wall comprises four layers: one primary and three secondary layers mainly made of lignin, hemicellulose, and cellulose. The middle lamella is a polysaccharide that binds cells together and provides stability to the fiber. The strength and stiffness in cellulose fibers are basically due to stable microfibril formation by hydrogen bonds [5]. Nanocellulose has a wide range of applications in both pure form and composite form, i.e., wound dressing, food, cosmetics, textiles, pharmaceuticals, energy, optoelectronics, bioprinting, tissue engineering, sensing, and environmental remediation [6].

Nanocellulose has developed a great interest, especially for the reinforcement in a polymer matrix for the formation of biocomposites because of its abundance, sustainability, mechanical properties—high stiffness, high tensile strength, high flexibility, and high surface-to-volume ratio—and excellent thermal and electrical properties [7]. This chapter summarizes nanocellulose in terms of its relationship with fiber morphology and mechanical properties.

3.2 Structural Properties of Nanocellulose

Depending on the extraction sources and treatment, there are different structures of nanocellulose exhibiting distinct characteristics. To acquire desired properties, different strategies are being used to produce nanocellulose: grinding, high-pressure homogenization, and enzymatic and chemical treatments [8]. These methods are mainly employed to isolate nanocrystalline and nanofibrillated cellulose (Fig. 3.1), whereas bacterial nanocellulose is produced naturally from different organisms [9].

Nanocrystalline cellulose is also known as whiskers and crystallites. These rod-like nanocrystals are obtained from various sources such as mulberry bark, wood, tunicin, straw, hemp, flax, and wheat. Their diameter is in the range of 5–70 nm, and length is 100–250 nm (from plant cellulose); they are extracted mainly by chemical treatment in which an amorphous portion of the cellulose is hydrolyzed by acid. The morphology of these produced cellulose nanocrystals (CNCs) is rod-like, observed under transmission electron microscopy (TEM). Before acid treatment, raw cellulose is treated with NaOH or bleaching powder to remove existing impurities, i.e., wax, lignin, esters, and hemicelluloses. Subsequently, the purified cellulose material is treated with acid for several hours at a high temperature that hydrolyzes the amorphous cellulose section, and then upon centrifugation and dialysis of suspension, purified CNCs are obtained.

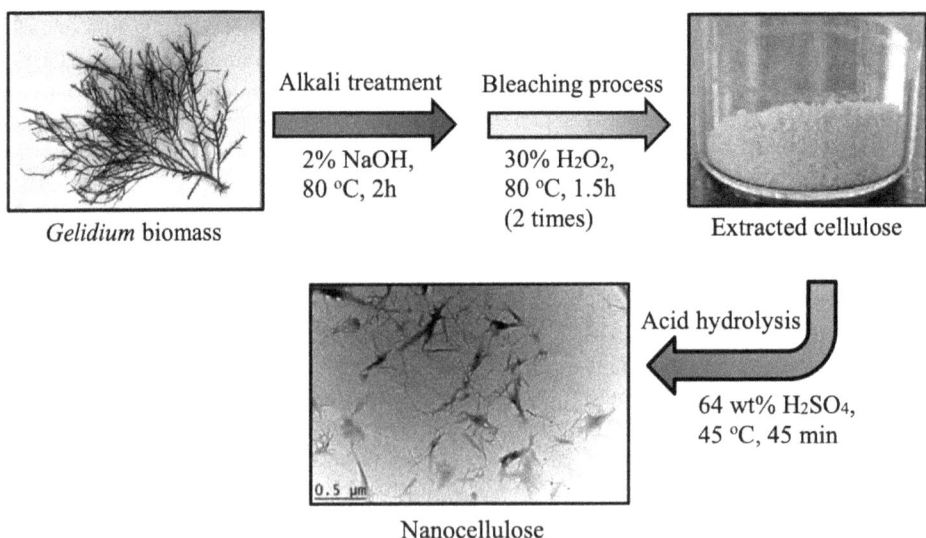

FIGURE 3.1   Schematic of extraction of nanocrystalline cellulose, nanofibrillated cellulose, and nanocrystalloids from cellulose fiber (Reproduced with permission from Ref. [13]. Copyright ©Elsevier 2016).

Depending on the source, the acid used, and the time and temperature of hydrolysis the diameter and length of produced CNCs vary. The CNCs extracted from wood pulp by sulfuric acid hydrolysis were found 100–300 nm in length and 3–5 nm in diameter [10]. CNCs extracted from fibers of banana pseudostems were used to fill a nanocomposite in polyvinyl alcohol (PVA) matrix. The CNCs obtained by acid hydrolysis were 135.0 ± 12.0 nm in length and 7.2 ± 1.9 nm in diameter [11]. The discarded cigarette filters (DCFs) were used as a source of CNCs. In this extraction process, CNCs are obtained by ethanolic extraction, followed by hydrolysis with sulfuric acid, and needles of nanocrystals were obtained with an average length of 143 nm [12]. Nanocellulose was extracted from red algae by chemical treatment, i.e., alkalization, bleaching treatment, and hydrolysis with acid, with a length of 21.8 ± 11.1 nm and a diameter of 547.3 ± 23.7 nm (Fig. 3.1) [13].

The dimensions of CNCs also depend on the acids used in the treatment process. For instance, hydrolysis of cotton with sulfuric acid at 65 °C produced CNCs with a diameter in the range of 10–30 nm and length in the range of 200–300 nm [14]. Whereas by HCl hydrolysis, CNCs with length 100–200 nm and diameter 10 nm were produced [15]. Treatment can be carried out in the presence of oxidants, such as tetramethyl-piperidin-1-oxyl (TEMPO) for the production of CNCs of different dimensions [16]. The approximate length of TEMPO-oxidized cellulose nanofibrils (CNFs) can be determined with polymerization by viscosity because of their linear relationship. The values of polymerization by viscosity decrease with the increasing amounts of NaClO during the oxidation process and can also be decreased by mechanical fibrillation in water [17]. Similarly, by controlling microbial hydrolysis of cotton fibers, cellulose nanowhiskers can be produced with no surface modification [18].

Cellulosic nanofibers are prepared by mechanical treatment of cellulose fibers, i.e., homogenization, ultrasonication, and grinding with no chemical changes in cellulose molecular structure [19, 20]. Besides mechanical treatment, sometimes mild chemical treatment can also produce CNFs [21], and sometimes both chemical and mechanical strategies can be applied simultaneously to make individual CNFs [22]. Wood is the main raw material for CNFs production. However, other raw materials, such as cotton fibers and tunicate, have also been used [23]. Needles of CNFs were extracted by acid hydrolysis of seeds of *Citrullus colocynthis* [24]. The diameter of CNFs varies from 2 nm to 50 nm, depending on the raw material used in the CNF production [25]. A very high aspect ratio of CNFs makes it suitable for reinforcing polymer materials.

Cellulosic nanofibers are extracted from wood, sugar beet, flax, and potato tuber by mechanical treatment with enzymatic and chemical treatments and are 60 nm in length. Microfibrils appear as long bundles of cellulose molecules interconnected by hydrogen bonding that stabilize it as microfibrils. These are positioned in a different cell layer with many directions, textures, and densities. The modulus of each thread of microfibril is observed to be in good agreement with the native cellulose. Its structure is very stable with outstanding Young's modulus and low thermal expansion [26], making them suitable for reinforcement in nanocomposites. It is microfibrils that attribute the natural stiffness and strength in cellulose fibers. Bio-based renewable citrus wastage on treatment with different technology produces nanofibers of crystallinity 55% and diameter 10 nm and length 458 nm [27]. The CNCs and CNFs of different morphologies can be produced from green and recyclable sources [28].

Bacterial nanocellulose (BNC) is another type of nanocellulose that is obtained from bacterial genera such as *Rhizobium, Acetobacter, Aerobacter, Agrobacterium, Azotobacter, Escherichia, Salmonella*, and *Sarcina* [29, 30] and also from systems with no cells [31]. It is produced inside the bacterial cell with β-1-4 glucan chains and is then isolated into cell membranes as protofibrils that crystallize and appear ribbon-shaped microfibrils and subsequent pellicle formation [32]. It is 20–100 nm in diameter and its mechanical properties depend on factors such as microbial strain type, culture condition, and synthesis method [33, 34]. Bacterial nanocellulose proves to be very useful for applications in a biomedical field such as regenerative medicines, 3D printing biomaterials, drug delivery, enzyme immobilization, and tissue engineering [35].

Bacterial nanocellulose is also known as microbial cellulose or bio-cellulose. It is highly crystalline and pure, and it does not require pretreatment of microbial strain in the extraction process to remove noncellulosic components. It has excellent water retention, high tensile strength and good biocompatibility, biodegradability, high crystallinity, optical transparency, hydrophilicity, and moldability into 3D structures [36] polyfunctionality. But its production cost is very high, which limits its use on a large scale.

Another important property of nanocellulose fibrils is its high aspect ratio and is always higher than that of the crystals. The fibrils' surface area is also very high, and these properties make them suitable to act as reinforcing agents.

3.3 Mechanical Properties of Nanocellulose

Nanofibrils present in nanocellulose play a key role in their excellent mechanical properties such as high tensile strength, low elongation breakpoint, and high elastic modulus. These properties make nanocellulose an amazing reinforcement agent for polymer matrix, and these properties can be easily modified by using different strategies. These specific properties are discussed in detail.

3.3.1 Tensile Strength

Tensile strength is a sign of fiber strength that depends on fiber length, fiber strength and bonding, molecular weight, orientation degree, crystallinity, and distribution of fibers' defects. The tensile strength of nanocellulose microfibril is about five times stronger than steel, and its weight is approximately one-fifth of that of steel. A sheet or solution of CNCs/CNFs is transparent toward visible light and not scattered by it. Nanocellulose has functional groups on their surface, so surface modification and chemical loading are more straightforward. The coefficient of thermal expansion is found compared to quartz glass. The tensile strength of the composite can be enhanced by modifying the surface of CNFs with NaOH that prohibits fibrils' aggregation and increases their dispersion within the polymer and hence improves reinforcing property in composites [37]. Bacterial nanocellulose upon dehydration followed by rehydration results in an increase of its tensile strength and makes it suitable for a wide range of biomedical field applications. The theoretical and experimental tensile properties of BNC have been discussed by many researchers and have proved to be a perfect reinforcing material for biocomposite formation [38].

3.3.2 Elastic Modulus

The Young's modulus, which defines the relationship between stress and strain in the linear elasticity region of a uniaxial deformation, measures the stiffness of nanocellulose. The value of Young's modulus of nanocellulose varies with the source and its treatment method. Nanocellulose extracted from tunicate has the value of longitudinal modulus as high as 151 GPa. The range of values of Young's modulus is between 58 and 180 GPa. These values can be increased by using different raw materials and preparation methods [39, 40]. The study was done using the X-ray diffraction method and Raman spectroscopy elastic modulus of tunicate cellulose fiber. A four-point bending was tested to observe the stress in the material by analyzing minor changes in the Raman band's characteristics. A slight change in intensity was observed when the sample is parallel to the direction of polarization of the laser and perpendicular, which indicates the two-dimensional structure and random arrangement of fibers in the tunicate sample. The elastic modulus value was found to be very high, at about 143 GPa [40].

The tensile modulus of microfibrillated cellulose (MFC) films depends on the duration of the mechanical treatment of cellulose fiber, and it increases with increasing duration. Pectin, naturally present in the cellulose, plays a key role in a binding agent between CMFs and increases material cohesion. When a film is subjected to any kind of mechanical stress, it helps to improve the mechanism of load transfer. But the elimination of pectin is decreased in a dry atmosphere by tensile modulus that has been observed for sugar beet pulp [41]. MFC has been prepared by reinforcing nanofibrils with the polymer matrix by Svagan et al. [42]. An increase in Young's modulus was investigated that was 13 GPa and tensile strength 180 MPa. However, in humid environments, the value of tensile modulus decreases due to pectin's hydrophilic nature.

3.3.3 Elongation at Break

It shows plant fibers' capability to resist shape changes and can be measured by taking the ratio of changed length and actual length. Fibers having high elongation values have lower strength and Young's modulus value. Therefore, if plant fibers are impregnated in thermoplastics such as polythene, the value of elongation at break decreases, and the value of Young's modulus increases.

Three-dimensional networks are formed within cellulose/CNFs because of hydroxyl groups present in the nanocellulose hydroglucose unit. There are three hydroxyl groups present in a cellulose unit. Out of them, one is primarily at carbon-6, and the other two are at the positions of carbon-2 and carbon-3. This network immobilizes the natural rubber chains, which decreases the elongation at the break of the composite.

Synthetic fibers' mechanical and physical properties are better than natural fibers, but the percentage of elongation at break and value of specific modulus of natural fibers are better. Some chemical treatment elongation at break can be enhanced by modifying mechanical treatments. For example, MFC extracted from sugar beet chips by high-pressure homogenizer showed better results than unhomogenized cellulose. The mechanical performance was improved by strengthening the CNFs with a phenol-formaldehyde matrix, with an elongation at break of 2.9%, a tensile strength of 127 MPa, and a modulus of elasticity of 9.5 GPa [41].

3.4 Structural Properties and Mechanical Structure-Property Relationship

The dimensions of nanocellulose produced from wood less than 100 nm are used for reinforcement, depending on their different morphologies such as rod-like, spherical, and platelet. It is only nanofibrils that provide extraordinary mechanical strength and other high performance to plants, and the crystalline and amorphous region of the molecule can characterize this mechanical strength. The chain molecules in the amorphous region provide plasticity and flexibility to the material, and molecules in the crystalline region are the main contributors to the material's elasticity. Potentially, it is stronger than steel because of its high modulus and its potential compared to Kevlar. Sakurada et al. [4] first experimentally studied its structure in 1962 through the crystal deformation of cellulose, and the elastic modulus values of 137 GPa were reported. Its structure highly affects the mechanical properties of the material in which it is used as a reinforcing agent because of its high E-module, high stiffness, and high strength.

The cell wall of plants differs in their composition and the orientation of cellulose fibrils. Cellulose fibrils are inclined to the main axis at some angles, which is known as the microfibrillar angle. In most cases, fiber strength increases with the increase in cellulose content and decreases with the spiral angle on the fiber axis. Nanofibrils can be considered the elementary cellulosic fibrils, having crystalline and amorphous segments of different length and diameter scales.

Nowadays, composite materials are being used in various applications. These are composed of polymer matrix and filler material for reinforcement. The filler materials with high performance are added to prepare biocomposite to enhance the mechanical strength of polymer matrix. As the nanocellulose content increases in the polymer matrix, the extent of the cellulose network decreases because of a decrease in micro-space at the filler interface and matrix. Thus, the strength of filler-matrix interfacial adhesion improves, increasing the specific surface area of the nanofiber exposed to the polymer matrix. Hence, the value of tensile strength increases. Synthetic fillers such as aramid fibers, glass, and carbon can be used for reinforcement. Their final disposal in the environment is a big issue because of their non-biodegradability or incomplete combustion. Cellulose could be an excellent substitute for this, as it is biodegradable and a cheaper source. CNFs obtained from animals show important reinforcement properties for their excellent compatibility, high modulus values, and high aspect ratio. Wakabayashi et al. [43] investigated that completely nanofibrillated TEMPO-oxidized cellulose had the highest tensile strength, transparency, and Young's modulus in comparison to the partially and non-fibrillated TEMPO-oxidized cellulose.

It enhances mechanical interlocking and interfacial bonding of nanofibers and polymer matrices because of a three-dimensional network that increases tensile strength and composite modulus. It increases the load transfer efficiency from the matrix to the filler fiber because of this interfacial bonding, which improves the composites' mechanical properties [44]. The modulus value of cellulose is in good agreement with polythene, which can be seen by considering the cross-sectional area of an individual molecule. This property makes it a perfect candidate to act as a reinforcing agent. The tensile strength of CNCs as reinforcing agent (7.5–7.7 GPa) is observed to be higher than that of Kevlar (3.5 GPa), steel wire (4.1 GPa), and carbon fiber (1.5–5.5 GPa) and less than that for carbon nanotubes (11–63 GPa). Simultaneously, elastic modulus was found to be lower than all these materials: (CNCs: 110–220 GPa, Kevlar: 124–130 GPa, steel wire: 210 GPa, carbon fiber: 150–500 GPa, carbon nanotubes: 270–950 GPa).

Various studies have been conducted to explore ways to enhance the mechanical properties of nanocomposites. It was investigated that if PVA is reinforced with the nanocellulose (extracted from bagasse) in the linear and crosslinked state, the tensile strength of linear PVA was observed to be 7.5 wt%, and it was 5 wt% for crosslinked PVA. By thermogravimetric analysis (TGA) studies, it was observed that thermal stability for linear PVA nanocellulose is higher than crosslinked PVA nanocomposite [45]. Nanocomposite synthesized from nanocellulose extracted from jute fibers and natural rubber was analyzed in detail. Nanocellulose with a diameter of 50 nm was reinforced into natural rubber using crosslinking agents to form a nanocomposite film. The elasticity modulus value was observed to increase from 1.3 to 3.8 GPa and was increasing with the increasing content of nanocellulose going forward. This is due to the restriction of mobility in adjoining polymer chains in the vicinity of the crosslinked cellulose and between the three-dimensional network of nanocellulose. It was also observed that the stress-strain curve is not affected by an increase in nanocellulose content because the effect of the crosslinking agent predominates over the reinforcing agent. The percentage of elongation is 884% for pure natural rubber and 415% for nanocomposites with 3 wt% nanocellulose. Therefore, it can be concluded that incorporating nanocellulose in natural rubber reduces the mobility of the matrix and increases the composite's stiffness [44].

CNFs have a more robust effect in polyurethane matrix than micrometer cellulose fibrils because of their small fiber dimensions and high aspect ratio. By mechanical and thermal studies, it has been observed that the value of tensile strength and the value of E-modulus for nanocellulose fibrils were higher than that for micrometer cellulose because of its better incorporating property in a polyurethane matrix. By combining 16.5 wt% of CNFs with a polyurethane matrix, it has been observed that the strength increased by about 500% and the stiffness increased by 3000% [46]. Ramie cellulose nanocrystallites were prepared by hydrolysis of ramie fibers with acid, and it was used as a filler for glycerol-plasticized starch for composite formation. The tensile strength value increases from 2.8 to 6.9 MPa, and Young's modulus value increases from 56 to 480 MPa because of the synergistic interaction of nanofillers with plasticized starch [47].

The mechanical properties of MFCs films (20–200 mm) have been reported by several authors and their high strength and their ability as reinforcing agents have also been analyzed. Several methods can be used for the synthesis of these films, such as casting evaporation of MFCs, filtration, or spin coating. The tensile modulus of MFC films depends on the duration of the cellulose's mechanical treatment, which increases with an increase in duration. Naturally occurring pectin cellulose plays a very important role in binding the cellulose microfibrils and increases the cohesion of material, which improves the load transfer mechanism in nanocellulose if subjected to mechanical stress. But the elimination of pectin in the dry atmosphere decreases the tensile modulus seen for the pulp of sugar beet [41]. However, due to pectin's hydrophilic nature, the value of tensile modulus decreases in a humid atmosphere.

Nanocomposite films have been prepared by reinforcing MFC with melamine formaldehyde. Dynamic mechanical thermal and tensile tests studied the prepared material, and the value of Young's modulus was observed to be 16.6 GPa, and the tensile strength value was 142 MPa [48].

A hydrogen bond is present between cellulose molecules, and it gives strong interaction between fibrils and fibers and within the fibrils. It provides flexibility to the cellulose polymer that enhances the polymer matrix's mechanical properties and thermal properties in which it is reinforced. Intramolecular hydrogen bonding is very important for determining the crystallite modulus and the mechanism of chain deformation. The elastic modulus values have been calculated for different polymorphs. It was interpreted that skeletal transformation takes place because of crystal transition, and intramolecular hydrogen bonding brings this skeletal contraction [49]. Intermolecular and intramolecular hydrogen bonding also plays a different role in elasticity. Cellulose fiber stretching with small amplitude affects elasticity, dynamic FTIR analyzed that strain distribution moves through the glucose ring containing carbon-oxygen-carbon linkage and seems less affected by intramolecular hydrogen bonding. Simultaneously, deformation at a right angle to the fiber axis affects more intermolecular hydrogen bonding [50].

The molecular simulation technique was used to calculate the value of elastic modulus and was derived between 124 and 155 GPa for the cellulose Ib crystal model, whose observed value was equal to 138 GPa.

Linear compressibility and anisotropy of Young's modulus in planes at right angles to the main axis were investigated, and the values of two moduli were 11 and 50 GPa, respectively [51].

All cellulose composites' morphology and mechanical characteristics are controlled by source, initial cellulose concentration, rate of precipitation, treatment, and dissolution time. Mechanical properties can be optimized by controlling the number of defects during precipitation. The tensile strength was observed to be 106 Mpa, and the tensile modulus was found to be 7.6 GPa in all cellulose composites.

Nanocellulose reinforced into PVA, and their mechanical and thermal properties were reported by chemical treatment derived from microcrystalline cellulose. This method obtained rod-like whiskers. The TEM technique was used to measure individual nanocelluloses and resulted in the dimensions of approximately 6.96 nm in diameter and 178 nm in length. It increased the thermal stability and tensile modulus of the nanocomposite. The crystallinity was also found to be increased compared to microcrystalline nanocellulose [52].

Iwamoto et al. extracted single microfibrils from tunicate by two methods. In one method, the tunic was oxidized by TEMPO, followed by mechanical disintegration with water. The value of the elastic modulus was 145.2 ± 31.3 GPa. In another, acid hydrolysis was carried out, and the value of elastic modulus was 150.7 ± 28.8 GPa. The morphology and dimension of microfibrils were studied by AFM, and the cross-sectional dimension was observed as 8×20 nm and few micrometers [39].

The reaction conditions change the degree of polymerization. This means that the hydrolysis of crystals increases with increasing reaction time and reduces the degree of polymerization. The concentration of the ratio of acid to the pulp also affects the morphology of the nanocellulose [13]. Along with the temperature and acid concentrations, the dimensions of the nanocellulose formed also depend on the intensity of ultrasonic irradiation applied to the dispersion of the particles of suspensions [53, 54]. The crystallinity can also be altered by changing reaction conditions [53].

The CNC has a special ability to organize a chiral liquid crystal phase arranged helically. This property introduces chirality and photonic properties in the materials and shows an excellent application in optical filters and sensors [55].

The transverse elastic modulus of the CNCs was calculated by comparing the experimentally determined force-distance curve with 3D finite element calculations of the tip indentation on CNCs [56]. The stiffness of CNCs is affected by comparing AFM results on the same nanoparticles under various humidity conditions. At 0.1% relative humidity (RH), the value of transverse modulus of an extracted CNC was calculated to be in the range of 18–50 GPa.

The relationship between molecular structure and elastic moduli has been studied for three different aromatic polyamides, poly(*m*-phenyl-eneisophthalamide), poly(*p*-phenylene terephthalamide), and poly-*p*-benzamide. The distribution of potential strain energy has been calculated in relation to the change in bond length, bond angles, and rotational angles [57].

Although the value of the elastic modulus of crystalline cellulose is not well-defined, several theoretical molecular models and many X-ray diffraction measurements have been done to measure the elastic modulus of crystalline cellulose. The mechanical properties of crystalline nanocellulose can compete with various current engineering materials such as steel and glass, especially considering density. A very high specific modulus of cellulose is due to the lower density of cellulose (1500 kg/m^3), allowing cellulose fibers to compete with current engineering reinforcement materials.

3.5 Conclusion

This chapter aims to study and understand the effect of structural properties on mechanical properties and how it changes with changes in treatment. The length scale of heterogeneity changes with and without the sonication of NC aqueous dispersion that can alter polymer nanocomposite's mechanical properties. Various mechanical properties depend on the degree of fibrillation of cellulose. There is a need to understand the basic molecular structure of fibers and their effect on mechanical properties for the economic and green use of cellulose at the nanoscale. The production of defect-free fibers or pure crystalline fibrils of different processing methods can improve the nanocrystalline cellulose and fibrils' mechanical properties. The mechanical properties of nanocellulose are excellent, but its practical use on an industrial scale is very difficult. It requires engineering and technology development to minimize production costs and the energy needed for its production. The mechanical degradation of cellulose in the nanoscale dimension requires an excessive amount of energy. It is still a big issue nowadays to study and understand its compatibility with other polymeric materials to form biocomposites. Chemical treatment for hydrolysis of cellulose is also time-consuming, hazardous, and requires a corrosion-resistant reactor. However, their mechanical properties can be enhanced by surface modification, the use of hydrophilic matrix, and crosslinking agents. The enzymatic and pre-oxidation treatment has been developed to minimize the energy needed for disintegration during mechanical treatment. It is useful in packaging, and the biomedical industries increased the interest of researchers in preparing nanocellulose-reinforced polymer composites. New opportunities can be opened by controlling the properties of nanocellulose during processing.

Acknowledgment

Meena Nemiwal is thankful to the Malaviya National Institute of Technology (MNIT), Jaipur, for its support.

REFERENCES

[1] Woodcock, Carrie, and Anatole Sarko. "Packing analysis of carbohydrates and polysaccharides: molecular and crystal structure of native ramie cellulose." *Macromolecules* 13 (1980): 1183–1187. doi:10.1021/ma60077a030.

[2] Sarko, Anatole, and Reto Muggli. "Packing analysis of carbohydrates and polysaccharides. III. Valonia cellulose and cellulose II." *Macromolecules* 7 (1974): 486–494. doi:10.1021/ma60040a016.

[3] Klemm, Dieter, Friederike Kramer, Sebastian Moritz, Tom Lindstroem, Mikael Ankerfors, Derek Gray, and Annie Dorris. "ChemInform abstract: Nanocelluloses: a new family of nature-based materials." *ChemInform* 42 (2011): doi:10.1002/chin.201138271.

[4] Sakurada, Ichiro, Yasuhiko Nukushina, and Taisuke Ito. "Experimental determination of the elastic modulus of crystalline regions in oriented polymers." *Journal of Polymer Science* 57 (1962): 651–660. doi:10.1002/pol.1962.1205716551.

[5] Abe, Kentaro, Shinichiro Iwamoto, and Hiroyuki Yano. "Obtaining cellulose nanofibers with a uniform width of 15 nm from wood." *Biomacromolecules* 8 (2007): 3276–3278. doi:10.1021/bm700624p.

[6] Nguyen, Long H., Sina Naficy, Rona Chandrawati, and Fariba Dehghani. "Nanocellulose for sensing applications." *Advanced Materials Interfaces* 6 (2019): 1900424. doi:10.1002/admi.201900424.

[7] Deepa, B., Eldho Abraham, Bibin M. Cherian, Alexander Bismarck, Jonny J. Blaker, Laly A. Pothan, Alcides L. Leao, Sivoney F. De Souza, and M. Kottaisamy. "Structure, morphology and thermal characteristics of banana nano fibers obtained by steam explosion." *Bioresource Technology* 102 (2011): 1988–1997. doi:10.1016/j.biortech.2010.09.030.

[8] Dahlem, Marcos A., Cleide Borsoi, Betina Hansen, and Andre L. Catto. "Evaluation of different methods for extraction of nanocellulose from yerba mate residues." *Carbohydrate Polymers* 218 (2019): 78–86. doi:10.1016/j.carbpol.2019.04.064.

[9] De Amorim, Julia D., Karina C. De Souza, Cybelle R. Duarte, Izarelle Da Silva Duarte, Francisco De Assis Sales Ribeiro, Girlaine S. Silva, and Patrícia M. De Farias, et al. "Plant and bacterial nanocellulose: production, properties and applications in medicine, food, cosmetics, electronics and engineering. A review." *Environmental Chemistry Letters* 18 (2020): 851–869. doi:10.1007/s10311-020-00989-9.

[10] Beck-Candanedo, Stephanie, Maren Roman, and Derek G. Gray. "Effect of reaction conditions on the properties and behavior of wood cellulose nanocrystal suspensions." *Biomacromolecules* 6 (2005): 1048–1054. doi:10.1021/bm049300p.

[11] Pereira, André L., Diego M. Nascimento, Men D. Souza Filho, João P. Morais, Niedja F. Vasconcelos, Judith P. Feitosa, Ana I. Brígida, and Morsyleide D. Rosa. "Improvement of polyvinyl alcohol properties by adding nanocrystalline cellulose isolated from banana pseudostems." *Carbohydrate Polymers* 112 (2014): 165–172. doi:10.1016/j.carbpol.2014.05.090.

[12] Ogundare, Segun A., Vashen Moodley, and Werner E. Van Zyl. "Nanocrystalline cellulose isolated from discarded cigarette filters." *Carbohydrate Polymers* 175 (2017): 273–281. doi:10.1016/j.carbpol.2017.08.008.

[13] Chen, You W., Hwei V. Lee, Joon C. Juan, and Siew-Moi Phang. "Production of new cellulose nanomaterial from red algae marine biomass *Gelidium elegans*." *Carbohydrate Polymers* 151 (2016): 1210–1219. doi:10.1016/j.carbpol.2016.06.083.

[14] Heux, L., G. Chauve, and C. Bonini. "Nonflocculating and chiral-nematic self-ordering of cellulose microcrystals suspensions in nonpolar solvents." *Langmuir* 16 (2000): 8210–8212. doi:10.1021/la9913957.

[15] Araki, Jun, Masahisa Wada, and Shigenori Kuga. "Steric stabilization of a cellulose microcrystal suspension by poly(ethylene glycol) grafting." *Langmuir* 17 (2001): 21–27. doi:10.1021/la001070m.

[16] Da Silva Perez, Denilson, Suzelei Montanari, and Michel R. Vignon. "TEMPO-mediated oxidation of cellulose III." *Biomacromolecules* 4 (2003): 1417–1425. doi:10.1021/bm034144s.

[17] Shinoda, Ryuji, Tsuguyuki Saito, Yusuke Okita, and Akira Isogai. "Relationship between length and degree of polymerization of TEMPO-oxidized cellulose nanofibrils." *Biomacromolecules* 13 (2012): 842–849. doi:10.1021/bm2017542.

[18] Satyamurthy, Prasad, Prateek Jain, Rudrapatna H. Balasubramanya, and Nadanathangam Vigneshwaran. "Preparation and characterization of cellulose nanowhiskers from cotton fibres by controlled microbial hydrolysis." *Carbohydrate Polymers* 83 (2011): 122–129. doi:10.1016/j.carbpol.2010.07.029.

[19] Pääkkö, M., M. Ankerfors, H. Kosonen, A. Nykänen, S. Ahola, M. Osterberg, and J. Ruokolainen, et al. "Enzymatic hydrolysis combined with mechanical shearing and high-pressure homogenization for nanoscale cellulose fibrils and strong gels." *Biomacromolecules* 8 (2007): 1934–1941. doi:10.1021/bm061215p.

[20] Uetani, Kojiro, and Hiroyuki Yano. "Nanofibrillation of wood pulp using a high-speed blender." *Biomacromolecules* 12 (2011): 348–353. doi:10.1021/bm101103p.

[21] Saito, Tsuguyuki, Satoshi Kimura, Yoshiharu Nishiyama, and Akira Isogai. "Cellulose nanofibers prepared by TEMPO-mediated oxidation of native cellulose." *Biomacromolecules* 8 (2007): 2485–2491. doi:10.1021/bm0703970.

[22] Wågberg, Lars, Gero Decher, Magnus Norgren, Tom Lindstrom, Mikael Ankerfors, and Karl Axnäs. "The build-up of polyelectrolyte multilayers of microfibrillated cellulose and cationic polyelectrolytes." *Langmuir* 24 (2008): 784–795. doi:10.1021/la702481v.

[23] Sacui, Iulia A., Ryan C. Nieuwendaal, Daniel J. Burnett, Stephan J. Stranick, Mehdi Jorfi, Christoph Weder, E. J. Foster, Richard T. Olsson, and Jeffery W. Gilman. "Comparison of the properties of cellulose nanocrystals and cellulose nanofibrils isolated from bacteria, tunicate, and wood processed using acid, enzymatic, mechanical, and oxidative methods." *ACS Applied Materials & Interfaces* 6 (2014): 6127–6138. doi:10.1021/am500359f.

[24] Kouadri, Imane, and Hamid Satha. "Extraction and characterization of cellulose and cellulose nanofibers from *Citrullus colocynthis* seeds." *Industrial Crops and Products* 124 (2018): 787–796. doi:10.1016/j.indcrop.2018.08.051.

[25] Chen, Wenshuai, Haipeng Yu, Sang-Young Lee, Tong Wei, Jian Li, and Zhuangjun Fan. "Nanocellulose: a promising nanomaterial for advanced electrochemical energy storage." *Chemical Society Reviews* 47 (2018): 2837–2872. doi:10.1039/c7cs00790f.

[26] Nishino, Takashi, Ikuyo Matsuda, and Koichi Hirao. "All-cellulose composite." *Macromolecules* 37 (2004): 7683–7687. doi:10.1021/ma049300h.

[27] Mariño, Mayra, Lucimara Lopes da Silva, Nelson Durán, and Ljubica Tasic. "Enhanced materials from nature: nanocellulose from citrus waste." *Molecules* 20 (2015): 5908–5923. doi:10.3390/molecules20045908.

[28] Li, Panpan, Juho A. Sirviö, Bright Asante, and Henrikki Liimatainen. "Recyclable deep eutectic solvent for the production of cationic nanocelluloses." *Carbohydrate Polymers* 199 (2018): 219–227. doi:10.1016/j.carbpol.2018.07.024.

[29] Ullah, Muhammad W., Sehrish Manan, Sabella J. Kiprono, Mazhar Ul-Islam, and Guang Yang. "Synthesis, structure, and properties of bacterial cellulose." *Nanocellulose*, (2019): 81–113. doi:10.1002/9783527807437.ch4.

[30] Jozala, Angela F., Leticia C. de Lencastre-Novaes, André M. Lopes, Valéria De Carvalho Santos-Ebinuma, Priscila G. Mazzola, Adalberto Pessoa Jr., Denise Grotto, Marli Gerenutti, and Marco V. Chaud. "Bacterial nanocellulose production and application: a 10-year overview." *Applied Microbiology and Biotechnology* 100 (2016): 2063–2072. doi:10.1007/s00253-015-7243-4.

[31] Ullah, Muhammad W., Mazhar Ul-Islam, Shaukat Khan, Yeji Kim, and Joong K. Park. "Structural and physico-mechanical characterization of bio-cellulose produced by a cell-free system." *Carbohydrate Polymers* 136 (2016): 908–916. doi:10.1016/j.carbpol.2015.10.010.

[32] Iguchi, M., S. Yamanaka, and A. Budhiono. "Bacterial cellulose: a masterpiece of nature's arts." *Journal of Materials Science* 35 (2000): 261–270. doi:10.1023/a:1004775229149.

[33] Ul-Islam, Mazhar, Taous Khan, and Joong K. Park. "Nanoreinforced bacterial cellulose–montmorillonite composites for biomedical applications." *Carbohydrate Polymers* 89 (2012): 1189–1197. doi:10.1016/j.carbpol.2012.03.093.

[34] Castro, Cristina, Robin Zuluaga, Catalina Alvarez, Jean-Luc Putaux, Gloria Caro, Orlando J. Rojas, Inaki Mondragon, and Piedad Ganan. "Bacterial cellulose produced by a new acid-resistant strain of *Gluconacetobacter* genus." *Carbohydrate Polymers* 89 (2012): 1033–1037. doi:10.1016/j.carbpol.2012.03.045.

[35] Khan, Shaukat, Mazhar Ul-Islam, Muhammad Ikram, Salman U. Islam, Muhammad W. Ullah, Muhammad Israr, Jae H. Jang, Sik Yoon, and Joong K. Park. "Preparation and structural characterization of surface modified microporous bacterial cellulose scaffolds: a potential material for skin regeneration applications *in vitro* and *in vivo*." *International Journal of Biological Macromolecules* 117 (2018): 1200–1210. doi:10.1016/j.ijbiomac.2018.06.044.

[36] De France, Kevin J., Todd Hoare, and Emily D. Cranston. "Review of hydrogels and aerogels containing nanocellulose." *Chemistry of Materials* 29 (2017): 4609–4631. doi:10.1021/acs.chemmater.7b00531.

[37] Panesar, Daman, Ramsey Leung, Mohini Sain, and Suhara Panthapulakkal. "The effect of sodium hydroxide surface treatment on the tensile strength and elastic modulus of cellulose nanofiber." *Sustainable and Nonconventional Construction Materials Using Inorganic Bonded Fiber Composites* (2017): 17–26. doi:10.1016/b978-0-08-102001-2.00002-4.

[38] Lee, Koon-Yang, and Alexander Bismarck. "Bacterial nanocellulose as reinforcement for polymer matrices." *Bacterial Nanocellulose* (2016): 109–122. doi:10.1016/b978-0-444-63458-0.00006-8.

[39] Iwamoto, Shinichiro, Weihua Kai, Akira Isogai, and Tadahisa Iwata. "Elastic modulus of single cellulose microfibrils from tunicate measured by atomic force microscopy." *Biomacromolecules* 10 (2009): 2571–2576. doi:10.1021/bm900520n.

[40] Šturcová, Adriana, Geoffrey R. Davies, and Stephen J. Eichhorn. "Elastic modulus and stress-transfer properties of tunicate cellulose whiskers." *Biomacromolecules* 6 (2005): 1055–1061. doi:10.1021/bm049291k.

[41] Leitner, Johannes, Barbara Hinterstoisser, Marnik Wastyn, Jozef Keckes, and Wolfgang Gindl. "Sugar beet cellulose nanofibril-reinforced composites." *Cellulose* 14 (2007): 419–425. doi:10.1007/s10570-007-9131-2.

[42] Svagan, Anna J., My A. Azizi Samir, and Lars A. Berglund. "Biomimetic polysaccharide nanocomposites of high cellulose content and high toughness." *Biomacromolecules* 8 (2007): 2556–2563. doi:10.1021/bm0703160.

[43] Wakabayashi, Moe, Shuji Fujisawa, Tsuguyuki Saito, and Akira Isogai. "Nanocellulose film properties tunable by controlling degree of fibrillation of TEMPO-oxidized cellulose." *Frontiers in Chemistry* 8 (2020). doi:10.3389/fchem.2020.00037.

[44] Thomas, Martin G., Eldho Abraham, P. Jyotishkumar, Hanna J. Maria, Laly A. Pothen, and Sabu Thomas. "Nanocelluloses from jute fibers and their nanocomposites with natural rubber: preparation and characterization." *International Journal of Biological Macromolecules* 81 (2015): 768–777. doi:10.1016/j.ijbiomac.2015.08.053.

[45] Mandal, Arup, and Debabrata Chakrabarty. "Studies on the mechanical, thermal, morphological and barrier properties of nanocomposites based on poly(vinyl alcohol) and nanocellulose from sugarcane bagasse." *Journal of Industrial and Engineering Chemistry* 20 (2014): 462–473. doi:10.1016/j.jiec.2013.05.003.

[46] Özgür Seydibeyoğlu, M., and Kristiina Oksman. "Novel nanocomposites based on polyurethane and micro fibrillated cellulose." *Composites Science and Technology* 68 (2008): 908–914. doi:10.1016/j.compscitech.2007.08.008.

[47] Lu, Yongshang, Lihui Weng, and Xiaodong Cao. "Morphological, thermal and mechanical properties of ramie crystallites: reinforced plasticized starch biocomposites." *Carbohydrate Polymers* 63 (2006): 198–204. doi:10.1016/j.carbpol.2005.08.027.

[48] Henriksson, Marielle, and Lars A. Berglund. "Structure and properties of cellulose nanocomposite films containing melamine formaldehyde." *Journal of Applied Polymer Science* 106 (2007): 2817–2824. doi:10.1002/app.26946.

[49] Nishino, Takashi, Kiyofumi Takano, and Katsuhiko Nakamae. "Elastic modulus of the crystalline regions of cellulose polymorphs." *Journal of Polymer Science Part B: Polymer Physics* 33 (1995): 1647–1651. doi:10.1002/polb.1995.090331110.

[50] Hinterstoisser, Barbara, Margaretha Åkerholm, and Lennart Salmén. "Effect of fiber orientation in dynamic FTIR study on native cellulose." *Carbohydrate Research* 334 (2001): 27–37. doi:10.1016/s0008-6215(01)00167-7.

[51] Tanaka, Fumio, and Tadahisa Iwata. "Estimation of the elastic modulus of cellulose crystal by molecular mechanics simulation." *Cellulose* 13 (2006): 509–517. doi:10.1007/s10570-006-9068-x.

[52] Cho, Mi-Jung, and Byung-Dae Park. "Tensile and thermal properties of nanocellulose-reinforced poly(vinyl alcohol) nanocomposites." *Journal of Industrial and Engineering Chemistry* 17 (2011): 36–40. doi:10.1016/j.jiec.2010.10.006.

[53] Hamad, Wadood Y., and Thomas Q. Hu. "Structure-process-yield interrelations in nanocrystalline cellulose extraction." *The Canadian Journal of Chemical Engineering* (2010): doi:10.1002/cjce.20298.

[54] Tsunashima, Yoshisuke, and Kimihiko Hattori. "Substituent distribution in cellulose acetates: its control and the effect on structure formation in solution." *Journal of Colloid and Interface Science* 228 (2000): 279–286. doi:10.1006/jcis.2000.6952.

[55] Lagerwall, Jan P., Christina Schütz, Michaela Salajkova, JungHyun Noh, Ji Hyun Park, Giusy Scalia, and Lennart Bergström. "Cellulose nanocrystal-based materials: from liquid crystal self-assembly and glass formation to multifunctional thin films." *NPG Asia Materials* 6 (2014): e80–e80. doi:10.1038/am.2013.69.

[56] Lahiji, Roya R., Xin Xu, Ronald Reifenberger, Arvind Raman, Alan Rudie, and Robert J. Moon. "Atomic force microscopy characterization of cellulose nanocrystals." *Langmuir* 26 (2010): 4480–4488. doi:10.1021/la903111j.

[57] Tashiro, Kohji, Masamichi Kobayashi, and Hiroyuki Tadokoro. "Elastic moduli and molecular structures of several crystalline polymers, including aromatic polyamides." *Macromolecules* 10 (1977): 413–420. doi:10.1021/ma60056a033.

4

Surface Chemistry of Nanocellulose and Its Composites

Jorge Padrão and Andrea Zille
University of Minho, Guimarães, Portugal

CONTENTS

4.1 Introduction

Cellulose was first described in 1838 by the French chemist Anselme Payen [1]. Cellulose is indubitably the most abundant biopolysaccharide on planet Earth. It is exclusively composed of D-glucose units (elemental chemical composition: $C_6H_{10}O_5$) covalently bounded by acetal bonds formed between the C1 atom carbon and the equatorial OH group of C4 (commonly named as β-1,4 glycosidic links). This molecule is defined as cellobiose (elemental chemical composition: $C_{12}H_{22}O_{11}$) (Fig. 4.1) [2]. Cellulose comprises at least 90 repeating units of cellobiose (30 kDa), forming an insoluble polymer because of its polymerization degree superior to 6. Cellulose encompasses amorphous domains and highly ordered regions. The ratio between amorphous and crystal domains has considerable variations, particularly among organisms, species, synthesis conditions, and circumstantial damage events. However, the exact factors determining the occurrence and abundance of the amorphous cellulose regions are still widely unknown. The cellulose protofibrils' assembly processes into cellulose itself is yet to be unraveled [3]. Cellulose displays three crystallization forms, or allomorphs: cellulose I, cellulose II, and cellulose III. Cellulose I is the most common allomorph present, comprising metastable parallel cellulose polysaccharide chains, and encompasses two sub-allomorphs: I_α and I_β. The I_α structure is composed of a single-chain triclinic cell unit. On the other hand, I_β comprises a two-chain monoclinic cell unit, being synthesized in its pure form only by tunicates. I_β has more thermodynamic stability than I_α [1, 2]. The ratio between I_α and I_β in cellulose depends on the organism species, namely I_α, which is dominant in algae and bacteria. In contrast, I_β is the principal sub-allomorph in higher plants. Cellulose II is constituted by antiparallel polysaccharide chains, being even more stable because of one extra hydrogen bound per glucose residue [3]. Cellulose II synthesis in nature is uncommon, only synthesized by some algae and bacteria [2]. Cellulose III allomorph, similarly to cellulose I, displays parallel chains. The addition of liquid ammonia to cellulose I and cellulose II will generate

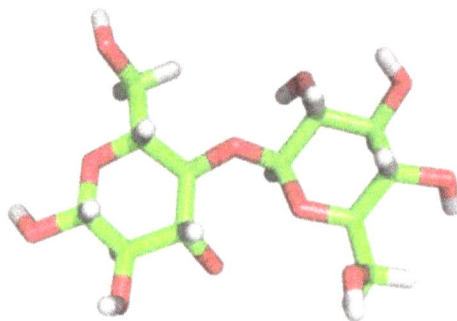

FIGURE 4.1 Cellobiose molecular structural formula. Atom color scheme: carbon—green, oxygen—red, and hydrogen—white. Image obtained using Pymol PyMOL™ software (DeLano Scientific LLC 2006) to highlight cellobiose present in the crystallographic image of cellobiose phosphorylase from *Cellulomonas uda* (Protein Data Bank file: 3S4A).

cellulose III$_I$ and III$_{II}$, respectively [4]. Despite the differences in hydrogen bonds, crystal structure, or chain orientation, allomorphs of cellulose comprise a rich surface of hydroxyl functional groups.

Anthropogenic applications containing cellulose date the most recondite of prehistorical times. Cellulose has been subjected to an interesting novel approach during the last five decades: nanosize cellulose, or nanocellulose. Nanosize materials (ranging from 1 nm to 100 nm in one dimension) encompass an undeniable set of advantages and enhanced properties since their shape and volume depict superior importance in their properties than just their absolute size [5]. Nanocellulose is commonly divided into five types: bacterial nanocellulose (BNC), cellulose nanocrystals (CNCs), cellulose nanofibrils (CNFs), electrospun nanocellulose (ESNC), and wet-spun nanocellulose (WSNC). In all cases, their nanosize considerably enhances the surface area. Thus, the availability of its surface hydroxyl groups is far superior. Despite its remarkable properties and wide range of applications, nanocellulose still possesses several limitations, which can be overcome by the selective combination with other materials. The combination of different materials generates a composite to achieve an end product with enhanced properties.

Composites are not novel, being the first known composites, the papyrus paper (4000 BC) and mud bricks reinforced using a straw (1300 BC). Interestingly, both composites contain cellulose in their formulation [6]. The first reports of nanocellulose used as a reinforcing material of a polymeric matrix depict the substantial improvement of the nanocomposite mechanical properties (up to twofold). To achieve similar results with regular macroscopic size cellulose as a reinforcing agent, considerably higher cellulose quantities had to be used, which would compromise the overall performance and viability of the composite [7]. For a more straight-forward comparison, Table 4.1 shows relevant properties of BNC, CNC, CNF, fibrous cellulose (cotton and wood), and para-aramid, better known through its commercial designation: Kevlar. It is observable that the different nanocelluloses are much closer to the aramid than to macroscopic cellulose. These notable features sparked a race for nanocomposite development encompassing nanocellulose from a plethora of different origins that contain plants, tunicates, algae, fungi, and bacteria. Each source generates nanocellulose with distinct properties and provides different surface modification levels and functionalization [8].

TABLE 4.1

Nanocellulose, Cellulose, and Para-Aramid Properties.

Cellulose Materials	Young's Modulus (GPa)	Density (g/cm³)	Crystallinity (%)	Degree of Polymerization	References
BNC	114	0.86	60–90	2000–8000	[1, 37, 60, 61]
CNC	130	Variable	71–96	250–350	[20, 62, 63]
CNF	100	Variable	77–86	750	[20, 64]
Para-aramid (Kevlar 49)	131	1.44	75	84	[65–68]
Cellulose fiber (cotton)	4	1.54	73	9000–15000	[69, 70]
Cellulose fiber (wood)	2	0.25	30	1190–1720	[69, 71]

4.2 Nanocellulose Surface and Nanocellulose Particularities

Cellulose and nanocellulose surface are ubiquitously densely populated with hydroxyl groups, independent of their origin or allomorph type, as depicted in cellulose monomer cellobiose (Fig. 4.1). In nanocellulose, the availability of the hydroxyl groups is extensively enhanced by the dramatic increase of the surface area compared to macroscopic cellulose. The availability of hydroxyl groups attached to carbon 1, 2, and 6 of each glucose monomer of cellobiose (named as O(2)H, O(3)H, and (O(6)H) depends on the extent of hydrogen bonding within the nanocellulose structure [9]. Therefore, cellulose allomorph and sub-allomorph types and abundance represent an important factor to consider not only in terms of hydroxyl group availability (where cellulose II has fewer hydroxyls available) but also for modification reactions complexity (cellulose I_α is the most easily modifiable) [10]. Cellulose I_α possesses a dislocation of cellulose sheets within the (110) lattice plane. Thus, cellulose I_α displaces +c/4 crystallographic point in each succeeding hydrogen-bonded cellulose sheet. At the same time, I_β sub-allomorph exhibits a displacement in the (200) lattice plane with hydrogen bonding sheets formed in alternating +c/4 and −c/4 crystallographic points [11]. Therefore, a brief description of BNC, CNC, and CNF is pivotal to discern the ideal nanocellulose to be used.

4.2.1 Bacterial Nanocellulose, Cellulose Nanocrystals, and Cellulose Nanofibrils Synthesis

BNC can be considered ideal for modifications. It does not require a delignification pretreatment and is mainly composed of sub-allomorph cellulose I_α, ranging from 70% to 60% whether it is synthesized in static or agitated culture conditions. BNC's detailed metabolomics is not yet fully unraveled due to its apparent complexity; several biosynthetic mechanisms were already soundly described [12, 13]. Briefly, BNC producing bacteria uses intracellular enzymatic machinery to produce the cellulose precursors uridine diphosphate glucose, subsequently assembled into cellulose in terminal enzymatic complexes usually at the outer envelope of the bacteria. Each BNC nanofiber produced presents a different thickness depending not only on the bacteria species but also on the culture environmental conditions, including temperature, culture medium formulation, dissolved oxygen concentration, light, and if the culture is performed in static or shaking conditions [13]. Therefore, BNC's width may vary between 10 nm and 100 nm and achieve a length of 50 nm to 100 μm [14]. The buildup of multiple fibrils eventually forms a macroscopic three-dimensional nanomesh containing pores ranging between 20 and 300 nm. BNC is often regarded as a hydrogel because of its high water holding capacity (99%) [15]. To remove bacterial cells, medium culture components, and other debris, BNC usually undergoes a mercerization treatment with sodium hydroxide, which reorganizes the packing of some BNC fibrils into allomorph cellulose type II and reduces the endotoxin level to approximately 1 EU/L. This value is comfortably below the threshold level defined by the Food and Drug Administration for materials that will contact with cerebrospinal fluid (60 EU/L) [16]. Therefore, BNC possesses staggering differences from plant cellulose in terms of purity, crystallinity, and size. BNC is comprised of 100% cellulose, without a single molecule of hemicellulose, lignin, or pectin in its formulation, and its crystallinity ranges between 75% and 90%. Due to its high crystallinity and nanometric architecture, BNC displays remarkable mechanical properties. All these features quickly caught the attention of many Research Center and manufacturers, and several applications were developed in a wide range of different areas (Table 4.2).

TABLE 4.2

BNC Applications.

Application Field	Commodity	Reference
Food	*Nata de coco*	[72]
Home appliances	High fidelity acoustic diaphragm	[73]
Medical	Artificial skin, artificial blood vessel	[74]
Wastewater treatment	Filtration membrane	[17]

The field of knowledge encompasses the most interesting medical field solutions since BNC allows an effective gas and nutrient diffusion, has no sensitization reports, and has notable biocompatibility. BNC structure semblances that of collagen may be a strong adjuvant for its biocompatibility [15]. BNC has also been recently used as a wastewater treatment filtration membrane, exhibiting the complete removal of soybean oil from a solution when BNC was incorporated in a polyvinyl chloride filtration system [17]. CNC can be roughly described as the collection and purification of crystalline cellulose sections as nanocellulose needles, or whiskers (because of similar dimensional size ratio) with a width of 3–25 nm and a length between 70 and 150 nm [18]. CNC appear as rigid needles, usually with very little agglomeration because of its synthesis process. CNC is obtained through the chemical hydrolysis of pure or delignified cellulose. Classical CNC production uses strong acids (particularly sulfuric acid or hydrochloric acid) to readily hydrolyze the amorphous cellulose sections and leave the crystalline regions intact solely because of their higher resistance to acid digestion [19]. Sulfuric acid digestion for periods superior to 30 minutes impedes CNC's clustering by providing a negative surface charge to the CNC. Otherwise, the abundant presence of hydroxyl groups would lead to their prompt aggregation [20] because of its high surface area. Strong acid use inevitably results in health safety and wastewater treatment issues, particularly on an industrial scale. Therefore, additional CNC production processes are now well established, namely, the use of enzymes, 2,2,6,6-tetramethylpiperidine-1-oxyl (TEMPO) oxidation, ionic liquids, and subcritical water [21]. The reported CNC properties include low polymerization due to their short size. Its high crystallinity index knows CNC. However, Tan and coworkers reported an impressive 96% crystallinity index when digested microcrystalline cellulose with an ionic liquid (3-methylimidazolium hydrogen sulfate). Besides, CNC solutions display liquid crystal properties, when assembled in a paper through vacuum filtration, they exhibited a visible light transmittance up to 90% [18]. Finally, to date, all studies focused on assessing the cytotoxicity of CNC shown negligible values and no genotoxicity [22]. In opposition to CNC that are synthesized through chemical treatment, CNF is obtained through using a mechanical process (with or without the combination of a chemical or enzymatic process). After cellulose delignification and bleaching, there is a vast plethora of processes to produce CNF, which include supermasscolloider, ball milling, blending, cryocrushing, extrusion, grinding, homogenization, microfluidization, steam explosion, and ultrasonication [23]. CNF produced commonly exhibit width between 20 and 60 nm and a length of 500 μm to more than 1 μm. The larger length of CNF in comparison to CNC makes CNF appear as an entangled mesh of nanofibers. All the referred CNF-producing methods consume relevant quantities of energy (more than 200 W/g of CNF), and some require the expenditure of considerable quantities of additional products, namely, liquid nitrogen for cryocrushing [22]. An additional major limitation of CNF production is the solid content yield, which leads to important limitations by reducing the feasibility of some processes and requiring larger storage areas and implying greater transportation costs. Only the twin-screw extruder can produce CNF with a high solid content (20%), whereas the remaining processes display a problematic low CNF solid content of less than 5% [24, 25]. Several strategies have been applied to reduce the energy requirements, which can be resumed as distinct pretreatments with a common goal: to improve the availability of the crystalline groups. These pretreatments include alkaline, enzymatic, and chemical pretreatments. Each has a particular limitation that must be pondered. Alkaline pretreatments usually generate hazardous wastewaters, and enzymatic treatments require long periods, and the most successful chemical pretreatment is mediated by TEMPO, which implies considerable purchase costs. However, CNF exhibits higher plasticity than CNC and enhanced rheological performance and good optical properties [26]. Interestingly, CNF displays a highly viscous behavior, even with low solid content (between 1 and 2%), making it a very effective edible viscosifying agent (without calories) for the food industry [22, 27]. Despite these impressive values per se, when used as reinforcing material in nanocomposites, its performance may drop to less than 0.05% and 4% of its Young's modulus and tensile strength, respectively [28]. Lack of nanofiber orientation may be a major parameter influencing these key mechanical properties. This is an important issue that has been recently approached during desynthesis of ESNC [29]. This chapter does not cover ESNC and WSNC and their composites. Despite their promising properties, the relevant CNC and CNF production costs currently drive their manufacture solely to high-end applications [30]. In addition, the optimization of their properties is still in its infancy. The authors would like to recommend a recent review on the subject [31].

4.2.2 Nanocellulose Surface Science

Each different type of nanocellulose displays interestingly distinct surface properties. As previously referred, BNC has more probability of exhibiting a higher concentration of hydroxyl groups at its surface. Atomic force microscopy (AFM) depicts the smooth and even fibrils of BNC with some occasional spanning, in opposition to the irregular surfaces of CNC and CNF [32–34]. In terms of surface area, using Barrett-Emmett-Teller (BET) method, at first glance, BNC, CNF, and CNC display a relatively low surface area: 5 m^2/g, 8 m^2/g, 0.4 m^2/g [35, 36]. This is most likely due to the drying process which removes the plasticizing effect of the water molecules and promotes the proximity of neighbor nanocellulose fibrils, endorsing additional hydrogen bonds, forming a tight structure [37]. Therefore, to perceive a more realistic surface area, nanocellulose aerogels must be analyzed. In simplistic terms, aerogels can be achieved through the sublimation of nanocellulose solution. Aerogel BNC surface, using BET, reached 160 m^2/g, CNC surface was estimated to be as high as 600 m^2/g [32, 38]. CNF aerogel displays a surface area of approximately 300 m^2/g [39]. Therefore, the difference between the surface area of the dried and aerogel nanocellulose is dramatic, being obvious that the aerogel architecture provides a value closer to reality. The total pore volume for each nanocellulose aerogel was estimated to be approximately the following: 0.2 cm^3/g for BNC, 10 cm^3/g, 0.9 cm^3/g for CNF [39–41]. The tenfold higher total pore volume of CNC is due to the large number of macropores [41]. However, surface areas and pore volumes of each value may present differences between works due to distinct production and measurement methodologies. All types of nanocellulose present a highly hydrophilic surface, with water contact angles ranging from approximately 20° (BNC) to 30° (CNC and CNF) [42–44].

Despite all the referred remarkable properties, nanocellulose still presents limited applicability due to its unanimous hydroxyl groups and hydrophilic surface. Therefore, not surprisingly, intense research and development have been being performed to modify the nanocellulose surface in innovative and ever more effective ways.

4.3 Nanocellulose Surface Modification and Nanocellulose Composites

Independent of each type of nanocelluloses, the ubiquitous chemical structure of their surface compiles a series of transversal surface modifications. These modifications may be directed to the hydroxyl groups or to break the β-D-anhydroglucopyranose (AUG) rings. Reported nanocellulose modifications include chemical oxidation [44], plasma treatment [45], acetylation [46], carboxymethylation [16], phosphorylation [47], cationization, ozonation, and sulfoethylation (Fig. 4.2) [24, 48].

4.3.1 Acid Hydrolysis

Acid hydrolysis may also be ascribed as nanocellulose modification due to the generation of several impurities (xylobiose, 3,4,5-trimethoxyphenol, 1,6-anhydroglucose, and vanillic acid) during sulfuric acid treatment for CNC production. This leads to the presence of sulfate esters, which considerably hinder the reproducibility of surface modifications [48].

4.3.2 Oxidation

Chemical oxidation is usually undertaken to obtain aldehyde and carboxylic groups on the cellulose surface. Periodate oxidation opens the AUG ring, forming an aldehyde in C2 and C3. These vicinal aldehydes are prompt to generate, among others, carbinolamines through nucleophilic attack of ε-NH_2 present in lysine of proteins. Therefore, this oxidation strategy represents a straightforward production of nanocellulose functionalized with proteins [44]. After periodate oxidation, chlorite oxidation may oxidize further the aldehydes formed, enhancing their stability. Ozonation may represent a less toxic procedure to generate aldehydes into nanocellulose and is easily scalable.

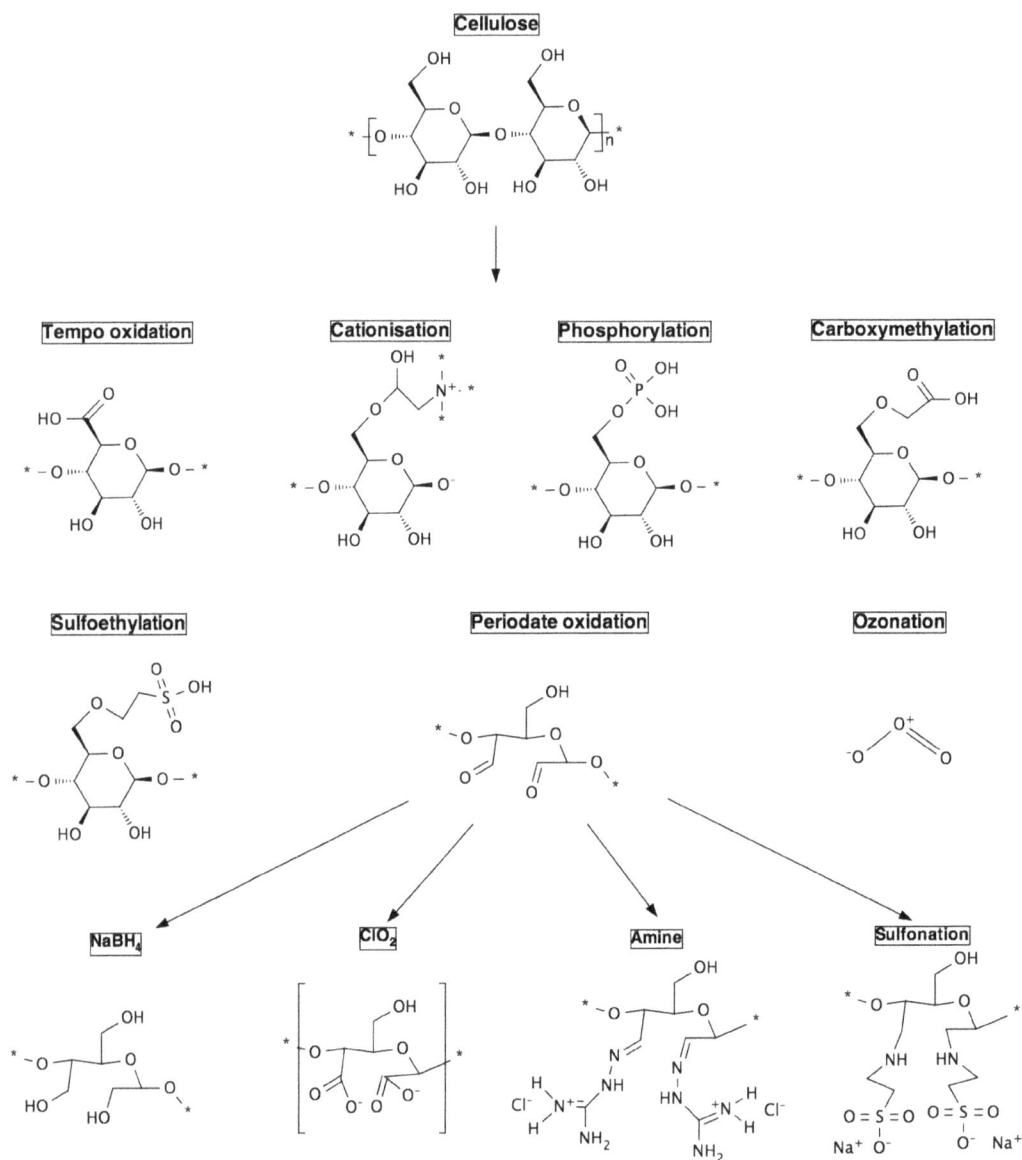

FIGURE 4.2 Nanocellulose surface modifications.

Plasma treatment also generates ozone which, at least partially, is responsible for nanocellulose oxidation. Ozone directly oxidizes nanocellulose surface or through the action of its radicals, and has two major limitations: it is difficult to precisely control its oxidation degree and usually results in an unsuitable loss of the degree of polymerization. TEMPO-mediated oxidation introduces a carboxylic group at C6 conferring a negative charge to the nanocellulose surface, it represents one of the most common nanocellulose treatment. This modification has been targeted for several optimizations to prevent the initial considerable loss of polymerization degree and to use cheaper and more environment-friendly compounds. The attempts to scale up this oxidation to industrial scale are still unforeseeable.

4.3.3 Sulfonation and Phosphorylation

The introduction of sulfite groups into nanocellulose surfaces is possible through the reaction of aldehydes produced during periodate oxidation and metabisulfite or taurine through sulfonation [49]. Phosphorylation of nanocellulose surface was also successfully reported using diammonium phosphate. Sulfonation and phosphorylation are two examples of esterification of nanocellulose and represent strategies to negatively charge the surface of nanocellulose [24, 48]. In particular, phosphorylation is applied to provide fire-retardant properties [47]. In opposition, cationization provides a positive charge to nanocellulose surface, conferring its antimicrobial properties. However, cationization still requires highly toxic reactants, and its process is highly complex; thus, several optimizations are still warranted.

4.3.4 Carboxymethylation

Carboxymethylation was first applied at the beginning of the twentieth century. However, it requires several complex processes and multiple hazardous chemicals. It is a reliable method to introduce a carboxyl group into the nanocellulose surface. It was successfully implemented at a pilot scale [24].

All the referred modifications generate anions or cations throughout the nanocellulose surface, promoting interplane repulsive forces, loosening nanocellulose hydrogen bonds and van der Waals interactions, improving the nanocellulose swelling, and further enhancing nanocellulose surface area. However, these charge-generated forces are subjugated to ambient variations, namely, to pH and medium ionic strength. This may limit the application or further treatments of the modified nanocellulose. Sulfoethylation of nanocellulose depicted an enhanced buffering capability to these important factors due to the sulfonate group's low pKa values. Moreover, sulfoethylation exhibits a good scale-up potential [24, 48]. The referred nanocellulose surface modifications represent the doorway to a plethora of further modifications able to provide an impressive number of novel capabilities to nanocellulose. Surface-modified nanocellulose may be functionalized through the adsorption of active molecules, polymer grafting and molecules grafting, and novel nanocomposite development. BNC, CNC, and CNF display several composite applications as observable in Tables 4.3–4.5, respectively.

BNC composites cover a wide range of applications. Nevertheless, its production costs until the beginning of the century promoted a focused development of biomedical applications, where the application-justifiable higher prices could easily cover the production costs. Recently, BNC production costs have been considerably lowered by successfully using low-cost formulations comprising ethanol, steep corn liquor, molasses, and ammonium sulfate and still maintain a high yield (approximately 6.5 g/L), and by using supplemented banana peel medium [50–52]. In addition, the wastewater generated during the production of BNC, the remaining culture medium after fermentation, and the washing wastewaters depicted an attractive biogas production potential, which can further mitigate production costs [53]. The BNC lower production costs provided sustainable development of textile and filtration applications, which possess less strict requirements than biomedical devices. BNC exhibited "waterproof" without hampering breathability when saturated with a common fabric hydrophobizer, and was successfully used as a matrix of a nanocomposite able to mimic leather [54, 55]. To authors' knowledge, BNC use as a filtration device started roughly 20 years ago [56, 57]. Actelayated pulverized BNC was used as a membrane for removing copper ions through polyelectrolyte-enhanced ultrafiltration, and the BNC-graphene nanocomposite membrane was able to selectively permeate inorganic ions [57, 58]. Finally, BNC has recently been used to synthesize Voronoi-nanonets when welded to electrospun nanonets of polyacrylonitrile, displaying a rejection efficiency of nearly 100% titanium oxide microparticles of 0.3 μm [59]. These ultrathin membranes possessed high porosity and exhibited promising antifouling properties and reusability, all highly relevant features for the effective and sustainable water filtration process. As for CNC, it is quite clear that CNC is commonly used as a reinforcing material of construction, packaging, and wearable devices, displaying a clear shift from biomedical applications to a common and widespread commodities application. Nevertheless, the major limitation of CNC is

TABLE 4.3

Nanocellulose Composites.

Application	Nanocellulose	Matrix	Filler	Nanocellulose Surface Modification	Objective	Reference
Electronic	*Ex-situ*	BNC	Multiwalled carbon nanotubes and polyaniline	None	Flexible, supercapacitor	[75]
Electronic device	*Ex-situ*	BNC	Nano-graphite	None	Conductive and enhanced thermal properties	[76]
Electronic device	*Ex-situ*	BNC	Palladium	None	Electrode	[77]
Electronic device	*Ex-situ*	BNC	Multiwalled carbon nanotubes and laccase	Cyanoethylation	Electrode capable of degrading recalcitrant pollutants and generate electricity	[78]
Filtration membrane	*Ex-situ*	BNC	Graphene	None	Selectively permeation	[58]
Filtration membrane	*Ex-situ*	BNC	Polyvinyl alcohol and biochar with nanosilver	None	Potential filtration membrane with bacteriostatic properties	[79]
Filtration membrane	*In-situ*	BNC	Palladium-decorated mesoporous polydopamine	None	Dye removal	[80]
Food film	*Ex-situ*	BNC	Lactoferrin	None	Antibacterial activity	[37]
Medical device	*In-situ /ex-situ*	BNC	Silver nanoparticles	None	Wound dressing with anti-bacterial activity	[81]
Medical device	*Ex-situ*	BNC	Catalase	None	Wound dressing with anti-bacterial activity	[82]
Medical device	*Ex-situ*	BNC	Benzalkonium chloride	None	Antibacterial activity	[83]
Medical device	*Ex-situ*	BNC	Chitosan and carboxymethyl cellulose	Carboxymethy-lation	Corneal regeneration	[16]
Medical device	*Ex-situ*	BNC	Urinary bladder matrix	Acetylation	Corenal regeneration	[46]
Medical device	*Ex-situ*	BNC	Hydroxyapatite	None	Bone regeneration	[84]
Pharmaceutical	*Ex- situ*	BNC	Ibuprofen	None	Controlled drug release	[85]
Paper	*Ex-situ*	Cotton line pulp	BNC pulp	None	High-quality paper	[72]
Paper	*Ex-situ*	BNC	Europium	None	Resilient florescent paper	[86]
Paper	*In-situ*	BNC	Phosphate	None	Fire retardant	[47]
Textile	*Ex-situ*	BNC	Acrylated epoxidized soybean oil	None	Leather replacement	[55]
Textile	*Ex-situ*	BNC	Polydimethylsiloxane and perfluorocarbon	None	Hydrophobization	[54]

TABLE 4.4

Recent CNC Nanocomposites.

Application	Matrix	Filler	Nanocellulose Surface Modification	Objective	Reference
Biodegradable polymer	Poly(vinyl alcohol)	CNC	None	Petrochemical-based plastic replacement	[87]
Packaging and construction	Poly(butylene succinate)	CNC	Acetylation	Enhanced thermal insulation	[88]
Packing material	CNC	Nanochitin and heptadecafluoro-1,1,2,2-tetrahydrodecyl dimethylchlorosilane-modified nano-SiO$_2$	TEMPO oxidation	Superhydrophobic and transparent	[89]
Packaging material	Poly(vinyl alcohol)	CNC	Citric acid	Food-active film	[90]
Bioremediation	CNC	Polyvinylamine	Periodate oxidation	Pesticide adsorption	[91]
Agronomy	Starch, poly (acrylic acid (AA)-*co*-acrylamide and polyvinyl alcohol	CNC	None	Superabsorbent hydrogel for agronomical applications	[92]
Pharmaceutical	CNC	Glycerol and polyethylene glycol	None	Drug capsule	[93]
Electronic device	Pig skin	CNC and carbon nanotubes	TEMPO oxidation	Flexible sensor and supercapacitor electrode	[94]
Pharmaceutical	Alginate	CNC	None	Drug carrier	[95]
Textile and electronic devices	Poly(vinyl alcohol)	CNC	None	Wearable devices	[96]

TABLE 4.5

Recent CNC Nanocomposites.

Application	Matrix	Filler	CNF Production Equipment	Nanocellulose Surface Modification	Objective	Reference
Food	Beeswax-in-water	CNF and carboxymethyl chitosan	High-pressure homogenizer	None	Edible food coating	[97]
Food	Mayonnaise	CNF	Microfluidizer	None	Viscosifying agent	[27]
Packaging	Poly(vinyl alcohol)	CNF and graphene oxide	High-pressure homogenizer	TEMPO oxidation	Ultraviolet shield	[98]
Bioremediation	CNF	L-Methionine	Steam explosion	Methionine grafting	Mercury adsorption	[99]
Active packaging	CNF	Nisin	Disk refiner nano-grinder	None	Antibacterial and low oxygen and vapor permeability	[100]

(Continued)

TABLE 4.5 *(Continued)*

Recent CNC Nanocomposites.

Application	Matrix	Filler	CNF Production Equipment	Nanocellulose Surface Modification	Objective	Reference
Security paper	CNF	Cadmium/ selenium/sulfur quantum dots	Ultrasonicator	None	1D ink application and 2D film preparation	[101]
Smart packaging	CNF	Chitosan	High-pressure homogenizer	None	Enhanced barrier and antibacterial properties	[102]
Medical device	CNF	Carbon spheres containing silver nanoparticles	Not available (commercially purchased)	None	Antibacterial paper	[103]

its brittleness, which remains to be surpassed. CNF displays an important role in food industry as a viscosifying agent or a food package reinforcement, and additional emerging applications are focused on low cost-effective bioremediation.

4.4 Conclusion

Cellulose is ubiquitous and a key material of the circular economy in a world that urgently struggles to reduce its dependence on petrochemical-based commodities. Nanocellulose, independent of its origin, modification, or production process, will play an important role in this transition. CNC can summarize the existing main nanocellulose limitations i.e. low yield and hazardous chemical production. CNF and BNC have limitations such as high production costs and inadequate production time for its industrial demand, respectively, which must be progressively mitigated. Despite CNC and CNF possess an additional advantage over BNC, which corresponds to the already established industrial infrastructures, completely adapted to process their main sources. The biotechnological potential of BNC should not be underestimated.

The estimates for the market size before the COVID-19 pandemic depicted the overwhelming expansion of nanocellulose: it was estimated to increase its production over 35 million metric tons per year and reach a value of more than half a billion euros by 2023. Naturally, this value requires a revision on its estimate; however, it may not represent a downturn. Environmental awareness of general population is constantly increasing, and sustainable products for everyday commodities and bioremediation and pollution control are ever more commendable. These facts allied to the current overwhelming need for highly efficient filtration materials for personal protective equipment, air purification units, and wastewater treatment without further compromising the environment can be a strong drive to widen further the massive application of nanocellulose per se and in the composite formulation. The particular development of nanocomposites with Voronoi-nanonets design using BNC exhibits a high water permeability flux and can be designed to achieve an impressive small particle rejection.

REFERENCES

[1] Klemm, Dieter, Brigitte Heublein, Hans-Peter Fink, and Andreas Bohn. "Cellulose: fascinating biopolymer and sustainable raw material." *ChemInform* 36 (2005): 3358–3393. doi:10.1002/chin.200536238.

[2] Saxena, Inder M., and R. Lcolm Brown. "Cellulose biosynthesis: current views and evolving concepts." *Annals of Botany* 96 (2005): 9–21. doi:10.1093/aob/mci155.

[3] Brown, R. M. "Cellulose structure and biosynthesis: what is in store for the 21st century?" *Journal of Polymer Science Part A: Polymer Chemistry* 42 (2003): 487–495. doi:10.1002/pola.10877.

[4] Wada, Masahisa, Henri Chanzy, Yoshiharu Nishiyama, and Paul Langan. "Cellulose III crystal structure and hydrogen bonding by synchrotron X-ray and neutron fiber diffraction." *Macromolecules* 37 (2004): 8548–8555. doi:10.1021/ma0485585.

[5] Paul, Donald R., and Lioyd M. Robeson. "Polymer nanotechnology: nanocomposites." *Polymer* 49 (2008): 3187–3204. doi:10.1016/j.polymer.2008.04.017.

[6] Herakovich, Carl T, "Mechanics of composites: a historical review." *Mechanics Research Communications* 41 (2012): 1–20. doi:10.1016/j.mechrescom.2012.01.006.

[7] Favier, V., G. R. Canova, J. Y. Cavaille, H. Chanzy, A. Dufresne, and C. Gauthier. "Nanocomposite materials from latex and cellulose whiskers." *Polymers for Advanced Technologies* 6 (1995): 351–355. doi:10.1002/pat.1995.220060514.

[8] Shak, Katrina Pui Yee, Yean Ling Pang, and Shee Keat Mah. "Nanocellulose: recent advances and its prospects in environmental remediation." *Beilstein Journal of Nanotechnology* 9 (2018): 2479–2498. doi:10.3762/bjnano.9.232.

[9] Rowland, Stanley P., and Phyllis S. Howley. "Hydrogen bonding on accessible surfaces of cellulose from various sources and relationship to order within crystalline regions." *Journal of Polymer Science Part A: Polymer Chemistry* 26 (1988): 1769–1778. doi:10.1002/pola.1988.080260708.

[10] Ling, Zhe, Sheng Chen, Xueming Zhang, Keiji Takabe, and Feng Xu. "Unraveling variations of crystalline cellulose induced by ionic liquid and their effects on enzymatic hydrolysis." *Scientific Reports* 7 (2017): 10230. doi:10.1038/s41598-017-09885-9.

[11] Poletto, Matheus, Vinícios Pistor, and Ademir J Zattera, "Structural characteristics and thermal properties of native cellulose." In *Cellulose: fundamental aspects*, eds. T. van de Ven and L. Godbout, 45–68. London, UK: IntechOpen. 2013.

[12] Rehm, Bernd. "Bacterial cellulose production: biosynthesis and applications." In *Microbial production of biopolymers and polymer precursors: applications and perspectives*, ed. B. Rehm, Norfolk: Caister Academic Press. 2009.

[13] Jacek, Paulina, Fernando Dourado, Miguel Gama, and Stanisław Bielecki. "Molecular aspects of bacterial nanocellulose biosynthesis." *Microbial Biotechnology* 12 (2019): 633–649. doi:10.1111/1751-7915.13386.

[14] Nagashima, Aya, Tsubasa Tsuji, and Tetsuo Kondo. "A uniaxially oriented nanofibrous cellulose scaffold from pellicles produced by *Gluconacetobacter xylinus* in dissolved oxygen culture." *Carbohydrate Polymers* 135 (2016): 215–224. doi:10.1016/j.carbpol.2015.08.077.

[15] Torres, Fernando, Solene Commeaux, and Omar Troncoso. "Biocompatibility of bacterial cellulose based biomaterials." *Journal of Functional Biomaterials* 3 (2012): 864–878. doi:10.3390/jfb3040864.

[16] Gonccalves, Sara, Jorge Padrao, Inees Patricio Rodrigues, Joao Pedro Silva, Vítor Sencadas, Senentxu Lanceros-Mendez, Henrique Giraao, Fernando Dourado, and Ligia R. Rodrigues. "Correction to bacterial cellulose as a support for the growth of retinal pigment epithelium." *Biomacromolecules* 16 (2015): 4032–4032. doi:10.1021/acs.biomac.5b01556.

[17] Galdino Jr, Claudio Jose S., Alexandre D. Maia, Hugo M. Meira, Thais C. Souza, Julia DP Amorim, Fabiola CG Almeida, Andrea FS Costa, and Leonie A. Sarubbo. "Use of a bacterial cellulose filter for the removal of oil from wastewater." *Process Biochemistry* 91 (2020): 288–296. doi:10.1016/j.procbio.2019.12.020.

[18] George, Johnsy, and Sabapathi S N. "Cellulose nanocrystals: synthesis, functional properties, and applications." *Nanotechnology, Science and Applications* 8 (2015): 45–54. doi:10.2147/nsa.s64386.

[19] Habibi, Youssef, Lucian A. Lucia, and Orlando J. Rojas. "Cellulose nanocrystals: chemistry, self-assembly, and applications." *Chemical Reviews* 110 (2010): 3479–3500. doi:10.1021/cr900339w.

[20] Dufresne, Alain. "Nanocellulose: a new ageless bionanomaterial." *Materials Today* 16 (2013): 220–227. doi:10.1016/j.mattod.2013.06.004.

[21] Novo, Lisias Pereira, Julien Bras, Araceli Garcia, Naceur Belgacem, and Antonio Aprigio da Silva Curvelo. "A study of the production of cellulose nanocrystals through subcritical water hydrolysis." *Industrial Crops and Products* 93 (2016): 88–95. https://doi.org/10.1016/j.indcrop.2016.01.012

[22] Ventura, Celia, Fatima Pinto, Ana F. Lourenco, Paulo J. Ferreira, Henriqueta Louro, and Maria J. Silva. "On the toxicity of cellulose nanocrystals and nanofibrils in animal and cellular models." *Cellulose* 27 (2020): 5509–5544. doi:10.1007/s10570-020-03176-9.

[23] Nagano, Takao, Yuya Arai, Hiromi Yano, Takafumi Aoki, Shin Kurihara, Rika Hirano, and Katsuyoshi Nishinari. "Improved physicochemical and functional properties of okara, a soybean residue, by nanocellulose technologies for food development: a review." *Food Hydrocolloids* (2020): 105964. doi:10.1016/j.foodhyd.2020.105964.

[24] Rol, Fleur, Mohamed N. Belgacem, Alessandro Gandini, and Julien Bras. "Recent advances in surface-modified cellulose nanofibrils." *Progress in Polymer Science* 88 (2019): 241–264. doi:10.1016/j.progpolymsci.2018.09.002.

[25] Trigui, Khadija, Clement De Loubens, Albert Magnin, Jean-Luc Putaux, and Sami Boufi. "Cellulose nanofibrils prepared by twin-screw extrusion: effect of the fiber pretreatment on the fibrillation efficiency." *Carbohydrate Polymers* 240 (2020): 116342. doi:10.1016/j.carbpol.2020.116342.

[26] Kandhola, Gurshagan, Angele Djioleu, Kalavathy Rajan, Nicole Labbe, Joshua Sakon, Danielle Julie Carrier, and Jin-Woo Kim. "Maximizing production of cellulose nanocrystals and nanofibers from pre-extracted loblolly pine kraft pulp: a response surface approach." *Bioresources and Bioprocessing* 7 (2020): 1–16. doi:10.1186/s40643-020-00302-0.

[27] Heggset, Ellinor B., Ragnhild Aaen, Trinelise Veslum, Marielle Henriksson, Sebastien Simon, and Kristin Syverud. "Cellulose nanofibrils as rheology modifier in mayonnaise: a pilot scale demonstration." *Food Hydrocolloids* 108 (2020): 106084. doi:10.1016/j.foodhyd.2020.106084.

[28] Hernandez, S., C. A. Brebbia, and W.P. D. Wilde. *High performance and optimum design of structures and materials II.* Ashurst Lodge: WIT Press. 2016.

[29] Kalantari, Mahsa, Mark McDermott, Cagri Ayranci, and Yaman Boluk. "Fabrication of oriented electrospun cellulose nanocrystals–polystyrene composite fibers on a rotating drum." *Journal of Applied Polymer Science* 137 (2020): 48942. doi:10.1002/app.48942.

[30] Teixeira, Marta A., Maria C. Paiva, M. T. Amorim, and Helena P. Felgueiras. "Electrospun nanocomposites containing cellulose and its derivatives modified with specialized biomolecules for an enhanced wound healing." *Nanomaterials* 10 (2020): 557. doi:10.3390/nano10030557.

[31] Niinivaara, Elina, and Emily D. Cranston. "Bottom-up assembly of nanocellulose structures." *Carbohydrate Polymers* 247 (2020): 116664. doi:10.1016/j.carbpol.2020.116664.

[32] Brinkmann, Andreas, Maohui Chen, Martin Couillard, Zygmunt J. Jakubek, Tianyang Leng, and Linda J. Johnston. "Correlating cellulose nanocrystal particle size and surface area." *Langmuir* 32 (2016): 6105–6114. doi:10.1021/acs.langmuir.6b01376.

[33] Guhados, Ganesh, Wankei Wan, and Jeffrey L. Hutter. "Measurement of the elastic modulus of single bacterial cellulose fibers using atomic force microscopy." *Langmuir* 21 (2005): 6642–6646. doi:10.1021/la0504311.

[34] Hai, Le V., Hyun C. Kim, Abdullahil Kafy, Lindong Zhai, Jung W. Kim, and Jaehwan Kim. "Green all-cellulose nanocomposites made with cellulose nanofibers reinforced in dissolved cellulose matrix without heat treatment." *Cellulose* 24 (2017): 3301–3311. doi:10.1007/s10570-017-1333-7.

[35] Abu-Danso, Emmanuel, Varsha Srivastava, Mika Sillanpaa, and Amit Bhatnagar. "Pretreatment assisted synthesis and characterization of cellulose nanocrystals and cellulose nanofibers from absorbent cotton." *International Journal of Biological Macromolecules* 102 (2017): 248–257. doi:10.1016/j.ijbiomac.2017.03.172.

[36] Chandrasekaran, Prathna T., Naimat K. Bari, and Sharmistha Sinha. "Enhanced bacterial cellulose production from *Gluconobacter xylinus* using super optimal broth." *Cellulose* 24 (2017): 4367–4381. doi:10.1007/s10570-017-1419-2.

[37] Padrao, Jorge, Sara Gonçalves, Joao P. Silva, Vitor Sencadas, Senentxu Lanceros-Mendez, Ana C. Pinheiro, Antonio A. Vicente, Ligia R. Rodrigues, and Fernando Dourado. "Bacterial cellulose-lactoferrin as an antimicrobial edible packaging." *Food Hydrocolloids* 58 (2016): 126–140. doi:10.1016/j.foodhyd.2016.02.019.

[38] Sai, Huazheng, Rui Fu, Li Xing, Junhui Xiang, Zhenyou Li, Fei Li, and Ting Zhang. "Surface modification of bacterial cellulose aerogels' web-like skeleton for oil/water separation." *ACS Applied Materials & Interfaces* 7 (2015): 7373–7381. doi:10.1021/acsami.5b00846.

[39] Rafieian, Fatemeh, Maleksadat Hosseini, Mehdi Jonoobi, and Qingliang Yu. "Development of hydrophobic nanocellulose-based aerogel via chemical vapor deposition for oil separation for water treatment." *Cellulose* 25 (2018): 4695–4710. doi:10.1007/s10570-018-1867-3.

[40] Liang, Hai-Wei, Zhen-Yu Wu, Li-Feng Chen, Chao Li, and Shu-Hong Yu. "Bacterial cellulose derived nitrogen-doped carbon nanofiber aerogel: an efficient metal-free oxygen reduction electrocatalyst for zinc-air battery." *Nano Energy* 11 (2015): 366–376. doi:10.1016/j.nanoen.2014.11.008.

[41] Heath, Lindy, and Wim Thielemans. "Cellulose nanowhisker aerogels." *Green Chemistry* 12 (2010): 1448. doi:10.1039/c0gc00035c.

[42] Dankovich, Theresa A., and Derek G. Gray. "Contact angle measurements on smooth nanocrystalline cellulose (I) thin films." *Journal of Adhesion Science and Technology* 25 (2011): 699–708. doi:10.1163/016942410x525885.

[43] Jonoobi, Mehdi, Jalaluddin Harun, Aji P. Mathew, Mohd Z. Hussein, and Kristiina Oksman. "Preparation of cellulose nanofibers with hydrophobic surface characteristics." *Cellulose* 17 (2009): 299–307. doi:10.1007/s10570-009-9387-9.

[44] Padrao, Jorge, Sylvie Ribeiro, Senentxu Lanceros-Mendez, Ligia R. Rodrigues, and Fernando Dourado. "Effect of bacterial nanocellulose binding on the bactericidal activity of bovine lactoferrin." *Heliyon* 6 (2020): e04372. doi:10.1016/j.heliyon.2020.e04372.

[45] Bhanthumnavin, W., P. Wanichapichart, W. Taweepreeda, S. Sirijarukula, and B. Paosawatyanyong. "Surface modification of bacterial cellulose membrane by oxygen plasma treatment." *Surface and Coatings Technology* 306 (2016): 272–278. doi:10.1016/j.surfcoat.2016.06.035.

[46] Goncalves, Sara, Inês Patricio Rodrigues, Jorge Padrao, Joao Pedro Silva, Vitor Sencadas, Senentxu Lanceros-Mendez, Henrique Girao, Francisco M. Gama, Fernando Dourado, and Lígia R. Rodrigues. "Acetylated bacterial cellulose coated with urinary bladder matrix as a substrate for retinal pigment epithelium." *Colloids and Surfaces B: Biointerfaces* 139 (2016): 1–9. doi:10.1016/j.colsurfb.2015.11.051.

[47] Basta, A. H., and H. El-Saied. "Performance of improved bacterial cellulose application in the production of functional paper." *Journal of Applied Microbiology* 107 (2009): 2098–2107. doi:10.1111/j.1365-2672.2009.04467.x.

[48] Eyley, Samuel, and Wim Thielemans. "Surface modification of cellulose nanocrystals." *Nanoscale* 6 (2014): 7764–7779. doi:10.1039/C4NR01756K.

[49] Sirvio, Juho A., Aleksi Kolehmainen, Miikka Visanko, Henrikki Liimatainen, Jouko Niinimaki, and Osmo E. Hormi. "Strong, self-standing oxygen barrier films from nanocelluloses modified with regioselective oxidative treatments." *ACS Applied Materials & Interfaces* 6 (2014): 14384–14390. doi:10.1021/am503659j.

[50] Rodrigues, Ana Cristina, Ana Isabel Fontao, Aires Coelho, Marta Leal, Francisco A. G. Soares da Silva, Yizao Wan, Fernando Dourado, and Miguel Gama. "Response surface statistical optimization of bacterial nanocellulose fermentation in static culture using a low-cost medium." *New Biotechnology* 49 (2019): 19–27. doi:10.1016/j.nbt.2018.12.002.

[51] Sijabat, Edwin K., Ahmad Nuruddin, Pingkan Aditiawati, and Bambang S. Purwasasmita. "Synthesis and characterization of bacterial nanocellulose from banana peel for water filtration membrane application." *Journal of Physics: Conference Series* 1230 (2019): 012085. doi:10.1088/1742-6596/1230/1/012085.

[52] Sijabat, Edwin K., Ahmad Nuruddin, Pingkan Aditiawati, and Bambang Sunendar Purwasasmita. "Optimization on the synthesis of bacterial nano cellulose (BNC) from banana peel waste for water filter membrane applications." *Materials Research Express* 7 (2020): 055010. doi:10.1088/2053-1591/ab8df7.

[53] Da Silva, Francisco A., Joao V. Oliveira, Catarina Felgueiras, Fernando Dourado, Miguel Gama, and M. M. Alves. "Study and valorisation of wastewaters generated in the production of bacterial nanocellulose." *Biodegradation* 31 (2020): 47–56. doi:10.1007/s10532-020-09893-z.

[54] Fernandes, Marta, Miguel Gama, Fernando Dourado, and Antonio P. Souto. "Development of novel bacterial cellulose composites for the textile and shoe industry." *Microbial Biotechnology* 12 (2019): 650–661. doi:10.1111/1751-7915.13387.

[55] Fernandes, Marta, Antonio P. Souto, Miguel Gama, and Fernando Dourado. "Bacterial cellulose and emulsified AESO biocomposites as an ecological alternative to leather." *Nanomaterials* 9 (2019): 1710. doi:10.3390/nano9121710.

[56] Tiongson, J. M., E. C. Bugante, and E. J. Del Rosario. "Development of ultrafiltration membranes from bacterial cellulose (nata de coco) for the separation of mango volatile organic compounds." *Philippine Agricultural Scientist* 85 (2002): 256–265.

[57] Espiritu, R. D., R. R. Navarro, and E. J. Del Rosario. "Development of membranes from bacterial cellulose (nata de coco) for the separation of copper (II) ions using polyelectrolyte-enhanced ultrafiltration." *Philippine Agricultural Scientist* 87 (2004): 87–95.

[58] Fang, Qile, Xufeng Zhou, Wei Deng, Zhi Zheng, and Zhaoping Liu. "Freestanding bacterial cellulose-graphene oxide composite membranes with high mechanical strength for selective ion permeation." *Scientific Reports* 6 (2016): 33185. doi:10.1038/srep33185.

[59] Tang, Ning, Shichao Zhang, Yang Si, Jianyong Yu, and Bin Ding. "An ultrathin bacterial cellulose membrane with a Voronoi-net structure for low pressure and high flux microfiltration." *Nanoscale* 11 (2019): 17851–17859. doi:10.1039/c9nr06028f.

[60] Hsieh, Y.C., H. Yano, M. Nogi, and S. J. Eichhorn. "An estimation of the Young's modulus of bacterial cellulose filaments." *Cellulose* 15 (2008): 507–513. doi:10.1007/s10570-008-9206-8.

[61] Tabuchi, Mari. "Nanobiotech versus synthetic nanotech?" *Nature Biotechnology* 25 (2007): 389–390. doi:10.1038/nbt0407-389.

[62] Tan, Xiao Y., Sharifah B. Abd Hamid, and Chin W. Lai. "Preparation of high crystallinity cellulose nanocrystals (CNCs) by ionic liquid solvolysis." *Biomass and Bioenergy* 81 (2015): 584–591. doi:10.1016/j.biombioe.2015.08.016.

[63] Li, Xia, Jun Li, Jie Gong, Yishan Kuang, Lihuan Mo, and Tao Song. "Cellulose nanocrystals (CNCs) with different crystalline allomorph for oil in water pickering emulsions." *Carbohydrate Polymers* 183 (2018): 303–310. doi:10.1016/j.carbpol.2017.12.085.

[64] Correa, Ana C., Eliangela De Morais Teixeira, Luiz A. Pessan, and Luiz H. Mattoso. "Cellulose nanofibers from curaua fibers." *Cellulose* 17 (2010), 1183–1192. doi:10.1007/s10570-010-9453-3.

[65] Kopeliovich, Dmitri. "Kevlar (aramid) fiber reinforced polymer." https://www.substech.com/dokuwiki/doku.php?id=kevlar_aramid_fiber_reinforced_polymers Last modified May 30, 2020.

[66] Kong, Haijuan, Qian Xu, and Muhuo Yu. "Microstructural changes of aramid fiber due to reaction with toluene 2,4-diisocyanate under tension in ScCO$_2$." *Polymers* 11 (2019): 1110. doi:10.3390/polym11071110.

[67] Timm, D. C., A. J. Ayorinde, C. H. Lee, L. F. Steele, and N. C. Plass. "Kevlar 49 composite performance: dependence on thermoset resin microstructure." *Polymer Engineering and Science* 24 (1984): 930–935. doi:10.1002/pen.760241113.

[68] Young, H. H., *Kevlar aramid fiber*. Hoboken, NJ: Wiley, 1999.

[69] Kompella, M. K, and J. Lambros. "Micromechanical characterization of cellulose fibers." *Polymer Testing* 21 (2002): 523–530. doi:10.1016/s0142-9418(01)00119-2.

[70] Barnhardt. "Properties of cotton." https://barnhardtcotton.net/technology/cotton-properties/ Last modified May 30, 2020.

[71] Sweet, M. S., and J. E. Winandy. "Influence of degree of polymerization of cellulose and hemicellulose on strength loss in fire-retardant-treated southern pine." *Holzforschung* 53 (1999): 311–317. doi:10.1515/hf.1999.051.

[72] Yamanaka, S., K. Watanabe, N. Kitamura, M. Iguchi, S. Mitsuhashi, Y. Nishi, and M. Uryu. "The structure and mechanical properties of sheets prepared from bacterial cellulose." *Journal of Materials Science* 24 (1989): 3141–3145. doi:10.1007/bf00584917.

[73] Uryu, Masaru, and Noboru Kurihara. "Acoustic diaphragm and method for producing same." U.S. Patent, 1993.

[74] Schumann, Dieter A., Jens Wippermann, Dieter O. Klemm, Friederike Kramer, Daniel Koth, Hartwig Kosmehl, Thorsten Wahlers, and Schariar Salehi-Gelani. "Artificial vascular implants from bacterial cellulose: preliminary results of small arterial substitutes." *Cellulose* 16 (2009): 877–885. doi:10.1007/s10570-008-9264-y.

[75] Li, Shaohui, Dekang Huang, Bingyan Zhang, Xiaobao Xu, Mingkui Wang, Guang Yang, and Yan Shen. "Electrodes: flexible supercapacitors based on bacterial cellulose paper electrodes." *Advanced Energy Materials* 4 (2014). doi:10.1002/aenm.201470050.

[76] Erbas Kiziltas, Esra, Alper Kiziltas, Kevin Rhodes, Nuri W. Emanetoglu, Melanie Blumentritt, and Douglas J. Gardner. "Electrically conductive nano graphite-filled bacterial cellulose composites." *Carbohydrate Polymers* 136 (2016): 1144–1151. doi:10.1016/j.carbpol.2015.10.004.

[77] Evans, Barbara R., Hugh M. O'Neill, Valerie P. Malyvanh, Ida Lee, and Jonathan Woodward. "Palladium-bacterial cellulose membranes for fuel cells." *Biosensors and Bioelectronics* 18 (2003): 917–923. doi:10.1016/s0956-5663(02)00212-9.

[78] Li, Xin, Pengfei Lv, Yixin Yao, Quan Feng, Alfred Mensah, Dawei Li, and Qufu Wei. "A novel single-enzymatic biofuel cell based on highly flexible conductive bacterial cellulose electrode utilizing pollutants as fuel." *Chemical Engineering Journal* 379 (2020): 122316. doi:10.1016/j.cej.2019.122316.

[79] Zhang, Liang, Sen Zheng, Zhihui Hu, Lvling Zhong, Yao Wang, Xiaomin Zhang, and Juanqin Xue. "Preparation of polyvinyl alcohol/bacterial-cellulose-coated biochar–nanosilver antibacterial composite membranes." *Applied Sciences* 10 (2020): 752. doi:10.3390/app10030752.

[80] GholamiDerami, Hamed, Prashant Gupta, Rohit Gupta, Priya Rathi, Jeremiah J. Morrissey, and Srikanth Singamaneni. "Palladium nanoparticle-decorated mesoporous polydopamine/bacterial nano-cellulose as a catalytically active universal dye removal ultrafiltration membrane." *ACS Applied Nano Materials* 3 (2020): 5437–5448. doi:10.1021/acsanm.0c00787.

[81] Eardley, William GP, Sarah A. Watts, and Jon C. Clasper. "Extremity trauma, dressings, and wound infection: should every acute limb wound have a silver lining?" *The International Journal of Lower Extremity Wounds* 11 (2012): 201–212. doi:10.1177/1534734612457028.

[82] Sampaio, Liliana M. P., Jorge Padrao, Jorge Faria, Joao P. Silva, Carla J. Silva, Fernando Dourado, and Andrea Zille. "Laccase immobilization on bacterial nanocellulose membranes: antimicrobial, kinetic and stability properties." *Carbohydrate Polymers* 145 (2016): 1–12. doi:10.1016/j.carbpol.2016.03.009.

[83] Wei, Bin, Guang Yang, and Feng Hong. "Preparation and evaluation of a kind of bacterial cellulose dry films with antibacterial properties." *Carbohydrate Polymers* 84 (2011): 533–538. doi:10.1016/j.carbpol.2010.12.017.

[84] Tazi, Neftaha, Ze Zhang, Younès Messaddeq, Luciana Almeida-Lopes, Lisinéia M. Zanardi, Dennis Levinson, and Mahmoud Rouabhia. "Hydroxyapatite bioactivated bacterial cellulose promotes osteoblast growth and the formation of bone nodules." *Amb Express* 2 (2012): 1–10. doi:10.1186/2191-0855-2-61.

[85] Jiji, Swaminathan, Sukumar Thenmozhi, and Krishna Kadirvelu. "Comparison on properties and effi-ciency of bacterial and electrospun cellulose nanofibers." *Fibers and Polymers* 19 (2018): 2498–2506. doi:10.1007/s12221-018-8527-y.

[86] Zhang, Mingquan, Xiao Wu, Zhenhua Hu, Zhouyang Xiang, Tao Song, and Fachuang Lu. "A highly efficient and durable fluorescent paper produced from bacterial cellulose/Eu complex and cellulosic fibers." *Nanomaterials* 9 (2019): 1322. doi:10.3390/nano9091322.

[87] Sonker, Amit K., Naveen Tiwari, Rajaram K. Nagarale, and Vivek Verma. "Synergistic effect of cel-lulose nanowhiskers reinforcement and dicarboxylic acids crosslinking towards polyvinyl alcohol prop-erties." *Journal of Polymer Science Part A: Polymer Chemistry* 54 (2016): 2515–2525. doi:10.1002/pola.28129.

[88] Yin, Dexian, Jianguo Mi, Hongfu Zhou, Xiangdong Wang, and Huafeng Tian. "Fabrication of branch-ing poly(butylene succinate)/cellulose nanocrystal foams with exceptional thermal insulation." *Carbohydrate Polymers* 247 (2020): 116708. doi:10.1016/j.carbpol.2020.116708.

[89] Xu, Junfei, Xiaolei Deng, Yunyuan Dong, Zhaozhong Zhou, Yuliang Zhang, Jianping Yu, Jianchen Cai, and Yuanxiang Zhang. "High-strength, transparent and superhydrophobic nanocellulose/nanochitin membranes fabricated via crosslinking of nanofibers and coating F-SiO$_2$ suspensions." *Carbohydrate Polymers* 247 (2020): 116694. doi:10.1016/j.carbpol.2020.116694.

[90] Yang, Weijun, Xiaoyan He, Francesca Luzi, Weifu Dong, Ting Zheng, Jose M. Kenny, Debora Puglia, and Piming Ma. "Thermomechanical, antioxidant and moisture behaviour of PVA films in presence of citric acid esterified cellulose nanocrystals." *International Journal of Biological Macromolecules* 161 (2020). doi:10.1016/j.ijbiomac.2020.06.082.

[91] Yang, Jie, Chengxiao Ma, Junhong Tao, Junfeng Li, Keqing Du, Zhen Wei, Cuizhong Chen, Zhaoyang Wang, Chun Zhao, and MingGuo Ma. "Optimization of polyvinylamine-modified nanocellulose for chlorpyrifos adsorption by central composite design." *Carbohydrate Polymers* 245 (2020): 116542. doi:10.1016/j.carbpol.2020.116542.

[92] Olad, Ali, Fatemeh Doustdar, and Hamed Gharekhani. "Fabrication and characterization of a starch-based superabsorbent hydrogel composite reinforced with cellulose nanocrystals from potato peel waste." *Colloids and Surfaces A: Physicochemical and Engineering Aspects* 601 (2020): 124962. doi:10.1016/j.colsurfa.2020.124962.

[93] Zhang, Yapei, Qing Zhao, Huashan Wang, Xingyu Jiang, and Ruitao Cha. "Preparation of green and gelatin-free nanocrystalline cellulose capsules." *Carbohydrate Polymers* 164 (2017): 358–363. doi:10.1016/j.carbpol.2017.01.096.

[94] Wu, Ying, Simiao Sun, Aobo Geng, Linjie Wang, Chi Song, Lijie Xu, Chong Jia, Jiangtao Shi, and Lu Gan. "Using TEMPO-oxidized-nanocellulose stabilized carbon nanotubes to make pigskin hydrogel conductive as flexible sensor and supercapacitor electrode: inspired from a Chinese cuisine." *Composites Science and Technology* 196 (2020): 108226. doi:10.1016/j.compscitech.2020.108226.

[95] Zhao, Junchai, Shichao Li, Yujing Zhao, and Zheng Peng. "Effects of cellulose nanocrystal polymorphs and initial state of hydrogels on swelling and drug release behavior of alginate-based hydrogels." *Polymer Bulletin* 77 (2019): 4401–4416. doi:10.1007/s00289-019-02972-z.

[96] Wu, Xinfeng, Yuan Gao, Hang Yao, Kai Sun, Runhua Fan, Xiaofeng Li, Yan An, Yanhua Lei, and Yuliang Zhang. "Flexible and transparent polymer/cellulose nanocrystal nanocomposites with high thermal conductivity for thermal management application." *Journal of Applied Polymer Science* 137 (2020): 48864. doi:10.1002/app.48864.

[97] Xie, Bing, Xingzhong Zhang, Xiaogang Luo, Yixiang Wang, Yan Li, Bin Li, and Shilin Liu. "Edible coating based on beeswax-in-water pickering emulsion stabilized by cellulose nanofibrils and carboxymethyl chitosan." *Food Chemistry* (2020): 127108. doi:10.1016/j.foodchem.2020.127108.

[98] Jia, Yuanyuan, Chunrui Hu, Peidong Shi, Qianqian Xu, Wenjing Zhu, and Rui Liu. "Effects of cellulose nanofibrils/graphene oxide hybrid nanofiller in PVA nanocomposites." *International Journal of Biological Macromolecules* 161 (2020): 223–230. doi:10.1016/j.ijbiomac.2020.06.013.

[99] Bisla, Vivek, Gaurav Rattan, Sonal Singhal, and Anupama Kaushik. "Green and novel adsorbent from rice straw extracted cellulose for efficient adsorption of Hg (II) ions in an aqueous medium." *International Journal of Biological Macromolecules* 161 (2020): 194–203. doi:10.1016/j.ijbiomac.2020.06.035.

[100] Yang, Yang, Handong Liu, Min Wu, Jinxia Ma, and Peng Lu. "Bio-based antimicrobial packaging from sugarcane bagasse nanocellulose/nisin hybrid films." *International Journal of Biological Macromolecules* 161 (2020): 627–635. doi:10.1016/j.ijbiomac.2020.06.081.

[101] Zhao, Yadong, and Jiebing Li. "Unique and outstanding quantum dots (QD)/tunicate cellulose nanofibrils (TCNF) nanohybrid platform material for use as 1D ink and 2D film." *Carbohydrate Polymers* 242 (2020): 116396. doi:10.1016/j.carbpol.2020.116396.

[102] Wu, Meiyan, Prakit Sukyai, Dong Lv, Fang Zhang, Peidong Wang, Chao Liu, and Bin Li. "Water and humidity-induced shape memory cellulose nanopaper with quick response, excellent wet strength and folding resistance." *Chemical Engineering Journal* 392 (2020): 123673. doi:10.1016/j.cej.2019.123673.

[103] Jiang, Qimeng, Bichong Luo, Zhengguo Wu, and Xiaoying Wang. "Antibacterial composite paper with corn stalk-based carbon spheres immobilized AgNPs." *Materials Science and Engineering: C* 113 (2020): 111012. doi:10.1016/j.msec.2020.111012.

5

Chemical Modification of Nanocellulose for Water Purification

Subrata Mondal
National Institute of Technical Teachers' Training and Research (NITTTR), Kolkata, India

CONTENTS

5.1 Introduction

Water plays an important role in every aspect of human activities; however, it is becoming scarce in various parts of the planet [1]. Nowadays, water pollution by various toxic contaminants is an alarming global concern. Rapid industrialization, significant growth of the worldwide population, and extended droughts have been the main causes of water pollution. For both human and industrial uses, clean water is essential [2, 3]. At present, the world faces significant challenges in meeting the growing demand of clean water for human consumption, industrial applications, and agricultural applications. Industrial processes require an enormous amount of clean water. Simultaneously, many industries produce large amounts of effluent contaminated with various pollutants, such as toxic heavy metals, color, order, organic materials, etc. New materials and technology need to be developed and scale up to meet the growing demand for purified water. Advances in nanotechnology research provide an excellent opportunity to develop cost-effective and environment-friendly water purification materials [4].

Nanoscale materials have significantly altered mechanical, optical, electrical, magnetic, and other functional properties compared with their conventional counterpart [5]. The excellent properties of nanomaterials, including high specific surface area and higher reactivity—hence less quantity required to achieve desired functionality—have enabled them for a wide range of applications, including various water treatment applications [5–9]. Applications of nanomaterials for water treatment are promising. They can be used in adsorption, catalysts, nanomembrane filtration, and nanosensors to detect contaminants in the polluted water. Water quality can be greatly improved by using nanotechnology approaches to purify water [6, 8, 10, 11]. Recently, research interest in preparing nanomaterials from renewable resources has been growing.

Cellulose is one of the most available natural organic polymers on Earth. There are four major sources of nanocellulose (NC) in nature: plants, bacteria, algae, and marine animals [12]. Cellulose molecule contains units of β-D-glucopyranose, which are joined by 1-4-β-glycosidic bonds [13]. Cellulosic nanomaterials can be commonly categorized into three classes: (i) nanofibrillated cellulose (NFC), (ii) cellulose nanocrystal (CNC), and (iii) bacterial nanocellulose (BNC) [14–17]. NFC is the primary constituent of parent cellulose. It can be manufactured from the lignocellulosic biomass by mainly two steps: pretreatment of lignocellulosic biomass and defibrillation of pretreated lignocellulose by mechanical means, e.g., by using a high-pressure homogenizer. CNC can be obtained from the partially controlled hydrolysis of CNF by removing the amorphous region [18–22]. Therefore, CNC has a lower aspect ratio as compared with NFC. Chauhan et al. reported CNC by sulfuric acid hydrolysis of microcrystalline cellulose from cotton linter. Cotton linters were treated with 64% sulfuric acid at 45 °C for 45 minutes. The average width and length of synthesized nanocellulose were 39 ± 7 nm and 150 ± 53 nm, respectively [22]. BNC can be synthesized by specific bacterial strains in an artificial culture environment [23]. BNC can be synthesized by the cultivation of *Gluconacetobacter xylinus* in a nutrient medium [24–29]. Bacteria-synthesized nanocelluloses are chemically pure and structurally significantly different as compared with CNF and CNC [13].

Nanomaterials derived from natural resources, e.g., lignocellulosic biomass, have unique advantages [30]. Nanocellulose has several advantages: low cost compared with other nanomaterials, renewability, wide availability, lightweight compared to metal-based nanoparticles, presence of hydroxyl groups on its surface, etc. [31]. Plenty of hydroxyl groups on the nanocellulose surface enable their surface modification with suitable molecules to alter the surface properties [13]. Applications of nanocellulose for water purification are quite a new research area. Native nanocellulose and surface-functionalized nanocellulose can remove various toxic contaminants from an aqueous medium [30]. The unique surface properties of nanocellulose offer the material scientists enormous possibilities to modify the nanocellulose with a wide range of molecules to tailor the properties. An appropriate selection of molecules can alter the surface properties of nanocellulose for surface modification. The tailored surface properties of nanocellulose can allow selective adsorption of molecules from the polluted water to remove various pollutants. Further, surface-functionalized nanocellulose can fabricate novel membranes for water purification. This chapter presents an overview of surface-modified nanocellulose for water purification. Chemical modifications of nanocellulose with hydrophilic and hydrophobic functional groups are discussed. Surface-modified nanocellulose as adsorption and membrane-based separations are taken into account.

5.2 Chemical Modifications of Nanocellulose

Nanomaterials have at least one dimension up to 100 nm. As per the definition of nanomaterials, nanocellulose has at least one dimension less than or equal to 100 nm. Some important features of nanocellulose include high aspect ratio (length/width ratio), high specific contact surface due to the nanodimension, the existence of −OH groups on the nanocellulose surface, excellent mechanical properties, biodegradability, low cost, availability, and sustainability [16]. Nanocellulose has Young's modulus from 100 to 130 GPa with a surface area of a few hundred m^2/g [32], and NC shows higher crystallinity than the source lignocellulosic biomass. Chemical modification of nanocellulose with various molecules depends on surface functional hydroxyl groups, surface morphology, and shape/size of nanocellulose. Acid hydrolysis time can influence the morphology of CNC. Jiang and Hsieh reported duration of sulfuric acid treatment on morphology of CNC. Transmission electron microscopic (TEM) images revealed that short-duration (e.g., 15 minutes) acid hydrolysis showed more heterogeneous morphology. With increasing hydrolysis time, more uniform CNC can be obtained. Fig. 5.1 shows the surface morphology of CNCs produced by using different acid hydrolysis duration (with sulfuric acid). Increasing acid hydrolysis time also reduces the width and length of nanocellulose [33].

The surface of pristine nanocellulose is incompatible with many organic and/or inorganic contaminants in wastewater. Chemical compatibility of nanocellulose and various organic and/or inorganic contaminants from the wastewater is important. Surface properties of nanocellulose can be

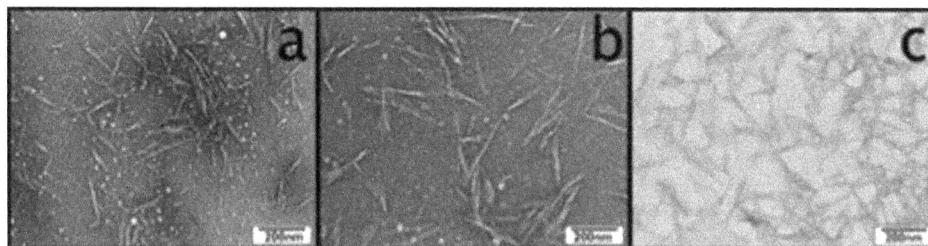

FIGURE 5.1 Transmission electron microscopic images of CNCs produced by three different durations of acid hydrolysis. (a) 15 minutes. (b) 30 minutes. (c) 45 minutes (Reproduced with permission from Ref. [33]. © Elsevier 2013).

engineered by attachments of various molecules on the nanocellulose surface by various direct reactions involving surface hydroxyl groups on nanocellulose. The existence of enormous —OH groups on the nanocellulose surface provides the material scientist an opportunity to functionalize nanocellulose with a wide range of molecules. Each unit of cellulose contains three —OH groups on its surface. Nanocellulose can be surface modified by oxidation (to introduce carboxyl or aldehyde functionality), esterification, amidation (to form amide products), carbamation (to modify with isocyanates), silylation, acetylation, acylation/alkanoylation, etherification, polymer grafting, etc. [31, 32, 34, 35]. Among these various functional groups, nanocellulose surfaces are widely modified with carboxylic and amino groups to be used for the water purification process. Nanocellulose has a high specific surface area, low density (~1.6 g/cc), and numerous reactive hydroxyl groups that enable grafting of various molecules to tailor the surface properties. Surface functionality of nanocellulose can be broadly categorized into three groups: (i) native surface chemistry as a result of nanocellulose extraction, (ii) physical (such as van der Waals force, electrostatic attraction, etc.) adsorption to the nanocellulose surface, and (iii) covalent attachment of specific molecules to the nanocellulose surface [36]. Covalent surface functionalization of nanocellulose involved attachments of various molecules on the nanocellulose surface by direct reactions involving the hydroxyl surface functional groups [12]. Nanocellulose can be surface functionalized by grafting of desirable molecule, e.g., acrylamide, on the nanocellulose surface [37].

5.2.1 Carboxylic Surface Functionalization of Nanocellulose

Espino-Pérez et al. presented an eco-friendly process of CNC modification with carboxylic functionality by solvent-free esterification. Two non-toxic carboxylic acids (CAs), phenylacetic acid and hydrocinnamic acid, were used for the surface functionalization. In this approach, CA not only acts as surface grafting but also as a solvent media above their melting point. The reaction mechanism and process are depicted in Fig. 5.2. Carboxylic surface functionalization did not much affect the properties of base nanocellulose [38].

5.2.2 Surface Functionalization of Nanocellulose with Amino Groups

Amino groups can be attached with nanocellulose surface via reaction with surface hydroxyl groups and various amino-functional groups containing molecules. Vivod et al. reported surface functionalization of cellulose nanofibers (CNFs) with hexamethylenediamine (HMDA). HMDA can be attached with CNF by two steps: sodium periodate oxidation of CNF to connected aldehyde functionality, followed by reaction with HMDA by a Schiff-base reaction to form CNF-ald-HMDA [39]. Khanjanzadeh et al. modified the surface of CNCs with 3-aminopropyltriethoxysilane (APTES) by an eco-friendly method. In this method, APTES was hydrolyzed into water and absorbed on the CNC surface by hydrogen bonding. Finally, the molecule was attached to the CNC surface via Si—O—C bonds through a chemical condensation reaction. The whole process is schematically shown in Fig. 5.3 [40]. Robles et al. reported surface functionalization of CNFs with APTES. Amino groups of APTES react with hydroxyl functionality of CNFs to make the nanocellulose surface hydrophobic [41].

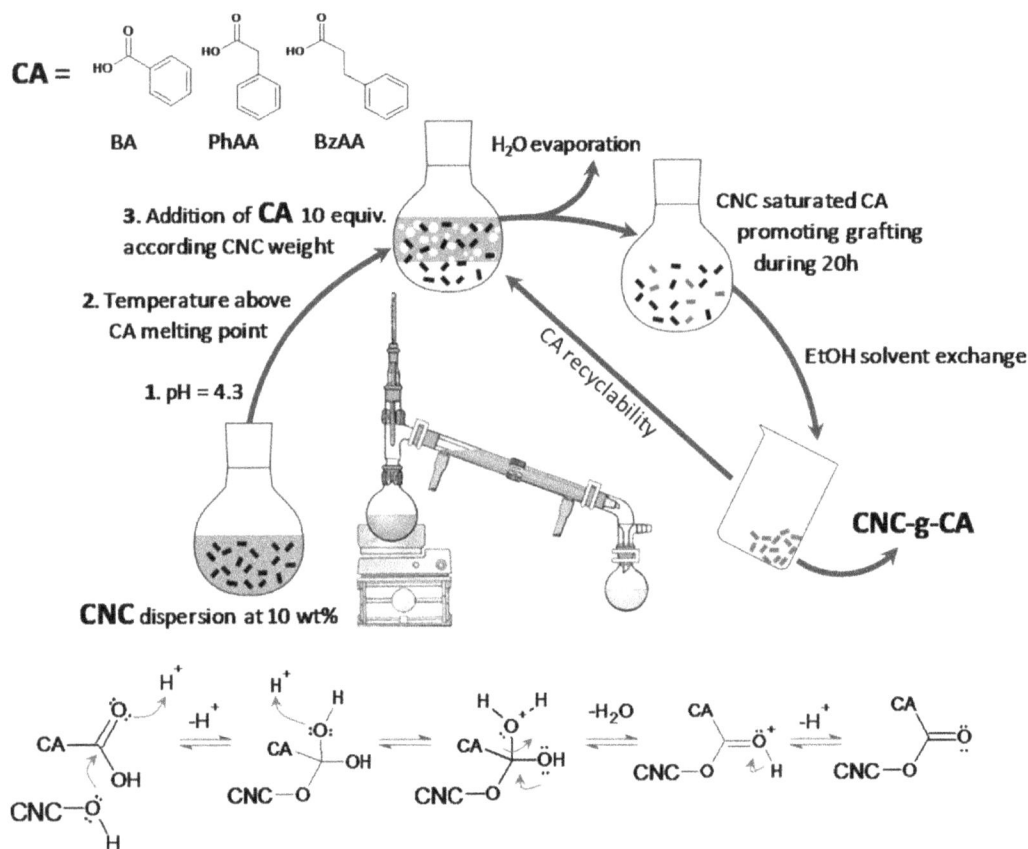

FIGURE 5.2 Schematic showing surface functionalization of cellulose nanocrystal with carboxylic acid by an eco-friendly approach (Reproduced with permission from Ref. [38]. © American Chemical Society 2014).

5.2.3 Surface Functionalization of Nanocellulose with Other Molecules

Nanocelluloses are also required to functionalize surfaces with other desirable molecules to remove specific contaminants from the wastewater. Phosphorylation reaction on the nanocellulose surface can be carried out by esterification of −OH group of nanocellulose with phosphoric acid molecules, and the phosphoric acid can be attached to the cellulose chains via phosphate group by one ester group [42]. Korhonen et al. reported surface-functionalized nanocellulose aerogel for the separation of oil from water-oil mixture. Nanofibrils of nanocellulose aerogel were functionalized with nanolayer of TiO_2 to

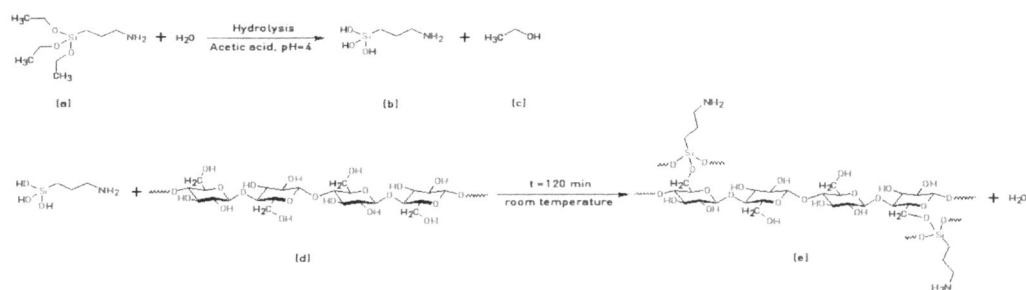

FIGURE 5.3 Schematic showing surface functionalization of cellulose nanocrystal by silane coupling compound. (a) APTES. (b) Silanol. (c) Ethanol. (d) CNC. (e) Modified CNC (Reproduced with permission from Ref. [40]. © Elsevier 2017).

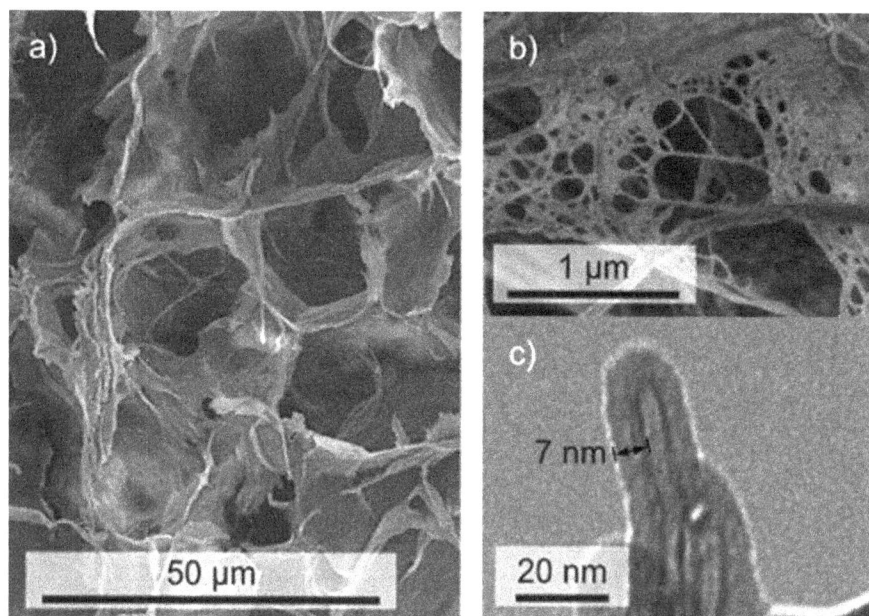

FIGURE 5.4 (a and b) Scanning electron microscopic (SEM) images of freeze-dried nanocellulose aerogels at different magnifications. (c) Transmission electron microscopic image of nanolayer TiO_2-coated fibrils (Reproduced with permission from Ref. [43]. © American Chemical Society 2011).

hydrophobic but oleophilic coatings. Fig. 5.4a and b shows the scanning electron microscopic images of aerogel, and Fig. 5.4c shows the TEM image of CNFs coated with TiO_2 nanolayer. Nanocellulose aerogels were formed by vacuum freeze-drying of nanocellulose hydrogels. Nanolayer of TiO_2 coated on CNFs was performed by atomic layer deposition technique [43].

5.3 Chemically Modified Nanocellulose for Water Purification

Nanocellulose is promising alternative adsorbents compared with other nanoadsorbents such as carbon-based nanoadsorbents, metal oxide nanoadsorbents, and polymer-based nanoadsorbents. Hydroxyl surface functional groups of nanocellulose attract a wide range of pollutants from contaminated water. The surface functional hydroxyl groups allow surface modifications of nanocellulose with various chemical moieties, enhancing the binding efficiency of contaminants with nanocellulose. Inherent hydrophilicity of nanocellulose can also be altered to improve the affinity of modified nanocellulose toward hydrophobic pollutants from the wastewater [44]. Surface-modified nanocellulose can be used for adsorption and membrane separation in water purification.

5.3.1 Chemically Modified Nanocellulose as Adsorbents

Adsorbents are widely used for the removal of various impurities from contaminated water. Adsorption is an accumulation of molecules or ions on a surface in contact with the water phase. Molecules or ions that are being adsorbed on the adsorbent surface is known as the adsorbate. During the adsorption process, adhesion of contaminants from liquid to the adsorbent surface would occur, and the process creates a thin layer of the adsorbate on the adsorbent surface. Therefore, adsorption is termed as a surface phenomenon [45, 46]. Unique surface properties of nanocellulose allow the material scientist to modify nanocellulose surface with various molecules. Surface-modified nanocellulose can be used as adsorbents for the removal of various contaminants from the wastewater. Original nanocelluloses adsorb a lower quantity of metal ions due to their less negative charge concentration. Carboxylation treatment enhanced the metal ion adsorption due to the oxidation treatment, which introduced carboxylate groups on the nanocellulose surface. Attached carboxylate groups provide negative sites for the electrostatic attraction of metal ions [47]. Yu et al. reported nanocellulose/graphene oxide composite by covalent bonding of nanocellulose with functionalized graphene oxide via an esterification reaction. Novel adsorbents were used for the removal of heavy metal ions such as nickel ion (Ni(II)).

Surface-modified adsorbent showed significant improvement of adsorption capacity as compared with unmodified nanocellulose-based adsorbent. Improvement of heavy metal adsorption capacity is due to the various functional groups (such as —C=O, —COOH, —OH) and adsorbent porosity [48]. Shahnaz et al. reported surface-modified nanocellulose with polypyrrole to remove chromium and Congo red dye from a binary mixture. Fig. 5.5 shows the mechanism of surface modification schematically for nanocellulose with polypyrrole and the adsorption mechanism of chromium ions and Congo red dye by modified nanocellulose. Hydroxyl groups of nanocellulose can provide active sites for polypyrrole reaction onto it [49]. Liu et al. reported influence of phosphate groups on nanocellulose surfaces for the effective removal of metal ions from the industrial effluent. Phosphate groups on nanocellulose surfaces can increase the sorption rate and capacity of metal ions removal. Phosphorylated nanocellulose removes nearly 100% of metal ions from the aqueous solution [50].

Zhang et al. reported silylated NFC sponge for the removal of oil. NFC was extracted from the suspension of cellulose pulp by high-speed homogenization at 20,000 rpm for 2 hours. NFC sponges were prepared by freeze-drying procedure. The oil removal property of the original NFC sponge and silylated NFC sponge were compared. Unmodified NFC sponge rapidly sunk after being saturated with dodecane and water, while silylated NFC sponge with 18.9 wt% of Si can selectively adsorb organic substance (such as dodecane) and floated on water [51]. TiO$_2$ nanolayer coated hydrophobic nanocellulose aerogels can absorb a significant amount of oil within the interior pores. Aerogel can absorb oil 30 times of its dry

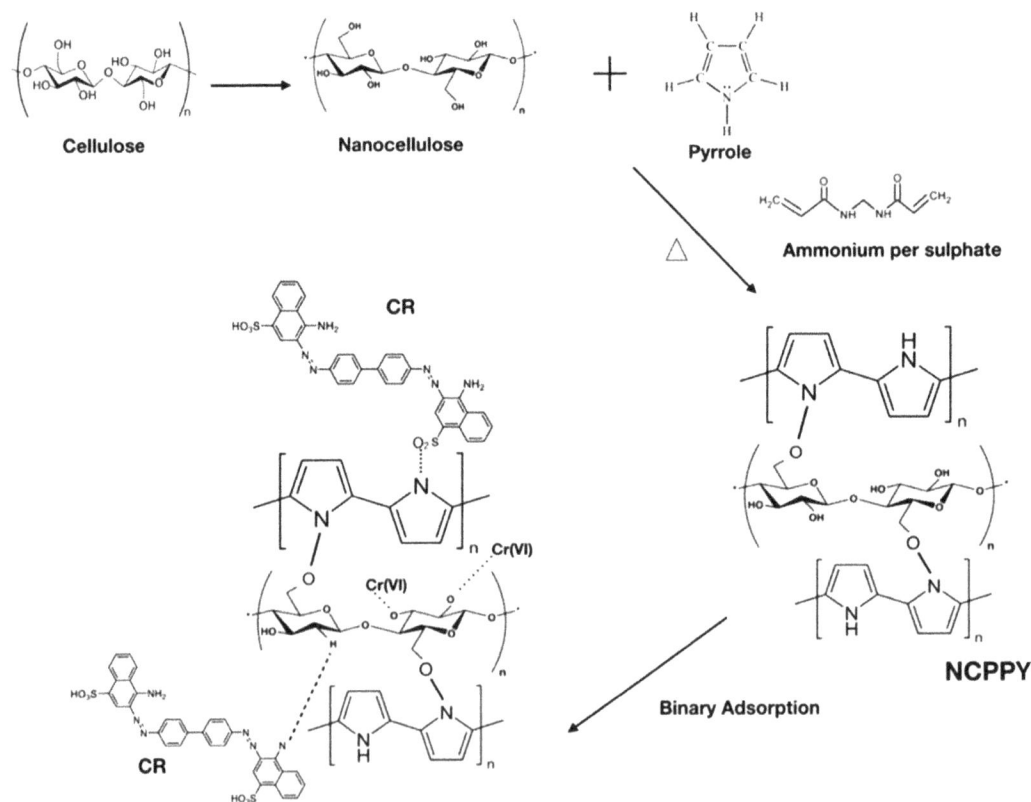

FIGURE 5.5 Schematic showing synthesis mechanism of nanocellulose modification with polypyrrole and its adsorption of Congo red dye and chromium ion (Reproduced with permission from Ref. [49]. © Elsevier 2020).

Unmodified

Silylated (18.9 wt% Si)

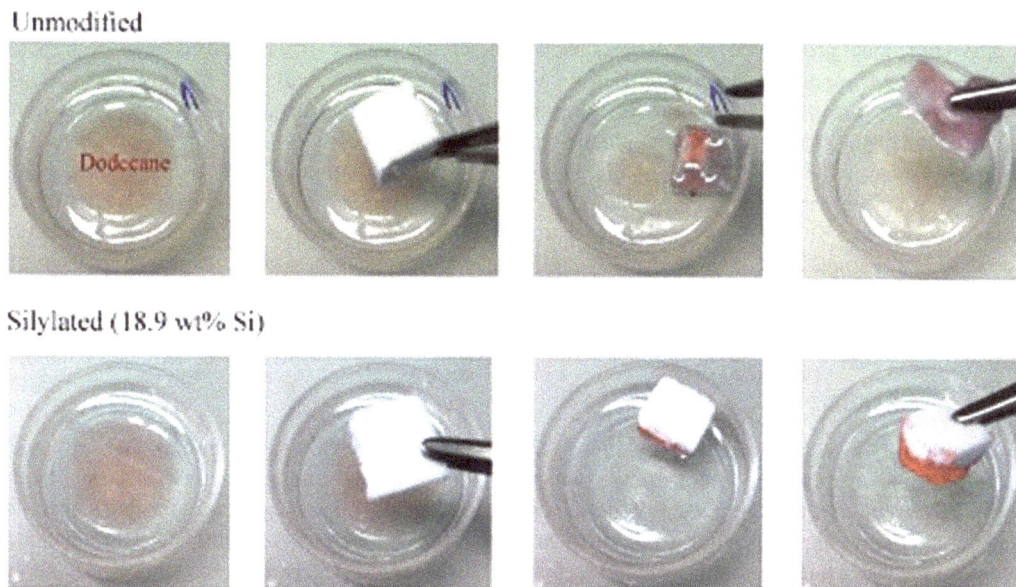

FIGURE 5.6 A comparison for the removal of dodecane spill from water by using unmodified nanofibrillated sponge and silylated NFC sponge (Reproduced with permission from Ref. [51]. © American Chemical Society 2014).

weight. Aerogel can be easily regenerated by simple washing in ethanol and reused [43]. Fig. 5.6 shows a comparison to remove dodecane spills from water using an unmodified nanofibrillated sponge and silylated NFC sponge.

5.3.2 Chemically Modified Nanocellulose for Membrane-Based Separation

Membrane-based separations are widely used for wastewater treatment because of its several advantages compared with chemical treatment of wastewater, such as no/little chemical cost, less space required to set up membrane-based treatment, low labor cost, faster process, no/less sludge production and therefore environment-friendly, etc. In the membrane-based water treatment process, the membrane acts as a selective barrier that allows a molecule, an ion, or other tiny particles to pass through but stops bigger molecules [52]. Nanocellulose can be used for membrane filtration to remove a wide range of contaminants from the wastewater because of its cost-effectiveness and sustainable and stable materials in an aqueous solution. Further, the abundance of surface hydroxyl groups provides material scientists opportunities to modify cellulosic membrane surface to remove target-specific contaminants from the wastewater [53, 54]. Rafieian et al. reported an amino-functionalized CNC-reinforced polyethersulfone (PES) matrix composite membrane to remove copper ion and red direct 16 dye from the wastewater. In the PES matrix, 1 wt% modified CNC loading can improve the copper ion removal from ~27% (for neat PES membrane) to 90% and direct red-16 reduction from 89% (for neat PES membrane) to 99% [55]. Karim et al. reported *in situ* TEMPO surface modification of nanocellulose membrane for metal ions removal from the aqueous solution. Nanocellulose-based membranes were *in situ* surface TEMPO functionalized with a thin functional layer of CNC to improve the metal ion adsorption capacity. *In situ* surface functionalization engineers the surface charge density without significantly affecting the bulk structure of base membrane. TEMPO oxidation introduces −COOH groups, which increases the hydrophilicity of the membrane. *In situ* TEMPO functionalization increases the metal ion adsorption capacity ~1.2- and 1.3-fold of Fe(II)/Fe(III) and Cu(II), respectively. The improved adsorption capacity of metal for the surface-functionalized membranes is due to the increased carboxylic group content on the membrane surface [30]. The deposition of a thin layer of graphene oxide on CNF membrane can significantly improve the pollutants removal capacity of the modified membrane due to the synergistic behavior of graphene oxide CNF membranes [56]. Park et al. reported poly(acryloyl hydrazide)-grafted CNC(PAH-CNC) based reverse osmosis membrane fabricated via layered interfacial polymerization technique. Plenty of reactive functional groups provided densely packed membrane matrix with improved water permeance and rejection of impurities. Significant boron-capturing capability of fabricated membranes is due to abundancy of hydroxyl and amino functional groups on the membrane surface [57]. Membrane fouling is one of the major problems of the membrane-based filtration process. Fouling is an undesirable inter-action of foulant molecules, viz., inorganic, colloidal, organic, or biofoulants, with membrane surface [58, 59]. Therefore, membrane fouling can be controlled by changing the surface characteristics of the membrane surface. The inherent hydrophilic nature of nanocellulose reduces the fouling of membrane surfaces. For example, higher hydrophilicity and greater negative charge on the membrane surface can significantly improve the organic foulant resistance [57].

5.4 Conclusions

The availability of clean water for human uses and industrial applications is a significant global issue. Adsorption and membrane filtration are the two most widely used conventional methods for water treat-ment. Increasing and emerging water contamination from multiple sources encourages researchers to find out alternative approaches for water purification. This chapter summarizes surface-modified nano-cellulose for water purification. Compared to other nanostructured materials, nanocellulose has many advantages: simple extraction process, low cost, environment-friendly natural nanomaterial. Besides, the availability of nanocellulose groups on the surface increases the absorption of contaminants. Also, it increases the opportunity to make nanocellulose functional with other desirable molecules to increase the removal of a wide range of pollutants from wastewater. All these excellent properties of nanocel-lulose make them ideal nanomaterials for water purification. The abundance of −OH groups on the nanocellulose surface provides material scientists the opportunity to functionalize nanocellulose with a wide range of molecules to accommodate a wide range of contaminants from the wastewater. Surface functional nanocellulose with different molecules can be used mainly in absorbing and membrane filtra-tion to remove pollutants from contaminated water. However, more research into education and industry requires functional counterparts for real-life applications in scale and commercialization of nanocellu-lose and their chemical water purification.

REFERENCES

[1] Theron, Jacques, J. A. Walker, and T. E. Cloete. "Nanotechnology and water treatment: appli-cations and emerging opportunities." *Critical Reviews in Microbiology* 34 (2008): 43–69. doi: 10.1080/10408410701710442.

[2] Santhosh, Chella, Venugopal Velmurugan, George Jacob, Soon Kwan Jeong, Andrews Nirmala Grace, and Amit Bhatnagar. "Role of nanomaterials in water treatment applications: a review." *Chemical Engineering Journal* 306 (2016):1116–1137. doi: https://doi.org/10.1016/j.cej.2016.08.053.

[3] Taghipour, Shabnam, Seiyed Mossa Hosseini, and Behzad Ataie-Ashtiani. "Engineering nanomaterials for water and wastewater treatment: review of classifications, properties and applications." *New Journal of Chemistry* 43 (2019): 7902–7927. doi: 10.1039/C9NJ00157C.

[4] Dhakras, Prathamesh A. "Nanotechnology applications in water purification and waste water treatment: a review." In *International Conference on Nanoscience, Engineering and Technology (ICONSET 2011)*, 285–291. IEEE, 2011.

[5] Lu, Haijiao, Jingkang Wang, Marco Stoller, Ting Wang, Ying Bao, and Hongxun Hao. "An overview of nanomaterials for water and wastewater treatment." *Advances in Materials Science and Engineering* 2016 (2016): 1–10. doi:10.1155/2016/4964828.

[6] Ghasemzadeh, Gholamreza, Mahdiye Momenpour, Fakhriye Omidi, Mohammad R. Hosseini, Monireh Ahani, and Abolfazl Barzegari. "Applications of nanomaterials in water treatment and environmen-tal remediation." *Frontiers of Environmental Science & Engineering* 8 (2014): 471–482. doi:10.1007/s11783-014-0654-0.

[7] Khan, Ibrahim, Khalid Saeed, and Idrees Khan. "Nanoparticles: properties, applications and toxicities." *Arabian Journal of Chemistry* 12 (2019): 908–931. doi:https://doi.org/10.1016/j.arabjc.2017.05.011.

[8] Savage, Nora, and Mamadou S. Diallo. "Nanomaterials and water purification: opportunities and challenges." *Journal of Nanoparticle Research* 7 (2005): 331–342. doi:10.1007/s11051-005-7523-5.

[9] Narayan, Roger, "Use of nanomaterials in water purification," *Materials Today* 13 (2010): 44–46. doi:https://doi.org/10.1016/S1369-7021(10)70108-5.

[10] Chaturvedi, Shalini, Pragnesh N. Dave, and N. K. Shah. "Applications of nano-catalyst in new era." *Journal of Saudi Chemical Society* 16 (2012): 307–325. doi:https://doi.org/10.1016/j.jscs.2011.01.015.

[11] Ahmed, Toqeer, Saba Imdad, Khwaja Yaldram, Noor Mohammad Butt, and Arshad Pervez. "Emerging nanotechnology-based methods for water purification: a review." *Desalination and Water Treatment* 52 (2014): 4089–4101. doi:10.1080/19443994.2013.801789.

[12] Alavi, Mehran. "Modifications of microcrystalline cellulose (MCC), nanofibrillated cellulose (NFC), and nanocrystalline cellulose (NCC) for antimicrobial and wound healing applications." *E-Polymers* 19 (2019): 103–119. doi:https://doi.org/10.1515/epoly-2019-0013.

[13] Lewandowska-Lancucka, Joanna, Anna Karewicz, Karol Wolski, and Szczepan Zapotoczny. "Surface functionalization of nanocellulose-based hydrogels." In *Cellulose-based superabsorbent hydrogels*, ed. Md. Ibrahim H. Mondal, 1–29. New York City: Springer International Publishing. 2019.

[14] Aitomaki, Yvonne, and Kristiina Oksman. "Reinforcing efficiency of nanocellulose in polymers," *Reactive and Functional Polymers* 85 (2014): 151–156. doi: https://doi.org/10.1016/j.reactfunctpolym.2014.08.010.

[15] Mabrouk, A. Ben, M. C. Brochier Salon, A. Magnin, M. N. Belgacem, and S. Boufi. "Cellulose-based nanocomposites prepared via mini-emulsion polymerization: understanding the chemistry of the nanocellulose/matrix interface." *Colloids and Surfaces A: Physicochemical and Engineering Aspects* 448 (2014): 1–8. doi: https://doi.org/10.1016/j.colsurfa.2014.01.077.

[16] Jorfi, Mehdi, and E. Johan Foster. "Recent advances in nanocellulose for biomedical applications." *Journal of Applied Polymer Science* 132 (2015). doi:10.1002/app.41719.

[17] Lee, Koon-Yang, Yvonne Aitomaki, Lars A. Berglund, Kristiina Oksman, and Alexander Bismarck. "On the use of nanocellulose as reinforcement in polymer matrix composites." *Composites Science and Technology* 105 (2014): 15–27. doi: https://doi.org/10.1016/j.compscitech.2014.08.032.

[18] El-Wakil, Nahla A., Enas A. Hassan, Ragab E. Abou-Zeid, and Alain Dufresne. "Development of wheat gluten/nanocellulose/titanium dioxide nanocomposites for active food packaging." *Carbohydrate Polymers* 124 (2015): 337–346. doi: https://doi.org/10.1016/j.carbpol.2015.01.076.

[19] Khan, Avik, Tanzina Huq, Ruhul A. Khan, Bernard Riedl, and Monique Lacroix. "Nanocellulose-based composites and bioactive agents for food packaging." *Critical Reviews in Food Science and Nutrition* 54 (2014): 163–174. doi: 10.1080/10408398.2011.578765.

[20] Kim, Joo-Hyung, Bong Sup Shim, Heung Soo Kim, Young-Jun Lee, Seung-Ki Min, Daseul Jang, Zafar Abas, and Jaehwan Kim. "Review of nanocellulose for sustainable future materials." *International Journal of Precision Engineering and Manufacturing: Green Technology* 2 (2015): 197–213. doi:10.1007/s40684-015-0024-9.

[21] Lani, N. S., N. Ngadi, A. Johari, and M. Jusoh. "Isolation, characterization, and application of nanocellulose from oil palm empty fruit bunch fiber as nanocomposites." *Journal of Nanomaterials* 2014 (2014). doi: 10.1155/2014/702538.

[22] Chauhan, Prashant, Caroline Hadad, Ana Herreros Lopez, Simone Silvestrini, Valeria La Parola, Enrico Frison, Michele Maggini, Maurizio Prato, and Tommaso Carofiglio. "A nanocellulose–dye conjugate for multi-format optical pH-sensing." *Chemical Communications* 50 (2014): 9493–9496. doi: 10.1039/C4CC02983F.

[23] Martinez, H. Avila, E. M. Feldmann, M. M. Pleumeekers, L. Nimeskern, W. Kuo, S. Schwarz, R. Muller et al. "Novel bilayer bacterial nanocellulose scaffold supports neocartilage formation *in vitro* and *in vivo*." *Biomaterials* 44 (2015): 122–133. doi: https://doi.org/10.1016/j.biomaterials.2014.12.025.

[24] Ahrem, Hannes, David Pretzel, Michaela Endres, Daniel Conrad, Julien Courseau, Hartmut Müller, Raimund Jaeger, Christian Kaps, Dieter O. Klemm, and Raimund W. Kinne. "Laser-structured bacterial nanocellulose hydrogels support ingrowth and differentiation of chondrocytes and show potential as cartilage implants." *Acta Biomaterialia* 10 (2014): 1341–1353. doi: https://doi.org/10.1016/j.actbio.2013.12.004.

[25] Berndt, Sabrina, Falko Wesarg, Cornelia Wiegand, Dana Kralisch, and Frank A. Muller. "Antimicrobial porous hybrids consisting of bacterial nanocellulose and silver nanoparticles." *Cellulose* 20 (2013): 771–783. doi: 10.1007/s10570-013-9870-1.

[26] Cheng, Jie, Minsung Park, and Jinho Hyun. "Thermoresponsive hybrid hydrogel of oxidized nanocellulose using a polypeptide crosslinker." *Cellulose* 21 (2014): 1699–1708. doi: 10.1007/s10570-014-0208-4.

[27] Farjana, Sadia, Farshad Toomadj, Per Lundgren, Anke Sanz-Velasco, Olga Naboka, and Peter Enoksson. "Conductivity-dependent strain response of carbon nanotube treated bacterial nanocellulose." *Journal of Sensors* 2013 (2013). doi:10.1155/2013/741248.

[28] Fu, Lina, Ping Zhou, Shengmin Zhang, and Guang Yang. "Evaluation of bacterial nanocellulose-based uniform wound dressing for large area skin transplantation." *Materials Science and Engineering: C* 33 (2013): 2995–3000. doi: https://doi.org/10.1016/j.msec.2013.03.026.

[29] Zhang, Shuo, Sandra Winestrand, Lin Chen, Dengxin Li, Leif J. Jönsson, and Feng Hong. "Tolerance of the nanocellulose-producing bacterium *Gluconacetobacter xylinus* to lignocellulose-derived acids and aldehydes." *Journal of Agricultural and Food Chemistry* 62 (2014): 9792–9799. doi: 10.1021/jf502623s.

[30] Karim, Zoheb, Minna Hakalahti, Tekla Tammelin, and Aji P. Mathew. "*In situ* TEMPO surface functionalization of nanocellulose membranes for enhanced adsorption of metal ions from aqueous medium." *RSC Advances* 7 (2017): 5232–5241. doi: 10.1039/C6RA25707K.

[31] Eyley, Samuel, and Wim Thielemans. "Surface modification of cellulose nanocrystals." *Nanoscale* 6 (2014): 7764–7779. doi: 10.1039/C4NR01756K.

[32] Dufresne, Alain. "Nanocellulose: a new ageless bionanomaterial." *Materials Today* 16 (2013): 220–227. doi: https://doi.org/10.1016/j.mattod.2013.06.004.

[33] Jiang, Feng, and You-Lo Hsieh. "Chemically and mechanically isolated nanocellulose and their self-assembled structures." *Carbohydrate Polymers* 95 (2013): 32–40. doi: https://doi.org/10.1016/j.carbpol.2013.02.022.

[34] Chin, Kwok-Mern, Sam Sung Ting, Hui Lin Ong, and Mf Omar. "Surface functionalized nanocellulose as a veritable inclusionary material in contemporary bioinspired applications: a review." *Journal of Applied Polymer Science* 135 (2018): 46065. doi:10.1002/app.46065.

[35] Sharma, Amita, Manisha Thakur, Munna Bhattacharya, Tamal Mandal, and Saswata Goswami. "Commercial application of cellulose nano-composites: a review." *Biotechnology Reports* 21 (2019): e00316. doi: https://doi.org/10.1016/j.btre.2019.e00316.

[36] Moon, Robert J., Ashlie Martini, John Nairn, John Simonsen, and Jeff Youngblood. "Cellulose nanomaterials review: structure, properties and nanocomposites." *Chemical Society Reviews* 40 (2011): 3941–3994. doi: 10.1039/C0CS00108B.

[37] Yang, Jie. "Liquid crystal of nanocellulose whiskers' grafted with acrylamide." *Chinese Chemical Letters* 23 (2012): 367–370. doi: https://doi.org/10.1016/j.cclet.2011.12.014.

[38] Espino-Pérez, Etzael, Sandra Domenek, Naceur Belgacem, Cecile Sillard, and Julien Bras. "Green process for chemical functionalization of nanocellulose with carboxylic acids." *Biomacromolecules* 15 (2014): 4551–4560. doi: 10.1021/bm5013458.

[39] Vivod, Vera, Branko Neral, Ales Mihelic, and Vanja Kokol. "Highly efficient film-like nanocellulose-based adsorbents for the removal of loose reactive dye during textile laundering." *Textile Research Journal* 89 (2019): 975–988. doi: 10.1177/0040517518760752.

[40] Khanjanzadeh, Hossein, Rabi Behrooz, Nader Bahramifar, Wolfgang Gindl-Altmutter, Markus Bacher, Matthias Edler, and Thomas Griesser. "Surface chemical functionalization of cellulose nanocrystals by 3-aminopropyltriethoxysilane." *International Journal of Biological Macromolecules* 106 (2018): 1288–1296. doi:10.1016/j.ijbiomac.2017.08.136.

[41] Robles, Eduardo, Inaki Urruzola, Jalel Labidi, and Luis Serrano. "Surface-modified nano-cellulose as reinforcement in poly(lactic acid) to conform new composites." *Industrial Crops and Products* 71 (2015): 44–53. doi: https://doi.org/10.1016/j.indcrop.2015.03.075.

[42] Mautner, Andreas, H. A. Maples, T. Kobkeatthawin, V. Kokol, Zoheb Karim, K. Li, and A. Bismarck. "Phosphorylated nanocellulose papers for copper adsorption from aqueous solutions." *International Journal of Environmental Science and Technology* 13 (2016): 1861–1872. doi: 10.1007/s13762-016-1026-z.

[43] Korhonen, Juuso T., Marjo Kettunen, Robin H. A. Ras, and Olli Ikkala. "Hydrophobic nanocellulose aerogels as floating, sustainable, reusable, and recyclable oil absorbents." *ACS Applied Materials & Interfaces* 3 (2011): 1813–1816. doi: 10.1021/am200475b.

[44] Carpenter, Alexis Wells, Charles-Francois de Lannoy, and Mark R. Wiesner. "Cellulose nanomaterials in water treatment technologies." *Environmental Science & Technology* 49 (2015): 5277–5287. doi: 10.1021/es506351r.

[45] Singh, N. B., Garima Nagpal, and Sonal Agrawal. "Water purification by using adsorbents: a review." *Environmental Technology & Innovation* 11 (2018): 187–240. doi: https://doi.org/10.1016/j.eti.2018.05.006.

[46] Sadegh, Hamidreza, A. M. Ali Gomaa, Vinod Kumar Gupta, Abdel Salam Hamdy Makhlouf, Ramin Shahryari-Ghoshekandi, Mallikarjuna N. Nadagouda, Mika Sillanpaa, and Elzbieta Megiel. "The role of nanomaterials as effective adsorbents and their applications in wastewater treatment." *Journal of Nanostructure in Chemistry* 7 (2017): 1–14. doi: 10.1007/s40097-017-0219-4.

[47] Sehaqui, Houssine, Uxua Perez de Larraya, Peng Liu, Numa Pfenninger, Aji P. Mathew, Tanja Zimmermann, and Philippe Tingaut. "Enhancing adsorption of heavy metal ions onto biobased nanofibers from waste pulp residues for application in wastewater treatment." *Cellulose* 21 (2014): 2831–2844. doi: 10.1007/s10570-014-0310-7.

[48] Yu, Haibiao, Shaobo Zhang, Yu Wang, Dingwen Yin, and Jintian Huang. "Covalent modification of nanocellulose (NCC) by functionalized graphene oxide (GO) and the study of adsorption mechanism." *Composite Interfaces* (2020): 1–14. doi:10.1080/09276440.2020.1731276.

[49] Shahnaz, Tasrin, V. C. Padmanaban, and Selvaraju Narayanasamy. "Surface modification of nanocellulose using polypyrrole for the adsorptive removal of Congo red dye and chromium in binary mixture." *International Journal of Biological Macromolecules* 151 (2020): 322–332. doi: 10.1016/j.ijbiomac.2020.02.181.

[50] Liu, Peng, Pere Ferrer Borrell, Mojca Bozic, Vanja Kokol, Kristiina Oksman, and Aji P. Mathew. "Nanocelluloses and their phosphorylated derivatives for selective adsorption of Ag^+, Cu^{2+} and Fe^{3+} from industrial effluents." *Journal of Hazardous Materials* 294 (2015): 177–185. doi: 10.1016/j.jhazmat.2015.04.001.

[51] Zhang, Zheng, Gilles Sèbe, Daniel Rentsch, Tanja Zimmermann, and Philippe Tingaut. "Ultralight weight and flexible silylated nanocellulose sponges for the selective removal of oil from water." *Chemistry of Materials* 26 (2014): 2659–2668. doi: 10.1021/cm5004164.

[52] Noble, Richard D. "An overview of membrane separations." *Separation Science and Technology* 22 (1987): 731–743. doi: 10.1080/01496398708068978.

[53] Gopakumar, Deepu A., Vishnu Arumughan, Daniel Pasquini, Shao-Yuan Ben Leu, Abdul Khalil HPS, and Sabu Thomas. "Nanocellulose-based membranes for water purification." In *Nanoscale materials in water purification*, ed. S. Thomas, D. Pasquini, S.Y. Leu, and D. A. Gopakumar, 59–85, Elsevier, Oxfordshire, United Kingdom, 2019.

[54] Mautner, Andreas. "Nanocellulose water treatment membranes and filters: a review." *Polymer International* 60 (2020): 741–751. doi: 10.1002/pi.5993.

[55] Rafieian, Fateme, Mehdi Jonoobi, and Qingliang Yu. "A novel nanocomposite membrane containing modified cellulose nanocrystals for copper ion removal and dye adsorption from water." *Cellulose* 26 (2019): 3359–3373. doi: 10.1007/s10570-019-02320-4.

[56] Valencia, Luis, Susanna Monti, Sugam Kumar, Chuantao Zhu, Peng Liu, Shun Yu, and Aji P. Mathew. "Nanocellulose/graphene oxide layered membranes: elucidating their behaviour during filtration of water and metal ions in real time." *Nanoscale* 11 (2019): 22413–22422. doi: 10.1039/C9NR07116D.

[57] Park, Chan Hyung, Sung Kwon Jeon, Sang-Hee Park, et al. "Cellulose nanocrystal-assembled reverse osmosis membranes with high rejection performance and excellent antifouling." *Journal of Materials Chemistry A* 7 (2019): 3992–4001. doi: 10.1039/C8TA10932J.

[58] Mondal, Subrata. "Polymeric membranes for produced water treatment: an overview of fouling behavior and its control." *Reviews in Chemical Engineering* 32 (2016): 611–628. doi: https://doi.org/10.1515/revce-2015-0027.

[59] Mondal, S., and S. R. Wickramasinghe. "Photo-induced graft polymerization of *N*-isopropyl acrylamide on thin film composite membrane: produced water treatment and antifouling properties." *Separation and Purification Technology* 90 (2012): 231–238. doi: https://doi.org/10.1016/j.seppur.2012.02.024.

6

Potential Differences between Cellulose Nanocrystal, Microfibrillated Cellulose, and Hairy Cellulose Nanocrystalloid in Water Purification

Kaushalya Bhakar
Central University of Gujarat, Gandhinagar, India

Meena Nemiwal
Malaviya National Institute of Technology, Jaipur, India

Dinesh Kumar
Central University of Gujarat, Gandhinagar, India

CONTENTS

6.1 Introduction

Cellulose is the most abundant biomaterial found on the planet, and its estimated production is about 7.5×10^{10} tons annually [1]. It has also been renamed bio-based or biopolymer. The term biopolymer is used for natural renewable polymers produced through living organisms such as plants or microbes rather than traditional chemical or physical methods. Naturally, cellulose is never found as an individual molecule, but as a number of cellulose units bonded together to make fibers. Cellulose is a major component present in the membrane of the cell wall of plants. The cell wall of a plant is a composite of various biopolymers, including cellulose, lignin, pectin, and hemicellulose, that provides strength and stiffness and prevents cell inflammation. It plays a vital role in increasing the mechanical strength of plants through cell walls [2]. The hierarchical structure of natural fibers, which focuses on their fundamental nanofibrillar elements, leads to different plant species' exceptional strength and superior performance properties. Various animals such as tunicates, bacteria, and algae are also known for cellulose production as structural polymers. Bacteria produce it as an additional cellular polysaccharide membrane [3]. The wooden

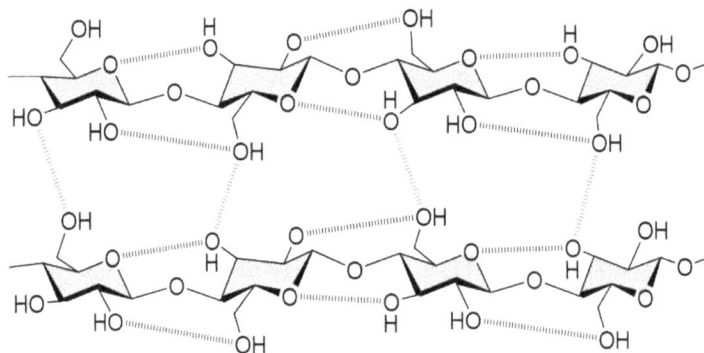

FIGURE 6.1 Intermolecular (——) and intramolecular (——) hydrogen bonding in cellulose structure. (Reproduced with permission from Ref. [10]. © KeAi Communications Co. Ltd. 2020).

structure contains about 30–40% cellulose. Cellulose is an excellent fibrous material used in the production of paper cotton, which is a relatively pure form of cellulose in plants. A polymer structure is made up of a repeat unit of anhydroglucose monomer, metabolized in the human body to survive. Two anhydroglucose units bonded together through β-1,4 linkage with a combination of hydrogen and hydroxyl groups and eliminate the water molecule to form a long series of cellulose polymers [4]. By the linkage of two sugar unit disaccharides, cellobiose is formed [5]. In the cellulose chain, glucose units are present as six-membered rings known as pyranoses. It is called polysaccharide because of cellulose formed by sugar monomer units. The hydroxyl group's definite chemical array in the beta-pyranose ring of cellulose made it a linear homo-polysaccharide. It increased the cellulose chain's capability to form inter- and intramolecular hydrogen bonds [6]. Hydroxyl group in the beta configuration of anhydroglucose present in the equatorial position, sidelong to the extended molecule, is available for comprehensive and hydrogen bonding. Due to hydrogen bonding, a highly systematic structure of cellulose is formed. Intermolecular hydrogen bonding between the cellulose chains gives extra strength to the cellulose fiber with insolubility in most solvents [7–9]. Intermolecular and intramolecular hydrogen bonding in cellulose fibers is illustrated in Fig. 6.1 [10].

Depending on cellulose fibers' origin, cellulose monomorphic is I, II, III, and IV, respectively [11, 12]. Cellulose I, a naturally occurring polymer inter-sheet, is present in parallel strands without hydrogen bonding and is treated with aqueous NaOH to obtain cellulose II. Cellulose II inter-sheet is present in the anti-parallel strand with hydrogen bonding, which makes it thermodynamically more stable than cellulose I. The treatment of cellulose I and II with amines can convert them into cellulose III. It is used to synthesize cellulose IV by glycerol at high temperatures [13]. When cellulose in small amounts reacts through proper conditions, it can be converted into different products. These can be further used to produce industrial products such as cellophane and rayon [14]. Cellulose is one of the most attractive ingredients that naturally occurs with some better features. These mostly depend on the material, size, and source of the extraction process used to extract the fiber. The source of cellulose is direct from nature, such as plant wood, and other sources include agricultural wastes such as rice husk and sugarcane bagasse [15]. Many hydroxyl groups and strong hydrogen bonding in cellulose enhance the physical and mechanical properties of nanofibers.

6.2 Nanocellulose

A polymer structure comprises two regions: the crystalline and amorphous regions. The presence of fractions of these areas controls the properties of a polymer. The crystalline area creates a highly ordered structure where all linear chains are arranged systematically. In the amorphous region, all the linear chains are arranged unsystematically to form a highly disordered region [16]. In cellulose fiber, the stiffness and strength resulted because of the crystalline area. The amorphous region gives flexibility to cellulose fiber [17]. The superior characteristics of nanocellulose, such as shape, size, highly ordered structure, density, porosity, etc., mainly depend upon cellulose fiber source [18–20]. The cellulose is produced by bio-organism like algae, bacteria is the purest form of the polymer because of the higher crystalline index compared to plants-synthesized cellulose. Nanocellulose is an organic fiber that can be separated from the plant's cell wall using the chemical or physical method. The size of nanocellulose is in the nanometer range, which is an important feature that distinguishes it from bulk materials. Nanofibers have a diameter in the range of 2–100 nm, while the length is between 100 and 2000 nm. The area of nanocellulose was first explored in early 1951. Initially, Bengt G. Ranby noted that cellulose fiber's treatment with strong acid formed nanoscale needle-shaped colloidal particles that are initially named micelle [21]. The mechanical and chemical treatment of cellulose fiber to synthesized nanocellulose resulted in a drastic change in the properties. Defibrillation of cellulose into nanofiber increases the surface area rapidly. Usually, the surface area of the soft pulp of cellulose is 1–4 m^2/g, which can change extensively at nanoscale around 500–600 m^2/g, depending upon the synthesis procedure [22, 23]. The mechanical properties of nanocellulose mostly depend upon the high surface area, which increases the availability of the hydroxyl group and hydrogen bonding. The surface modification of nanocellulose occurs through various chemical reactions: carboxylation, esterification, etherification, silylation, sulfonation, phosphorylation, and amidation. In these recent years, nanocellulose has become a material of interest to develop various eco-friendly nanocomposites. As compared to other polysaccharide nanomaterials, nanocellulose can be produced at a large scale with low cost using various preparation methods. Nanocellulose is a natural biopolymer that has the capability of being renewed. The extraordinary strength properties with being easily degradable by living organisms make it a superior candidate among all variant carbohydrate-based nanomaterials [24]. For instance, starch is one biopolymer found on Earth and treated chemically or physically to produce nanomaterial, containing crystalline and amorphous regions [25]. The yield of the product synthesized through chemical treatment is very low as compared to mechanical treatment. The poor mechanical strength and strong water attraction of starch films have led material researchers toward developing cellulose-based starch nanocomposites [26]. The continuous work for developing these novel nanofibers has been widely encouraged by both the academic world and industry in nanotechnology. Nanofibers' applications include effective catalysts, nanocellulose membrane, bio-adsorbent, electro-optical films, nanofiber-strengthened composites, microelectronics, gas-barrier films, cosmetics, fire-resistant tools, and other highly advanced and superior-performance materials [27–30]. Recently, the increasing concern for the environment and the establishment of renewable and sustainable organizations has guided the elementary research and application of bio-based nanofiber composites in this era. Fig. 6.2 shows the different properties of nanocellulose fiber.

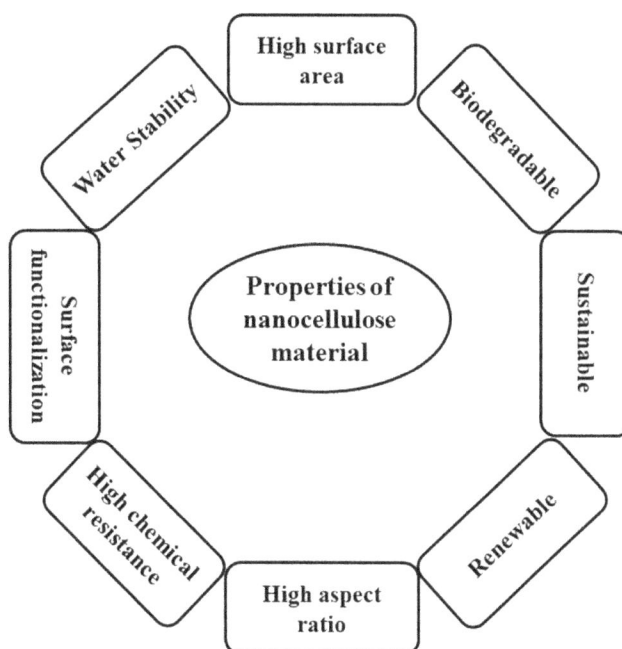

FIGURE 6.2   Different properties of nanocellulose fiber.

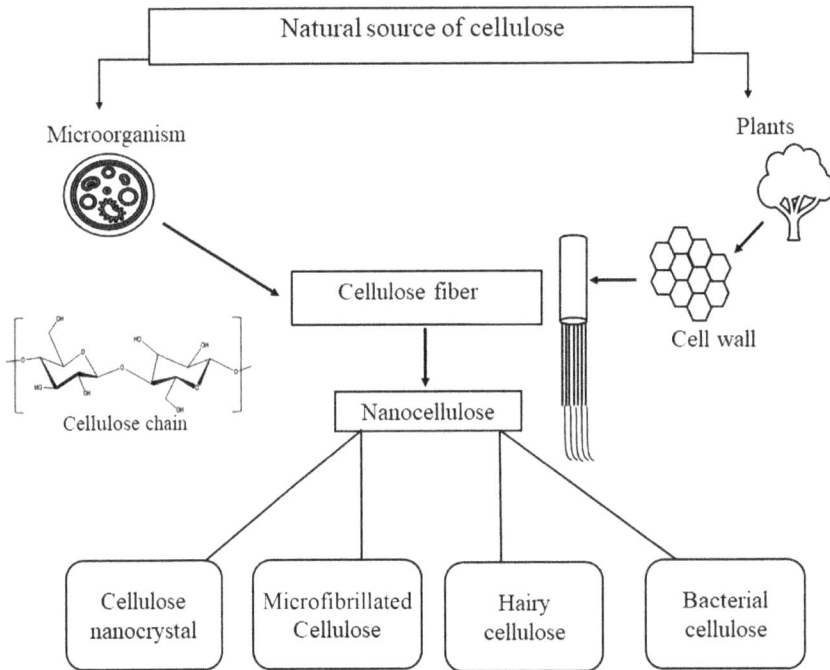

FIGURE 6.3 Extraction of different nanocelluloses from natural sources.

Cellulose nanofiber (CNF) is an active material that is used continuously for the removal of pollutants from drinking water. The superior mechanical properties with inherent fibrillar structure, chemical inertness, high surface area with hydrophilic nature, and high absorption capacity make it a next-generation filter and highly advanced filtration membrane [31].

Fig. 6.3 shows the extraction of different nanocelluloses forms natural sources. Nanocellulose can be categorized into four main types, depending upon their synthesis method: nanocrystalline cellulose (CNC), microfibrillated cellulose (MFC), hairy cellulose (HC), and bacterial nanocellulose (BC). Although all types of nanocellulose are similar in molecular arrangement, they contain different features—morphology, particle size, crystallinity, and other properties—because of the variation of sources of cellulose fiber and isolation methods [32–35].

Nanocellulose with negative or positive charge containing functional groups on the surface has been shown to eliminate harsh metallic contaminants from aqueous solutions. Nanocellulose having ionic and non-ionic surface moieties has also been employed as an adsorbent to remove organic pollutants, including dyes, oils, and pesticides, from aqueous solutions.

6.3 Types of Nanocellulose

6.3.1 Cellulose Nanocrystal (CNC)

Cellulose nanocrystal is a nanomaterial of cellulose, also known as cellulose nanowhisker, derived from natural sources like wood pulp or plant fiber. Cellulose nanocrystal isolation is mainly based on the hydrolysis of CNF using HCl, H_2SO_4, and H_3PO_4 [36]. Fig. 6.4 shows the isolation of cellulose nanocrystal from cellulose fiber by acid hydrolysis. The amorphous CNF region is hydrolyzed, while the crystalline region can maintain the structure during acid attack [37, 38]. During the acid hydrolysis, the hydrogen bonding disrupts, and cleavage of the disordered region takes palace. The cellulose nanocrystal displays similar morphology as native cellulose with a rod-like structure, high crystallinity,

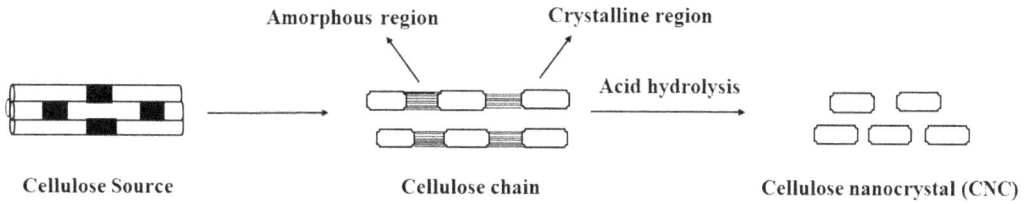

FIGURE 6.4 Representation of cellulose nanocrystal isolated from cellulose fiber by acid hydrolysis (Reproduced with permission from Ref. [10]. © KeAi Communications Co. Ltd. 2020).

and size between 10 and 20 nm in diameter and 100–500 nm in length, depending upon the extraction method. The isolation of CNC is controlled by various factors, including hydrolysis conditions such as temperature, time, and stirring. The CNC holds many desirable cellulose properties, namely, biodegradability, good stability, interesting optical properties, and high strength, while also owning a high specific surface area, which provides a large surface for functionalization [39]. These characteristics make it an ideal candidate for the replacement of carbon nanotube. Specifically, nanocellulose has several advantages as an adsorbent, including its structural stability in water, which diminishes the biofouling, and surface modification and functionalization, which make it available for various applications [40]. The nature of acid mainly affects the extraction procedure. By changing the acid, the surface functionalization of a nanofiber can be modified; for example, by acid hydrolysis using H_2SO_4, CNC with sulfate group surface is obtained via the hydroxyl group's esterification; while with HCl acid treatment, hydroxy group functionalized surface can be achieved. Cellulose nanocrystal stability developed using H_2SO_4 acid hydrolysis is more than using HCl and H_3PO_4 hydrolysis. This is due to the generation of negative charge on the CNC's surface, which keeps the structure by avoiding agglomeration using electrostatic forces [41]. Concerning the nanoparticle structure, during hydrolysis, if the mixture of HCl and H_2SO_4 is added, spherical-shaped cellulose nanocrystal is formed in place of rod-like nanocrystal under ultrasonic treatment [42]. The presence of fewer amounts of sulfate groups on their surface improves nanocrystal thermal stability [43].

The removal of contaminants from wastewater using a sorption mechanism is one of the major applications of cellulose nanocrystal. The negative density on the CNC's surface can be increased using 2,2,6,6-tetramethylpiperidine-1-oxyl (TEMPO)-mediated oxidation, which maximizes the binding capacity with water and absorption capacity of CNC-based adsorbent.

6.3.2 Microfibrillated Cellulose (MFC)

Microfibrillated cellulose is also called cellulose microfibril, microfibrillar cellulose, and recently used term nanofibrillated cellulose. Both microfibrillated and nanocrystal cellulose can be extracted from plant cell walls. MFC was first synthesized by Herrick et al. in 1983 [44], and Turbak et al. [45] first used wood pulp as a starting material in 1985. The rapid and unique procedure to synthesize a new type of cellulose was patent by Turbak et al. and called cellulose microfibrillated. The source of cellulose is fragmented many times in a high-pressure homogenizer and white aqueous suspension is thus obtained that comprise gel-like features at low concentrations [45]. It also can convert into transparent films once dried. These cellulose fibrils have length in the micrometer range and the diameter is between 10 and a few hundred nanometers depending upon the source. Various mechanical treatments like refining, grinding, and high-pressure homogenization can synthesize MFC, depending upon the desired MFC (Fig. 6.5).

The disintegration into cellulose fibrils is controlled by the different parameters set for pretreatment, including oxidation, enzymatic, and acidic treatment. The pretreatment of cellulose pulp through acid hydrolysis before homogenization results in fine structure of MFC compared to direct homogenization [33]. The chemical or enzymatic treatment, such as TEMPO oxidation, of wood pulp with mechanical refining can decrease energy during the refining process to develop MCF [46]. The MCF

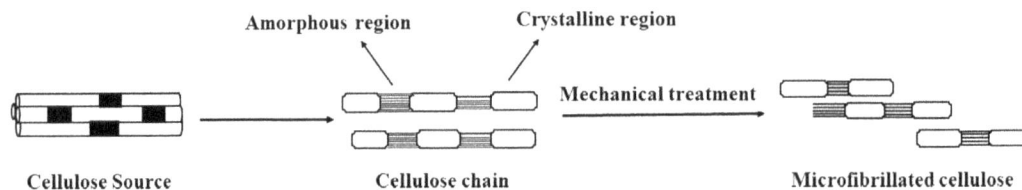

FIGURE 6.5 Representation of MCF isolated from cellulose fiber by mechanical treatment. (Reproduced with permission from Ref. [10]. © KeAi Communications Co. Ltd. 2020).

exhibits three-dimensional structure comprising both amorphous and crystalline regions with source chemical conformation [9]. This nanomaterial comprises high surface area, long microfibrils, and many hydroxyl groups that are readily available for surface grafting compared to cellulose nanocrystals. Cellulose-based nanomaterials are an effective alternative adsorbent because of their high aspect ratio, which favors strong hydrogen bonding to hold the network structure, low cost, and high degree of polymerization. The functional surface of CNF allows easy integration of various functional chemical moieties to improve the binding capacity of impurities for CNF. The surface functionalization of CNFs via carboxylation is one of the most investigated methods for increasing their sportive capacity [40].

6.3.3 Bacterial Cellulose (BC)

Although cellulose is mainly produced by plants, including cotton fiber, dried hemp, and wood. Cellulose is also found in microorganisms such as fungi, algae, and bacteria, known as bacterial cellulose. BC was first extracted by Brown in 1886 from bacteria *Acetobacter xylinum* [47]. The current research studies show that BC is also found in different Gram-negative (*Acetobacter, Azotobacter, Rhizobium, Agrobacterium, Pseudomonas, Salmonella, Alcaligenes*) and Gram-positive (*Sarcina ventriculi*) bacterial species [48]. Compared to plant-based cellulose, BC is the purest form of a highly ordered crystalline structure containing naturally occurring cellulose. It comprises only glucose monomer units, and all other types of non-cellulose polysaccharides such as hemicellulose and lignin are absent in BC with no chemical or mechanical treatment. All cellulose types obtained from various bacteria are exhibited with different compositions, physical and mechanical properties, and applications in many fields. BC shows remarkable physical and chemical properties, which make it more interesting among the scientific community. It has a high crystalline structure with a high degree of polymerization, high tensile strength, structural integrity, and high aspect ratio. It also has a high capacity for binding water molecules on the surface, biodegradability, and sustainability to form an ultrafine web-like structure that makes it more stable in water [49]. BC's spherical shape makes it a good adsorbent for the elimination of toxic metal ions, organic contaminants, and oil from water. To improve its adsorbing ability, a variety of nanoparticles are grafted on its surface. BC can combine with other materials to form nanocomposite such as spherical-shaped graphene/BC nanocomposite synthesized by Hu et al. (2014) [50]. It possesses the excellent capacity to adsorb metal ions, organic dyes, and oil from water. The highly porous, 3D interconnected honeycomb-like structure increases its stability and mechanical properties. Its superior characteristics have considerable applications in many fields, including tissue engineering, wound dressing, drug delivery, food packing material, sensor, adsorbent, optically transparent films, vacuum insulation, battery separator, and beauty products [51].

6.3.4 Hairy Cellulose Nanocrystalloids (HCNCs)

Hairy cellulose is the newest type in the nanocellulose family, synthesized by van de Ven and colleagues through periodate cellulose oxidation [52]. In the extraction process of nanocellulose, all chemical modifications occur in a highly disorganized field. If all cellulose chains present at the amorphous region

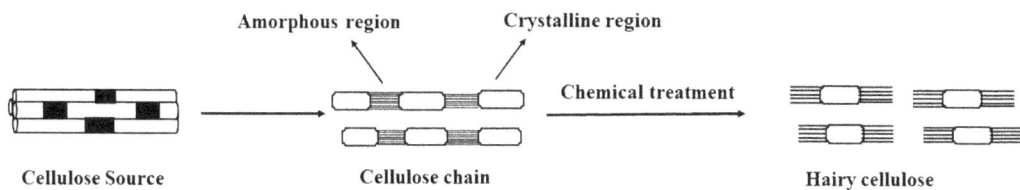

FIGURE 6.6 Representation of hairy cellulose isolated from cellulose fiber by chemical treatment (Reproduced with permission from Ref. [10]. © KeAi Communications Co. Ltd. 2020).

formed CNF, then the modification of the amorphous region takes place to disintegrate the disordered region. To make the amorphous areas soluble, it is vital to rupture the chain present in the amorphous area. This is an ideal method where the disordered cellulose region is solubilized, and enough number of cellulose chains also break down [52]. Hairy cellulose nanocrystalloid (HCNC) comprises crystalline and amorphous regions such as microfibrillated cellulose and also possesses high crystallinity with rod-shaped cellulose nanocrystals. Different hairy cellulose derivatives, including sterically and electrostatically stabilized nanocrystalline cellulose, can synthesize by using different chemical reactions [52–54], as shown in Fig. 6.6. The sterically stabilized nanocrystalline cellulose (SNCC) can be synthesized by chemical treatment of cellulose chain using periodate (IO_4^-) before heating, while the electrostatically stabilized nanocrystalline cellulose (ENCC) is developed by periodate and perchlorate (ClO_4^-) oxidation [52–54].

6.4 Potential Difference between Cellulose Nanocrystal (CNC), Microfibrillated Cellulose (MFC), and Hairy Cellulose Crystalloids in Water Filtration

Water is a valuable natural resource, and the availability of quality drinking water is essential for the survival of living beings on the planet. Potable water is considered an important indicator of the country's progress, and according to recent reports, around 663 million people worldwide do not have access to safe potable water. Coastal areas of the seas and oceans have also seen a dramatic increase in contaminants. The world's growing population and rapid industrialization are considered major causes of pollution. It is essential to treat drinking water and wastewater for the development of human society. Various approaches, including nanomaterial-based adsorbents, polymer membranes, and filters, have been developed for water purification. The natural abundance in nanoscale [55], non-hazardous characters, considers cellulose ideal due to stability, renewability, and high efficiency. Production of nanocellulose at a commercial level at a low cost enhances its application in various sectors. Nanocellulose-based materials CNC, CMF, and HC can be used as an adsorbent and filtration membrane to remove inorganic and organic pollutants from an aqueous system. From the viewpoint of construction of membrane for water filtration, CNFs are a more appropriate building block material because of their higher aspect ratio and higher interfibrillar connectivity because of fiber entanglement and small pore size membrane. Compared with CNC, MFC contains longer length fiber with high surface area and highly ambient hydroxyl groups, which are effortlessly available for surface functionalization. CNCs are being considered as more appropriate alternative for use as a functional nanofiller material. HC is synthesized through two steps of oxidation. HCNC-based adsorbents are a new class of nanocellulose with significant adsorption potential for water treatment. HC has a high surface area, non-toxicity, renewability, and easily approachable amorphous regions than the crystalline region, which allows surface modification, thereby making it a promising adsorbent for removing dyes and heavy metal ions like other nanocelluloses [56]. The charge density on ENCC is much higher than the traditional CNC, which increases its adsorption capacity and makes it more colloidally stable, and prevents its aggregation [57]. The synthesis procedure is also more facile than traditional CNC.

While the CNC is a good adsorbent for water treatment, during the synthesis of CNC, more harsh acid conditions are required, making it less suitable as a recyclable nanomaterial. The hairy cellulose with a positive and negative charged functional group can eliminate ionic species from an aqueous system. Nanocellulose-based adsorbents can be synthesized in different ways, such as cellulose surface functionalization by the various functional groups, e.g., aldehyde groups, carboxyl, and carbonyl groups, and cationic groups [58]. The surface modification also takes place through grafting and polymerization. All types of nanocellulose, including CNC, MFC, and HC, contain similar physical properties: size distribution, tensile strength, Young's modulus, and optical properties. The surface functionalization is one of the most important factors, which regulates the water treatment efficiency of nanocellulose. Nanocellulose-based membranes are also actively used for the removal of contaminants from water. Numerous types of nanocellulose-based membranes are synthesized continuously with customized structure and improved physical and chemical properties using various kinds of synthesis procedures and materials. The highly hydrophilic nature of nanocellulose enhances the antifouling activity, especially toward organic containments, which are intrinsic [55]. Much work is done to strengthen nanocellulose-based membranes for separation and antifouling capacity for commercial use in water purification. Various methods such as vacuum filtration, electrospinning, solution casting, etc. are used to develop membranes. Some data on CNC, MCF, and HC with regard to the elimination of toxins from water are provided in Tables 6.1 and 6.2.

TABLE 6.1

Adsorption of Heavy Metal Ions from Water by Nanocellulose.

Pollutant	Surface Modification	Nanocellulose	Maximum Efficiency (mg/g)	Reference
	Sulfonation	CNC	34	[59]
Ag(I)	None	MFC	14	[59]
	Phosphorylated and carboxylation	MFC	120	[59]
	None	CNC	47	[60]
Cu(II)	None	MFC	14	[60]
	Phosphorylated and carboxylation	MFC	114	[59]
	Oxidation	HC	185	[57]
Pb(II)	Carboxylation	SCNC	365.9	[61]
	Succination	SCNC	300	[61]
	Carboxylated CNC/alginate composite	CNC	280	[62]
	TEMPO oxidation	MFC	9.7	[63]
	Thiol modification	m-MFC	137.7	[64]
	Carboxylation via poly(methacrylic acid-*co*-maleic acid)	MFC	116	[65]
	Mercerization	MFC	192.7	[66]
UO_2(II)	Carboxylation	MFC	167	[67]
	Sulfonation	CNC	9.7	[68]
Cd(II)	Succination	CNC	152	[61]
	Carboxylation via poly(methacrylic acid-*co*-maleic acid)	MFC	134	[65]
	Amination	MFC	58.1	[69]
Zn(II)	Carboxylation via poly(methacrylic acid-*co*-maleic acid)	MFC	136	[65]
	TEMPO oxidation	MFC	66	[63]
	Sulfonation	CNC	8.55	[68]
Ni(II)	Carboxylation via poly(methacrylic acid-*co*-maleic acid)	MFC	115	[65]
	Amination	MFC	56.1	[69]
	Amination with diethylenetriamine	CNC	70	[70]
As(V)	Amination	CNC	9.8	[71]
	Magnetite modification	MFC	20-40	[72]
Hg(II)	Polyethyleneimine grafting	CNC	23.7	[73]
	Itaconic-acid-grafted magnetite modified with 2-mercaptobenzamide	CNC	69.7	[74]
	TEMPO oxidation/thiolation	MCF	145	[75]

TABLE 6.2

Adsorption of Dyes from Water by Nanocellulose.

Dye	Surface Modification	Nanocellulose	Adsorption Efficiency (mg/g)	Reference
	Sulfonation	CNC	107	[76]
Methylene blue	Carboxylation	CNC	233	[77]
	Polyacrylamide composite	CNC	19	[78]
	None	MCF	100	[79]
	Carboxylation	MCF	122.2	[70]
	ALG–ENCC beads	HC	1250	[80]
	Oxidation	HC	1400	[80]
	Hybrid hydrogel Composite	HC	300	[56]
	Amination	CNC	134.7	[81]
Congo red	PVAm grafting	CNC	199.5	[82]
	Quaternization	MCF	664	[83]
Crystal violet	Sulfonation	CNC	178	[76]
	Carboxylation	CNC	223	
Malachite green	Sulfonation	CNC	92	[76]
	Carboxylation	CNC	148	
Acid green	Quaternization	MFC	683	[81]
Acid red	Amination	CNC	134.7	[79]
	PVAm grafting	CNC	199.9	[80]
Reactive red 195	CoCO$_3$/FeCO$_3$	MCF	18	[84]
Toluidine blue	Carboxylation	MCF	4.15	[84]
Methyl orange	Hybrid hydrogel Composite	HC	270	[56]

6.5 Conclusion

Nanocellulose is a bio-adsorbent with flexible surface chemistry. Several facile methods are used to synthesize cellulose nanocrystal, microfibrillated cellulose, and hairy cellulose. These nanocellulose-based composites have a high capacity to remove heavy metal ions, dyes, oil, and pesticides from an aqueous system via adsorption, absorption electrostatic attraction, and the size exclusion mechanisms. Nanocellulose-based mechanically stable and surface-functionalized membrane with antifouling properties are also developed for water filtration. The researchers are continuously working toward large-scale commercial production of nanocellulose-based material for water purification application with improved properties.

Acknowledgment

Dinesh Kumar is thankful to DST, New Delhi, for providing financial support to this work (sanctioned vide project Sanction Order F. No. DST/TM/WTI/WIC/2K17/124(C)).

REFERENCES

[1] Habibi, Youssef. "Key advances in the chemical modification of nanocelluloses." *Chemical Society Reviews: Royal Society of Chemistry* 43 (2014): 1519–1542. doi:10.1039/c3cs60204d.

[2] Khandelwal, Mudrika, and Alan H. Windle. "Hierarchical organisation in the most abundant biopolymer-cellulose." *Materials Research Society Symposium Proceedings* 1504 (2013): 16–21. doi: https://doi.org/10.1557/opl.2013.379.

[3] Alpers, D. H. *Encyclopedia of food sciences and nutrition.* Cambridge, MA: Academic Press. 2003.

[4] Iguchi, M., S. Yamanaka, and A. Budhiono. "Bacterial cellulose: a masterpiece of nature's arts." *Journal of Materials Science* 35 (2000): 261–270. doi: https://doi.org/10.1023/A:1004775229149.

[5] Stecher, Paul G., Martha Windholz, Dolores S. Leahy, D. M. Bolton, and L. G. Eaton. *The Merck Index.* Rahway, NJ: Merck & Co Inc, 1968.

[6] Feldman, D. "Wood-chemistry, ultrastructure, reactions, by D. Fengel and G. Wegener, Walter de Gruyter, Berlin and New York, 1984." *Journal of Polymer Science: Polymer Letters Edition* 23 (1985): 601–602. doi:10.1002/pol.1985.130231112.

[7] Anwar, Zahid, Muhammad Gulfraz, and Muhammad Irshad. "Agro-industrial lignocellulosic biomass a key to unlock the future bio-energy: a brief review." *Journal of Radiation Research and Applied Sciences* 7 (2014): 163–173. doi:https://doi.org/10.1016/j.jrras.2014.02.003.

[8] Habibi, Youssef, Lucian A. Lucia, and Orlando J. Rojas. "Cellulose nanocrystals: chemistry, self-assembly, and applications." *Chemical Reviews* 110 (2010): 3479–3500. doi: https://doi.org/10.1021/cr900339w.

[9] Lavoine, Nathalie, Isabelle Desloges, Alain Dufresne, and Julien Bras. "Microfibrillated cellulose—its barrier properties and applications in cellulosic materials: a review." *Carbohydrate Polymers* 90 (2012): 735–764. doi: https://doi.org/10.1016/j.carbpol.2012.05.026.

[10] Phanthong, Patchiya, Prasert Reubroycharoen, Xiaogang Hao, Guangwen Xu, Abuliti Abudula, and Guoqing Guan. "Nanocellulose: extraction and application." *Carbon Resources Conversion* 1 (2018): 32–43. doi: https://doi.org/10.1016/j.crcon.2018.05.004.

[11] Baghaei, Behnaz, and Mikael Skrifvars. "All-cellulose composites: a review of recent studies on structure, properties and applications." *Molecules* 25 (2020). doi: 10.3390/molecules25122836.

[12] Shak, Katrina Pui Yee, Yean Ling Pang, and Shee Keat Mah. "Nanocellulose: recent advances and its prospects in environmental remediation." *Beilstein Journal of Nanotechnology* 9 (2018): 2479–2498. doi:10.3762/bjnano.9.232.

[13] Nunes, R. C. R. "Rubber nanocomposites with nanocellulose." In *Progress in rubber nanocomposites*, eds. S. Thomas and H. J. Maria, 463–494. Woodhead Publishing, 2017.

[14] Susheel Kalia, Alain Dufresne, Bibin Mathew Cherian, B. S. Kaith, Luc Averous, James Njuguna, and Elias Nassiopoulos. "Cellulose-based bio-and nanocomposites: a review." *International Journal of Polymer Science* (2011). doi:10.1155/2011/837875.

[15] Kumar Gupta, Praveen, Shreeya Sai Raghunath, Deepali Venkatesh Prasanna, Priyadharsini Venkat, Vidhya Shree, Chandrananthi Chithananthan, Shreya Choudhary, Krithika Surender, and Keerthana Geetha. "An update on overview of cellulose, its structure and applications." In *Cellulose* (2019). doi: 10.5772/intechopen.84727.

[16] Wasim, Maria, Aneela Sabir, Muhammad Shafiq, and Tahir Jamil. "Electrospinning: a fiber fabrication technique for water purification." In *Nanoscale materials in water purification*, eds. S. Thomas, D. Pasquini, S. Y. Leu, and D. A. Gopakumar, 289–308. Oxfordshire, UK: Elsevier. 2019.

[17] Lee, H. V., S. B. A. Hamid, and S. K. Zain. "Conversion of lignocellulosic biomass to nanocellulose: structure and chemical process." *Scientific World Journal* 2014 (2014). doi: https://doi.org/10.1155/2014/631013.

[18] Lin, Ning, Jin Huang, and Alain Dufresne. "Preparation, properties and applications of polysaccharide nanocrystals in advanced functional nanomaterials: a review." *Nanoscale* 4 (2012): 3274–3294. doi: https://doi.org/10.1039/c2nr30260h.

[19] Jonoobi, Mehdi, Reza Oladi, Yalda Davoudpour, Kristiina Oksman, Alain Dufresne, Yahya Hamzeh, and Reza Davoodi. "Different preparation methods and properties of nanostructured cellulose from various natural resources and residues: a review." *Cellulose* 22 (2015): 935–969. doi: https://doi.org/10.1007/s10570-015-0551-0.

[20] Bras, David Le, Maria Stromme, and Albert Mihranyan. "Characterization of dielectric properties of nanocellulose from wood and algae for electrical insulator applications." *Journal of Physical Chemistry B* 119 (2015): 5911–5917. doi: https://doi.org/10.1021/acs.jpcb.5b00715.

[21] Ranby, Bengt G. "III. Fibrous macromolecular systems: cellulose and muscle—the colloidal properties of cellulose micelles." *Discussions of the Faraday Society* 11 (1951): 158–164. doi: https://doi.org/10.1039/DF9511100158.

[22] Sehaqui, Houssine, Qi Zhou, Olli Ikkala, and Lars A. Berglund. "Strong and tough cellulose nanopaper with high specific surface area and porosity." *Biomacromolecules* 12 (2011): 3638–3644. doi: https://doi.org/10.1021/bm2008907.

[23] Sehaqui, Houssine, Uxua Perez de Larraya, Peng Liu, Numa Pfenninger, Aji P. Mathew, Tanja Zimmermann, and Philippe Tingaut. "Enhancing adsorption of heavy metal ions onto biobased nanofibers from waste pulp residues for application in wastewater treatment." *Cellulose* 21 (2014): 2831–2844. doi: https://doi.org/10.1007/s10570-014-0310-7.

[24] Klemm, Dieter, Friederike Kramer, Sebastian Moritz, Tom Lindstrom, Mikael Ankerfors, Derek Gray, and Annie Dorris. "Nanocelluloses: a new family of nature-based materials." *Angewandte Chemie: International Edition* 50 (2011): 5438–5466. doi: https://doi.org/10.1002/anie.201001273.

[25] Corre, Deborah Le, Julien Bras, and Alain Dufresne. "Starch nanoparticles: a review." *Biomacromolecules* 11 (2010): 1139–1153. doi: https://doi.org/10.1021/bm901428y.

[26] Savadekar, N. R., and S. T. Mhaske. "Synthesis of nano cellulose fibers and effect on thermoplastics starch-based films." *Carbohydrate Polymers* 89 (2012): 146–151. doi: https://doi.org/10.1016/j.carbpol.2012.02.063.

[27] Paul, D. R., and L. M. Robeson. "Polymer nanotechnology: nanocomposites." *Polymer* 49 (2008): 3187–3204. doi: https://doi.org/10.1016/j.polymer.2008.04.017.

[28] Isogai, Akira, Tsuguyuki Saito, and Hayaka Fukuzumi. "TEMPO-oxidized cellulose nanofibers." *Nanoscale* 3 (2011): 71–85. doi: https://doi.org/10.1039/c0nr00583e.

[29] Sturcova, Adriana, Geoffrey R. Davies, and Stephen J. Eichhorn. "Elastic modulus and stress-transfer properties of tunicate cellulose whiskers." *Biomacromolecules* 6 (2005): 1055–1061. doi: https://doi.org/10.1021/bm049291k.

[30] de Azeredo, Henriette M. C. "Nanocomposites for food packaging applications." *Food Research International* 42 (2009):1240–1253. doi: https://doi.org/10.1016/j.foodres.2009.03.019.

[31] Carpenter, Alexis Wells, Charles François De Lannoy, and Mark R Wiesner. "Cellulose nanomaterials in water treatment technologies." *Environmental Science and Technology* 49 (2015): 5277–5287. doi: https://doi.org/10.1021/es506351r.

[32] Nechyporchuk, Oleksandr, Mohamed Naceur Belgacem, and Julien Bras. "Production of cellulose nanofibrils: a review of recent advances." *Industrial Crops and Products* 93 (2016): 2–25. doi: https://doi.org/10.1016/j.indcrop.2016.02.016.

[33] Zimmermann, Tanja, Nicolae Bordeanu, and Esther Strub. "Properties of nanofibrillated cellulose from different raw materials and its reinforcement potential." *Carbohydrate Polymers* 79 (2010): 1086–1093. doi: https://doi.org/10.1016/j.carbpol.2009.10.045.

[34] Mishra, Raghvendra Kumar, Arjun Sabu, and Santosh K. Tiwari. "Materials chemistry and the futurist eco-friendly applications of nanocellulose: status and prospect." *Journal of Saudi Chemical Society* 22 (2018): 949–978. doi: 10.1016/j.jscs.2018.02.005.

[35] Abdul Khalil, H. P. S., Y. Davoudpour, Md Nazrul Islam, Asniza Mustapha, K. Sudesh, Rudi Dungani, and M. Jawaid. "Production and modification of nanofibrillated cellulose using various mechanical processes: a review." *Carbohydrate Polymers* 99 (2014): 449–464. doi: https://doi.org/10.1016/j.carbpol.2013.08.069.

[36] Yu, Houyong, Zongyi Qin, Banglei Liang, Na Liu, Zhe Zhou, and Long Chen. "Facile extraction of thermally stable cellulose nanocrystals with a high yield of 93% through hydrochloric acid hydrolysis under hydrothermal conditions." *Journal of Materials Chemistry A* 1 (2013): 3938–3944. doi: https://doi.org/10.1039/c3ta01150j.

[37] Dufresne, Alain. "Nanocellulose: potential reinforcement in composites." *Natural Polymers* 2 (2012): 1–32.

[38] Moon, Robert J., Ashlie Martini, John Nairn, John Simonsen, and Jeff Youngblood. "Cellulose nanomaterials review: structure, properties and nanocomposites." *Chemical Society Reviews* 40 (2011): 3941–3994. doi: https://doi.org/10.1039/c0cs00108b.

[39] van de Ven, T. G. M., and Amir Sheikhi. "Hairy cellulose nanocrystalloids: a novel class of nanocellulose." *Nanoscale* 33 (2013): 1–3.

[40] Mahfoudhi, Norhene, and Sami Boufi. "Nanocellulose as a novel nanostructured adsorbent for environmental remediation: a review." *Cellulose* 24 (2017): 1171–1197. doi: https://doi.org/10.1007/s10570-017-1194-0.

[41] Grishkewich, Nathan, Nishil Mohammed, Juntao Tang, and Kam Chiu Tam. "Recent advances in the application of cellulose nanocrystals." *Current Opinion in Colloid and Interface Science* 29 (2017): 32–45. doi: https://doi.org/10.1016/j.cocis.2017.01.005.

[42] Neng, Wang, Ding Enyong, and Cheng Rongshi. "Preparation and liquid crystalline properties of spherical cellulose nanocrystals." *Langmuir* 24 (2008): 5–8. doi: https://doi.org/10.1021/la702923w.

[43] Wang, Neng, Enyong Ding, and Rongshi Cheng. "Thermal degradation behaviors of spherical cellulose nanocrystals with sulfate groups." *Polymer* 48 (2007): 3486–3493. doi: https://doi.org/10.1016/j.polymer.2007.03.062.

[44] Siro, Istvan, and David Plackett. "Microfibrillated cellulose and new nanocomposite materials: a review." *Cellulose* 17 (2010): 459–494. doi: https://doi.org/10.1007/s10570-010-9405-y.

[45] Heiskanen, I., and A. Harlin. "Process for the production of microfibrillated cellulose in an extruder and microfibrillated cellulose produced according to the process." *US Patent* 1 (2012): 4.

[46] Einchhorn, S. J., A. Dufresne, M. M. Aranguren, J. R. Capadona, S. J. Rowan, C. Weder, and S. Veigel. "Review: current international research into cellulose nanofibres and composites." *Journal of Materials Science* 45 (2010): 1–33. doi: https://doi.org/10.1007/s10853-009-3874-0.

[47] Brown, R. Malcolm, J. H. M. Willison, and Carol L. Richardson. "Cellulose biosynthesis in *Acetobacter xylinum*: visualization of the site of synthesis and direct measurement of the *in vivo* process." *Proceedings of the National Academy of Sciences of the United States of America* 73 (1976): 4565–4569. doi: https://doi.org/10.1073/pnas.73.12.4565.

[48] Jung, Ho Il, O. Mi Lee, Jin Ha Jeong, et al.. "Production and characterization of cellulose by *Acetobacter* sp. V6 using a cost-effective molasses-corn steep liquor medium." *Applied Biochemistry and Biotechnology* 162 (2010): 486–497. doi: https://doi.org/10.1007/s12010-009-8759-9.

[49] Douglass, Eugene F., Huseyin Avci, Ramiz Boy, Orlando J. Rojas, and Richard Kotek. "A review of cellulose and cellulose blends for preparation of bio-derived and conventional membranes, nanostructured thin films, and composites." *Polymer Reviews* 58 (2018): 102–63. doi: https://doi.org/10.1080/15583724.2016.1269124.

[50] Luo, Honglin, Fangfang Feng, Fanglian Yao, Yong Zhu, Zhiwei Yang, and Yizao Wan. "Improved removal of toxic metal ions by incorporating graphene oxide into bacterial cellulose." *Journal of Nanoscience and Nanotechnology* 20 (2019): 719–730. doi: https://doi.org/10.1166/jnn.2020.16902.

[51] Campano, Cristina, Ana Balea, Angeles Blanco, and Carlos Negro. "Enhancement of the fermentation process and properties of bacterial cellulose: a review." *Cellulose* 23 (2016): 57–91. doi: https://doi.org/10.1007/s10570-015-0802-0.

[52] Gomez, I. Jennifer, Blanca Arnaiz, Michele Cacioppo, Francesca Arcudi, and Maurizio Prato. "Nitrogen-doped carbon nanodots for bioimaging and delivery of paclitaxel." *Journal of Materials Chemistry B* 6 (2018): 1–3.

[53] Yang, Han, Md Nur Alam, and Theo G. M. van de Ven. "Highly charged nanocrystalline cellulose and dicarboxylated cellulose from periodate and chlorite oxidized cellulose fibers." *Cellulose* 20 (2013): 1865–1875. doi: https://doi.org/10.1007/s10570-013-9966-7.

[54] Yang, Han, Dezhi Chen, and Theo G. M. van de Ven. "Preparation and characterization of sterically stabilized nanocrystalline cellulose obtained by periodate oxidation of cellulose fibers." *Cellulose* 22 (2015): 1743–1752. doi: https://doi.org/10.1007/s10570-015-0584-4.

[55] Wan Ngah, W. S., and M. A. K. M. Hanafiah. "Removal of heavy metal ions from wastewater by chemically modified plant wastes as adsorbents: a review." *Bioresource Technology* 99 (2008): doi: https://doi.org/10.1016/j.biortech.2007.06.011.

[56] Heidari Nia, Marzieh, Mandana Tavakolian, Ali Reza Kiasat, and Theo G. M. van de Ven. "Hybrid aerogel nanocomposite of dendritic colloidal silica and hairy nanocellulose: an effective dye adsorbent." *Langmuir* 26 (2020). doi: https://doi.org/10.1021/acs.langmuir.0c02090.

[57] Sheikhi, Amir, Salman Safari, Han Yang, and Theo G. M. Van De Ven. "Copper removal using electrosterically stabilized nanocrystalline cellulose." *ACS Applied Materials and Interfaces* 7 (2015): 11301–11308. doi: https://doi.org/10.1021/acsami.5b01619.

[58] Weijuan, Yixiang Wang, Zhiqiang Huang, Xiaolan Wang, Lingyun Chen, Yu Zhang, and Lina Zhang. "On-demand dissolvable self-healing hydrogel based on carboxymethyl chitosan and cellulose nanocrystal for deep partial thickness burn wound healing." *ACS Applied Materials and Interfaces* 10 (2018): 41076–41088. doi: 10.1021/acsami.8b14526.

[59] Liu, Peng, Houssine Sehaqui, Philippe Tingaut, Adrian Wichser, Kristiina Oksman, and Aji P. Mathew. "Cellulose and chitin nanomaterials for capturing silver ions (Ag⁺) from water via surface adsorption." *Cellulose* 21 (2014): 449–461. doi: https://doi.org/10.1007/s10570-013-0139-5.

[60] Ulmanu, Mihaela, Elena Maranon, Yolanda Fernandez, Leonor Castrillon, Ildiko Anger, and Daniela Dumitriu. "Removal of copper and cadmium ions from diluted aqueous solutions by low cost and

waste material adsorbents." *Water, Air, and Soil Pollution* 142 (2003): 357–373. doi: https://doi.org/10.1023/A:1022084721990.

[61] Linghu, Wen Sheng, and Chao Wang. "Adsorption of heavy metal ions from aqueous solution by chitosan." In *Advanced Materials Research*, 881–883 (2014): 570–573. doi: https://doi.org/10.4028/www.scientific.net/AMR.881-883.570.

[62] Hu, Zhao Hong, Ahmed Mohamed Omer, Xiao Kun Ouyang, and Di Yu. "Fabrication of carboxylated cellulose nanocrystal/sodium alginate hydrogel beads for adsorption of Pb (II) from aqueous solution." *International Journal of Biological Macromolecules* 108 (2018): 149–157. doi: https://doi.org/10.1016/j.ijbiomac.2017.11.171.

[63] Srivastava, Shalini, Abhishek Kardam, and Kumar Rohit Raj. "Nanotech reinforcement onto cellulosic fibers: green remediation of toxic metals." *International Journal of Green Nanotechnology: Biomedicine* 4 (2012): 46–53. doi: https://doi.org/10.1080/19430892.2012.654744.

[64] Yang, Rui, Katherine B. Aubrecht, Hongyang Ma, Ran Wang, Robert B. Grubbs, Benjamin S. Hsiao, and Benjamin Chu. "Thiol-modified cellulose nanofibrous composite membranes for chromium (VI) and lead (II) adsorption." *Polymer 55* (2014): 1167–1176. doi: https://doi.org/10.1016/j.polymer.2014.01.043.

[65] Maatar, Wafa, and Sami Boufi. "Poly(methacylic acid-*co*-maleic acid) grafted nanofibrillated cellulose as a reusable novel heavy metal ions adsorbent." *Carbohydrate Polymers* 126 (2015): 199–207. doi: https://doi.org/10.1016/j.carbpol.2015.03.015.

[66] Suopajarvi, Terhi, Henrikki Liimatainen, Osmo Hormi, and Jouko Niinimaki. "Coagulation-flocculation treatment of municipal wastewater based on anionized nanocelluloses." *Chemical Engineering Journal* 231 (2013): 59–67. doi: https://doi.org/10.1016/j.cej.2013.07.010.

[67] Ma, Hongyang, Benjamin S. Hsiao, and Benjamin Chu. "Ultrafine cellulose nanofibers as efficient adsorbents for removal of UO_2^{2+} in water." *ACS Macro Letters* 1 (2012): 213–216. https://doi.org/10.1021/mz200047q.

[68] Kardam, Abhishek, Kumar Rohit Raj, Shalini Srivastava, and M. M. Srivastava. "Nanocellulose fibers for biosorption of cadmium, nickel, and lead ions from aqueous solution." *Clean Technologies and Environmental Policy* 16 (2014): 385–393. doi: https://doi.org/10.1007/s10098-013-0634-2.

[69] Hokkanen, Sanna, Eveliina Repo, Terhi Suopajarvi, Henrikki Liimatainen, Jouko Niinimaa, and Mika Sillanpaa. "Adsorption of Ni(II), Cu(II) and Cd(II) from aqueous solutions by amino modified nanostructured microfibrillated cellulose." *Cellulose* 21 (2014): 1471–1487. doi: https://doi.org/10.1007/s10570-014-0240-4.

[70] Chen, Shiyan, Yu Zou, Zhiyong Yan, Wei Shen, Shuaike Shi, Xiang Zhang, and Huaping Wang. "Carboxymethylated-bacterial cellulose for copper and lead ion removal." *Journal of Hazardous Materials* 161 (2009): 1355–1359. doi: https://doi.org/10.1016/j.jhazmat.2008.04.098.

[71] Singh, Kiran, T. J. M. Sinha, and Shalini Srivastava. "Functionalized nanocrystalline cellulose: smart biosorbent for decontamination of arsenic." *International Journal of Mineral Processing* 139 (2015): 51–63. doi: https://doi.org/10.1016/j.minpro.2015.04.014.

[72] Dwivedi, Amarendra Dhar, Naresh D Sanandiya, Jitendra Pal Singh, Syed M Husnain, Keun Hwa Chae, Dong Soo Hwang, and Yoon Seok Chang. "Tuning and characterizing nanocellulose interface for enhanced removal of dual-sorbate (As^V and Cr^{VI}) from water matrices." *ACS Sustainable Chemistry and Engineering* 5 (2017): 518–528. doi: https://doi.org/10.1021/acssuschemeng.6b01874.

[73] Jin, Xuchen, Zhouyang Xiang, Qingguo Liu, Yong Chen, and Fachuang Lu. "Polyethyleneimine-bacterial cellulose bioadsorbent for effective removal of copper and lead ions from aqueous solution." *Bioresource Technology* 244 (2017): 844–849. doi: https://doi.org/10.1016/j.biortech.2017.08.072.

[74] Anirudhan, T. S., and F. Shainy. "Effective removal of Hg(II) ions from chlor-alkali industrial wastewater using 2-mercaptobenzamide modified itaconic acid-grafted-magnetite nanocellulose composite." *Journal of Colloid and Interface Science* 456 (2015): 22–31. doi: https://doi.org/10.1016/j.jcis.2015.05.052.

[75] Geng, Biyao, Haiying Wang, Shuai Wu, Jing Ru, Congcong Tong, Yufei Chen, Hongzhi Liu, Shengchun Wu, and Xuying Liu. "Surface-tailored nanocellulose aerogels with thiol-functional moieties for highly efficient and selective removal of Hg(II) ions from water." *ACS Sustainable Chemistry & Engineering* 5 (2017): 11715–11726. doi: https://doi.org/10.1021/acssuschemeng.7b03188.

[76] Yu, Hou Yong, Dong Zi Zhang, Fang Lu, and Juming Yao. "New approach for single-step extraction of carboxylated cellulose nanocrystals for their use as adsorbents and flocculants." *ACS Sustainable Chemistry and Engineering* 4 (2016): 2632–2643. doi: https://doi.org/10.1021/acssuschemeng.6b00126.

[77] Qiao, Han, Yanmei Zhou, Fang Yu, Enze Wang, Yinghao Min, Qi Huang, Lanfang Pang, and Tongsen Ma. "Effective removal of cationic dyes using carboxylate-functionalized cellulose nanocrystals." *Chemosphere* 141 (2015): 297–303. doi: https://doi.org/10.1016/j.chemosphere.2015.07.078.

[78] Zhou, Chengjun, Sunyoung Lee, Kerry Dooley, and Qinglin Wu. "A facile approach to fabricate porous nanocomposite gels based on partially hydrolyzed polyacrylamide and cellulose nanocrystals for adsorbing methylene blue at low concentrations." *Journal of Hazardous Materials* 263 (2013): 334–341. doi: https://doi.org/10.1016/j.jhazmat.2013.07.047.

[79] Chan, Chi Hoong, Chin Hua Chia, Sarani Zakaria, Mohd Shaiful Sajab, and Siew Xian Chin. "Cellulose nanofibrils: a rapid adsorbent for the removal of methylene blue." *RSC Advances* 5 (2015): 18204–18212. doi: https://doi.org/10.1039/x0xx00000x.

[80] Tavakolian, Mandana, Hannah Wiebe, Mohammad Amin Sadeghi, and Theo G. M. van de Ven. "Dye removal using hairy nanocellulose: experimental and theoretical investigations." *ACS Applied Materials and Interfaces* 12 (2020): 5040–5049. doi: https://doi.org/10.1021/acsami.9b18679.

[81] Jin, Liqiang, Weigong Li, Qinghua Xu, and Qiucun Sun. "Amino-functionalized nanocrystalline cellulose as an adsorbent for anionic dyes." *Cellulose* 22 (2015): 2443–2456. doi: https://doi.org/10.1007/s10570-015-0649-4.

[82] Jin, Liqiang, Qiucun Sun, Qinghua Xu, and Yongjian Xu. "Adsorptive removal of anionic dyes from aqueous solutions using microgel based on nanocellulose and polyvinylamine." *Bioresource Technology* 197 (2015): 348–355. doi: https://doi.org/10.1016/j.biortech.2015.08.093.

[83] Pei, Aihua, Nuria Butchosa, Lars A. Berglund, and Qi Zhou. "Surface quaternized cellulose nanofibrils with high water absorbency and adsorption capacity for anionic dyes." *Soft Matter* 9 (2013): 2047–2055. doi: https://doi.org/10.1039/c2sm27344f.

[84] Chan, C. H., C. H. Chia, S. Zakaria, M. S. Sajab, and S. X. Chin. "Cellulose nanofibrils: a rapid adsorbent for the removal of methylene blue." *RSC Advances* 5 (2015): 18204–18212. doi:10.1039/x0xx00000x.

7

Potential of Nanocellulose as a Sustainable Replacement of CNTs in Water Purification Methods

Ankita Dhillon
Banasthali Vidyapith, Banasthali, India

Meena Nemiwal
Malaviya National Institute of Technology, Jaipur, India

Dinesh Kumar
Central University of Gujarat, Gandhinagar, India

CONTENTS

7.1 Introduction

The major global environmental challenges include CO_2 emissions, plastic waste accumulation, ill effects of various pollution types on human health, and lack of non-renewable energy resources. There is an urge for advanced, eco-friendly, cost-effective, sensing, and adsorbent materials to resolve these issues. Nanocellulose (NC), mixed cellulosic materials, and carbon nanotubes (CNTs) may present great potential as they may be tailored to have many functionalities. However, to access their water purification applications, numerous challenges like efficiency, assessment, proper usage, fabrication methods, and price need to be addressed. Although CNTs have been extensively used in water purification, their applications and commercialization have been restricted to luxurious products [1, 2] because of complicated and costly manufacturing practices [3–5]. This is chiefly due to easy aggregation of CNTs in water, which reduces surface area and mechanical properties and deteriorates optical and electrical properties. Therefore, their dispersion is a key step in their manufacturing procedures [6–8]. On the contrary, NC is the critical component of plant cell walls, a renewable and natural polymer adsorbent worldwide. Over the years, NC has gained wide attention because of its exceptional properties like great strength, high surface area, eco-friendly, and cost-effective nature. This chapter aims to present a thorough interpretation of NC's capacity as a justifiable substituent of CNTs in water purification methods. In the end, the chapter includes NC's application in water.

7.2 Properties of CNTs and NC

This section includes classifications, proportions, and electronical, mechanical, optical, and thermal properties of NC and CNTs to compare them to demonstrate the advantages of NC over CNTs in water purification.

7.2.1 Properties of NC

NC is considered a fascinating class of bioadsorbents for a range of remediation applications [9, 10]. The variation in NC resources and the extraction procedures results in a major variation in morphology and other aspects of the NCs. A study reported that the acid hydrolysis extraction of NC from various sources resulted in extended, elastic, and intertwined nanocellulosic fibers. In contrast, similar banana rachis and kapok treatment resulted in more discrete and rod forms [11]. The lignocellulosic resources determine the crystallographic arrangement and extent of NC crystallinity [12]. These sources also determine the thermal degradation property of NC. It has been found that compared to the α-cellulose of original lignocellulosic biomass, NC extracted from natural cellulosic resources have improved thermal constancy (Fig. 7.1) [13]. This better thermal stability of NC is due to their greater crystallinity, structural flexibility, and lower content of lignin materials having low thermal stability [11].

7.2.2 Properties of CNTs

A wound graphene sheet composed of carbon atoms present as one-dimensional tubular materials is known as CNTs. The CNTs are classified as single-wall carbon nanotubes (SWCNTs), double-wall carbon nanotubes (DWCNTs), and multi-wall carbon nanotubes (MWCNTs) (Fig. 7.1). The radius of CNTs can be in the range of 0.5–150 nm [14]. SWCNTs, DWCNTs, and MWCNTs vary in their mechanical properties. More particularly, CNTs have outstanding mechanical properties [15–17], excellent mass density [18], good thermal conductivity [19, 20], enhanced coefficient of thermal expansion [21], and

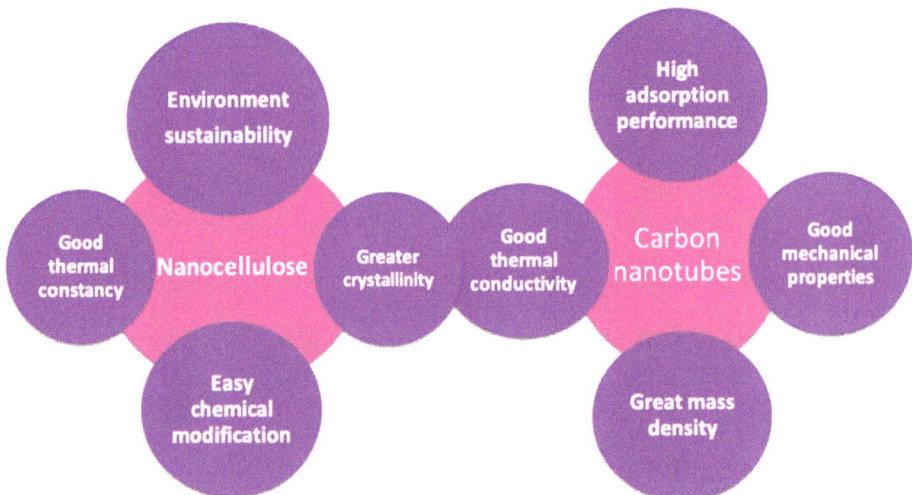

FIGURE 7.1 A schematic depiction of comparative properties of NC and CNTs.

excellent capacitance [22, 23]. Based on the graphene sheet coiling like chirality, SWCNTs are found to show both semiconducting and metallic properties [24, 25]. To use CNTs in innumerable applications, their dispersion plays an important role. Therefore, many organic substances and surfactants have been reported for the dispersion of CNTs [26]. The processes of dispersion of CNTs have been largely divided into physical alteration and chemical alteration. The introduction of covalent bonds during chemical treatment may negatively affect the superior properties of CNTs because of bond breaking within themselves. In contrast, the process of physical alteration can involve two different approaches. The first one involves hydrophobic interactions, for instance, solubilization of micelle, and the second one involves physical adsorption, for instance, π–π stacking [7]. A very stable physical adsorption of DNA molecules by π–π stacking has been carried out on SWCNTs [27, 28]. It is successfully utilized for numerous applications, for instance, in drug delivery and biosensors [29, 30]. Fig. 7.1 shows a schematic drawing of comparative properties of NC and CNTs.

7.3 Potential of NC over CNTs in Water Purification

Although CNTs are a marvelous example of nanotechnology research and advancement, they can have determinant impacts on the environment. Potential CNTs discharge into the environment can cause severe environmental issues. For example, an occupational inhalation of the nanosized carbon particles as suspended particulate matter can lead to an exposure threat when it persists in the atmosphere [31]. Similarly, there is an increased risk of leaching of CNTs in the water treatment plant afterward, a few cycles that can damage the treatment plant. If leached, microbial metabolic functions can disrupt in the treatment plant. Substantial harm to aquatic life can occur if CNTs pass into the environment [32]. They are also found responsible for the induction of lung cellular spread and pulmonary infection in rats [31]. Therefore, their persistence in the biological environment is a severe threat to water treatment plants' utilization. Yet, CNTs toxicity relies upon several factors like different CNTs, sizes, various impurities, method of synthesis, and physical extent of dispersion. The well-dispersed CNTs are less toxic than the agglomerated ones [33, 34]. Therefore, the agglomeration of nanotubes resulting from the weak cohesion between them and the matrix increases their toxicity. Interestingly, soluble CNTs have been found to have a non-toxic consequence of cell sustainability [35]. Similarly, agglomeration of pristine CNTs can occur quickly, and so are extra toxic compared to functionalized CNTs. The high price of CNTs limits their application at the field level [36]. In this regard, NC is good alternative to CNTs because of its various advantages over CNTs. NC has a negligible carbon footprint with cost-effective nature compared to CNTs [37, 38]. They are obtained from renewable resources like agricultural products and have presented exceptional nanocomposites [39]. NC has been characterized by its excellent properties like pronounced strength, distinctive morphology, superior tensile strength, high surface area, low thermal expansion coefficient, biodegradability, and cost-effective nature [40].

To be more precise, the development of NCs as novel eco-friendly solids having varied possible applications has boosted their applications as innovative bio-adsorbents. The modification of their surface to adapt various surface properties, such as reactivity, surface charge density, processability, and functionality, further enhances their applications for numerous pollutants. Compared to CNTs, the NC adsorbents exhibit many benefits, including environmental sustainability, cost-effective nature, good adsorption capacity, and adsorption selectivity by suitable functionalization of the surface. As compared to CNTs, NC eradicates the safety concerns as it is an innocuous material [41] with complete biodegradability and negligible deleterious effects on the environment [42].

These noteworthy benefits recommend NC as a sustainable replacement for CNTs. The apparent advancement and turning from CNTs to NC are due to the lower environmental, health, and safety concerns with NC compared to CNTs in water filtration membranes. Based on preparation procedures, NC is found to be less expensive than CNTs, which is due to the high energy requirement in case of CNTs for freezing of dry slurry material. Since NC have several biodegradation pathways, their chances of persistence in the environment are negligible compared with CNTs showing less concern in the lasting environmental exposures, bioaccumulation, and tropical transmission [43].

7.4 Comparative NC and CNTs' Studies in Water Purification Applications

7.4.1 Comparative NC and CNTs' Studies for Heavy Metal Ions

Earth's crust contains heavy metals as natural components. Although primary exposure to heavy metals occurs through water consumption, some of them can pass via bioaccumulation inside the body. Accumulation of these toxic ions can lead to serious human health risks, for instance, cancer and other threats.

The adsorption of heavy metals occurs by metal interaction with the adsorbent surface, having selected anionic functional groups that can be monodentate or bidentate [44, 45]. Fortunately, the color change during the interaction is a good sign of significant metal uptake [46, 47]. Likewise, Hu et al. developed beads of carboxylated cellulose nanocrystal (CNC)/sodium alginate hydrogel for the successful Pb(II) uptake from water. The same ions uptake reported by another sulfonated CNC in the same study shows that various anionic functional groups can be employed [48]. For the consecutive adsorption-desorption cycles, 0.1 M HCl was utilized as eluent with a decrease in capacity to about 66%.

Liu et al. fabricated succinic anhydride-modified CNC having a high density of carboxylate groups for Pb(II) and Cd(II) uptake from an aqueous medium [49]. Mautner et al. used the benefits of a natural adsorbent material and membrane filtration methods to develop cellulose nanofibrils (CNFs), which were further modified with phosphate groups [50]. The developed phosphorylated CNF nanopapers were then used as an ion exchanger to uptake the Cu(II) ions in dynamic filtration experiments. As compared to unmodified CNF nanopapers, phosphorylated CNF nanopapers showed reduced performance. The surface phosphate groups of the nanopaper carried out more copper uptake than the bulk functional groups inside the nanopapers. The presence of calcium ions in the same concentration had a negligible effect on copper adsorption capacity [50].

Similarly, Cu(II) ions uptake was achieved using novel chitosan-coated CNTs composites adsorbent [51]. The adsorbent was fabricated using combined Diels-Alder reaction and mercaptoacetic acid locking imine reaction. As compared to pristine CNTs with similar sulfonic groups, the formulated CNTs-CHO-CS composites adsorbent presented higher Cu(II) uptake with good adsorption efficiency at the pH range of 7–8. However, the adsorption performance was lower as compared to CNF nanopapers due to their higher surface area and bulk electrostatic interaction between active sites and heavy metal ions. In analogy to conventional CNTs, the nanopapers presented good regeneration without significant leaching [51]. The mechanical properties of adsorption filter were further improved by Li et al. by developing a new adsorption filter of carboxylated cellulose fabrics prepared by controlled 2,2,6,6-tetramethyl-1-piperidinyloxy (TEMPO)-mediated oxidation [52]. The adsorption filter's improved mechanical properties were guaranteed by woven structure of cellulose fabrics, whereas the functional sites localized on the outside. Carboxylated groups provided the inside of the fibers for the simultaneous uptake of heavy metal ions and dyes. The dynamic adsorption performance studies presented an unexpected filtration rate, which was tenfold greater than that of unmodified cellulose fabrics [52]. In a recent survey, Rashid et al. studied Pb(II) remediation applications using sulfonated polyethersulfone/multi-wall carbon nanotubes composite (SPES/MWCNTs) from wastewaters through ion exchange process [53]. In the fabrication process, MWCNTs were implanted with modified SPES. It was synthesized using polyethersulfone (PES) and concentrated H_2SO_4 solution refluxing for 3 hours at 60 °C. The developed chemisorbent presented the highest adsorption of 83.18% compared to parental materials. The adsorption-desorption studies showed a significant loss in chemisorption efficiency of the adsorbent for three successive cycles. Although developed adsorbents showed good absorbent behavior, the chemical stability and reusability of functional adsorbents were broader issues than nanocellulosic adsorbents [53].

Similarly, NC was chemically modified by Yu et al. by dissolving it in AmimCl ionic liquid and covalent interaction with functionalized graphene oxide (GO) to prepare nanocrystalline cellulose/graphene oxide (NCC/GO) composites adsorbent [54]. The esterification reaction involved a covalent modification of NC with functionalized GO that destabilized the molecular hydrogen bonding and increased the chemical reactivity. The functional groups' abundance and development of multiple adsorption sites like carboxylic, alcoholic, ketonic, etc. considerably increased because of introducing GO [54]. Huang et al.

fabricated a novel adsorbent system based on NC by upcycling sugarcane waste [55]. The approach resulted in the formation of a suspension of nanoscale adsorbents, which after electrostatic interactions with negatively charged pollutants resulted in NC's coagulation, easing the secondary contaminant remediation. The authors carried out metaperiodate oxidation of cellulose followed by cationizing to eventually develop cationic dialdehyde cellulose (C-DAC) nanofibers using Grignard's reagent. The resultant nanofibers were used for Cr(VI) ions uptake, with the maximum charge density having C-DAC demonstrated a maximum adsorption capacity of 80.5 mg/g [55]. For the first time, Kumar et al. fabricated a dual, competent, and eco-friendly composite, namely, polyaniline-impregnated NC composite (PANI-NC), for rapid uptake of Cr(VI) ions in wastewaters along with wastewater cleaning [56]. The PANI-NC was utilized in columns as pebbles and tea bags to easily separate the powdered adsorbent, making the complete process adaptable. The developed composite presented 96% adsorption efficiency toward chromium and 99% remediation of other industrial dyes. Therefore, it is not merely developed for lab-scale applications but also as an industrial wastewater cleaner [56]. In a study by Sambaza et al., MWCNTs were used to develop branched polyethyleneimine-multi-walled carbon nanotube (PEI-MWCNT) polymeric nanocomposite adsorbents to remove Cr(VI) ions [57]. Maximum sorption of Cr(VI) ions was achieved at pH $\leq$ 4, 60 minutes of contact time. Compared to NC-based adsorbents, the expensive nature, rapid agglomeration, and metallic impurities of MWCNTs hindered their large-scale applications [57]. Although numerous NC-based adsorbents carried out efficient heavy metal ions uptake from water, their successful recovery from solution after interactions is necessary practical applications that need to be addressed. The problem was successfully addressed by Wei et al. by developing hybrid magnetic aerogel [58]. The adsorbent was developed by integrating Fe_3O_4 nanoparticles with NC and then successfully utilizing physical adsorption for Cr(VI) ions. The magnetic adsorbent owned notable ferromagnetic property allowing efficient and controlled recovery of the aerogel. The developed adsorbent showed comparable adsorption performance for both Pb(II) and Cu(II) ions, signifying extended hybrid aerogel applications for additional heavy metal ions uptake [58]. Similarly, the capability of CNTs modified with four generations of poly-amidoamine dendrimer (PAMAM, G4) has been studied for the Pb(II) and Cu(II) ions uptake in single and binary component systems [59]. After the careful examination of various experimental factors, unique highest adsorption capacity of Pb(II), i.e., 4870 mg/g, and Cu(II), i.e., 3333 mg/g, was achieved at pH 7.0 in both single and binary phases [59]. Although the developed nanocomposite functioned as a superadsorbent for uniquely large quantities of Pb(II) and Cu(II) ions uptake compared to NC, the aggregation of CNTs suspension into an aqueous solution, environmental sustainability, and cost-ineffective nature are considerable issues. That is why a more potent functionalization is essentially required for improved sustainability [59]. In a study, a simplistic and competent crosslinking method was applied to develop NC-based amino-functionalized aerogel adsorbent for effectual uptake of Cu(II) ions [60]. Self-prepared adsorbent assembly of TEMPO-oxidized CNF (TO-CNF) crosslinked trimethylol propane-tris-(2-methyl-1-aziridine) propionate (TMPTAP) and polyethyleneimine (PEI). The TOCNF/TMPTAP/PEI aerogel developed abundant oxygen-containing and amino groups. The developed aerogel composite signified many distinguished novelties. First, the developed aerogel composite presented the highest adsorption capacity of 485.4 mg/g compared to other related cellulose-based biosorbent. Second, it presented outstanding recyclability lacking any noteworthy adsorption performance reduction above four cycles. Afterward, the crosslinking tactic used for adsorbent fabrication presented as a simplistic, eco-friendly, inexpensive, and competent approach. Finally, the developed TMPTAP moiety functioned as a crosslinker and presented coordination binding sites of Cu(II) ions [60]. Guo et al. used CNFs of high aspect ratio and carbon dots (CDs) high-performance luminescence to develop fluorescent nanocellulosic hydrogel, namely, carboxymethylated cellulose nanofibrils modified with carbon dots (CM-CNF-CDs) [61]. The fabricated material served both as an efficient adsorbent for the uptake of toxic metals and an optical sensor for heavy metals recognition. The synthesized hydrogel presented maximum adsorption capacities of 769, 212, 1246, and 2056 mg/g for Fe (III), Ba(II), Cu(II), and Pb(II), respectively. The excellent adsorption capacities owned to the 3D structures developed hydrogel and large functional groups like amide, hydroxyl, and carboxyl to facilitate the accumulation and uptake of heavy metal ions. The developed fluorescent hydrogel successfully enhanced fluorescent CDs stability and sensitivity to detect heavy metals' concentration [61].

In another study, GO and functionalized f-MWCNTs were investigated for their sorption efficiencies in treating mine drainage contaminated with Cu(II), Mn(II), Zn(II), Pb(II), Fe(II), and Cd(II) metal ions [62]. The authors used the modified Hummers' method and subsequent acid treatment of the MWCNTs for the development of several layered GO nanosheets and f-MWCNTs, thereby increasing the metal ion uptake. The characterization studies confirmed that the increased oxidation of developed adsorbent increased the adsorption capacities of the MWCNTs. Although the work presented an adsorbent with promising potential for metal ions uptake, further studies will be needed to find the modes of decreasing agglomeration process to improve adsorbent reusability [62] further.

7.4.2 Comparative NC and CNT's Studies for Dyes

The pharmaceutical, paper, pulp, paints, textile, and plastics industries commonly use dyes in their various applications. Because of their recalcitrant nature, their discharge into the water bodies imposes a serious risk to living bodies. Therefore, there is a need to remediate them from wastewaters effectively. NC having cationic moieties are used for anionic dyes remediation. Zhu et al. developed cationic NC via oxidation using sodium periodate; afterward, hyperbranched PEI reaction on dialdehyde functionalized cellulose powder was carried out [63]. The developed composite exhibited a great adsorption capacity of 2100 mg/g toward Congo red and 1860 mg/g toward basic yellow [63]. Another Congo red dye removal study was conducted by Arabi et al. from aqueous systems. For enhanced interaction between adsorbent and dye, the adsorbent surface was modified via surface oxidation of MWCNTs and zinc oxide (ZnO) [64]. The adsorbent presented improved adsorption efficiency with increasing adsorbent dosage, the reaction mixture's temperature, and contact time. Although the ZnO/MWCNTs nanocomposite showed the ability to adsorb Congo red dye, the agglomeration of CNTs can be extra toxic than NC materials. The high price of CNTs limits their application at the field level [64]. Jin et al. developed amino-functionalized nanocrystalline cellulose (ANCC) composite by following the subsequent steps. First, nanocrystalline cellulose (NCC) was extracted from fully bleached hardwood kraft pulp hydrolysis using sulfuric acid [65]. NCC oxidation was achieved by using sodium periodate, subsequently yielding C_2/C_3 dialdehyde NC (DANC). Finally, the formation of ANCC was achieved by ethylenediamine grafting via reductive amination treatment. During the treatment process, it was found that increases in ethylenediamine dosage increased the ANCC's primary amine groups. The resulting ANCC adsorbent then successfully removed anionic dyes like Congo red 4BS, acid red GR, and reactive light yellow K-4G from wastewater [65]. Cellulose nanoparticles, also known as hairy NC, which are easily synthesized and are eco-friendly, have been developed for dye removal. In this manner, Tavakolian et al. developed a negatively charged hairy NC electrostatically stabilized nanocrystalline cellulose (ENCC) for methylene blue uptake [66]. The developed ENCC adsorbent showed a much higher adsorption capacity of 1400 mg/g than other similar adsorbents reported in the past. Further, composite hydrogel beads of sodium alginate and ENCC (ALG-ENCC beads) were prepared to ease their handling during their employment on a larger dyes remediation scale. The developed beads still presented a high removal capacity of 1250 mg/g during the methylene blue dye [66]. For the first time, Bhakta et al. developed maghemite nanocrystals-adorned MWCNTs by IR irradiation, disabling major cons of the approaches for the CNTs decoration with maghemite nanoparticles [67]. Treatment with sodium hydroxide resulted in the purification of MWCNTs (p-MWCNTs). Subsequently, the developed p-MWCNTs treatment with maghemite nanocrystals (p-MWCNTs/MC) led to the final adsorbent formation. The p-MWCNTs/MC adsorbent showed good adsorption capacity toward methylene blue dye. Compared to other NC, they have a high carbon footprint with a costly nature, limiting their applications in domestic and eco-friendly dye removal methods [67]. In a study, selected H-ZSM-5 form of zeolites having various Si/Al ratios (P26, P371) were modified in various ways, and advanced water-stable films were prepared from CNFs having high surface area and good flexibility [68]. The fabricated films were used for the black anionic reactive dye uptake in textile wastewater. The structure and surface chemistry effects of zeolite on dye uptake were studied, and nearly 80% of dye uptake occurred within 20 minutes. The performance of filtration films was also evaluated, and it displayed an extremely high flux rate with 45% dye uptake efficacy, presenting their potential as a good filter membrane [68]. Maleki et al. addressed the negligible suspension of CNTs into an aqueous solution by developing amine-functionalized MWCNT-NH$_2$ [69]. The fabricated material was verified for the black anionic reactive dyes uptake in textile wastewater. The experimental studies directed stronger CNT-NH$_2$ interactions with black anionic reactive dyes, resulting in their higher adsorption in single and binary phases of dyes. Integrating CNTs and amine groups resulted in manifold increase in dye adsorption than unmodified ones, making it an appropriate adsorbent for dye uptake. However, since their chances of persistence in the environment are very high, NC shows deeper concern about lasting environmental exposures, bioaccumulation, and tropical transmission [69]. A facile electrospinning procedure was used to develop porous nanocomposite gels involving CNCs as crosslinker and heating based on partially hydrolyzed polyacrylamide (HPAM) [70]. The developed nanocomposite membranes achieve water insolubility because of crosslinking during esterification of HPAM and CNCs under thermal conditions. Further, the porous nanocomposite's rapid swelling was realized with an efficient methylene blue absorptivity from wastewater [70]. Similarly, incorporating CNC in keratin sponge matrix was achieved to develop efficient and biodegradable bioadsorbents for the dye's uptake from wastewater [71]. Adsorption studies presented excellent adsorption capacities of 1201 mg/g for Reactive Black (RB5) and 1070 mg/g for Direct Red 80 dye [71]. Similarly, RB5 adsorption studies were also conducted using MWCNTs from aqueous solutions [71]. Although experimental studies presented good adsorption capacity of RB5 on MWCNTs at solute concentration between 10 and 200 mg/L within 1 hour, the adsorbent has the drawbacks of high cost in both manufacturing and regeneration parts compared to NC adsorbents [72].

7.4.3 Comparative NC and CNT's Studies for Other Pollutants

Various industries, such as petrochemical, nutrition, and medicinal, generate many pollutants, primarily organic pollutants. Even their presence in trace concentration makes water unhealthy for various purposes. Therefore, it is imperative to remove them from the water bodies. Cellulose nanomaterials can be efficient candidates for many pollutants because of high surface area and functionality. Sehaqui et al. prepared cationic ammonium-modified CNFs for the uptake of anions like nitrate, fluoride, phosphate, and sulfate [73]. Since the nanofibers showed the highest nitrates adsorption, they prepared permeable nanopaper as adsorption membrane for nitrate uptake from wastewater [73]. Jodeh et al. studied the adsorption of nitrate ions by strong complexation properties, which was accomplished by developing MWCNTs functionalized with chitosan [74]. Although the adsorbent presented good adsorption behavior, they can negatively impact the environment [74]. On the other hand, NC materials have several existing pathways of biodegradation. Their chances of persistence in the environment are negligible compared with CNTs showing less concern in the lasting environmental exposures, bioaccumulation [43].

There has been a growing interest in developing remediation methods for diuron because of its negative effect on ecosystem integrity and living beings. Hence, Liu et al. prepared an efficient NC-based biocomposite having *Arthrobacter globiformis* D47 bacterium for the quick degradation of diuron [75]. The developed NC composites proved as an alternative bioremediator toward organic xenobiotics [75]. Similarly, the quick degradation of diuron pesticide was also achieved using MWCNTs [76]. The adsorbent showed highest performance toward diuron at 100 μg/L diuron concentration, 1 hour contact time, and pH 7.0. However, the increased removal efficiency of 90% was achieved still; the costly nature of CNTs adsorbents limits their large-scale applications [76].

NC nanocomposites are obtained from renewable resources like agricultural products, which are cheaper to use in large treatment plants [39]. A high-speed mixing and freeze-drying process was used for the synthesis of spherical magnetic superabsorbents using hydrophobized CNFs [77]. The developed adsorbent showed exceedingly high performance toward both oil and organic pollutants adsorption. The adsorbent's magnetic behavior made easy collection after using an external magnetic field [77]. A similar technique was used to synthesize magnetic NC aerogel (NCA) for both oil and organic pollutant remediation [78]. Both the adsorption property of oleic acid (OA) and magnetic property of Fe$_3$O$_4$ were utilized to develop NCA/OA/Fe$_3$O$_4$ aerogels, which showed 56.32, 68.06, and 33.24 g/g of adsorption capacities to ethyl acetate, cyclohexane, and vacuum pump oil, consequently [78]. Similar pollutants were remediated using MWCNTs decorated with magnetite MWCNTs modified with silicon oxide (MWCNTs/ magnetite/SiO$_2$) [79]. As a result of good magnetic separability, fast operation enhanced performance, and having both hydrophobic and hydrophilic properties, they are the right candidate in removing dispersant and spilled oil from seawater [79]. However, the agglomeration problem associated with CNTs makes them extra toxic compared to other biodegradable alternatives. The high price of CNTs limits their application at the field level [36]. Therefore, NC is a suitable alternative to CNTs because of having a negligible carbon footprint with cost-effective nature compared to CNTs.

7.5 Conclusion

NC signifies a novel sustainable adsorbents having exceptional optical, structural, and mechanical properties. In this chapter, NC's beneficial role is represented in water treatment because of their remarkable properties like high surface area, low toxicity, good mechanical strength, efficient functionality, and ecological viability. A comparative revision of both NC and CNTs involving physical, chemical, and mechanical properties, cost-effective analysis, usage, and disposal has been carried out to show NC's potential as a sustainable replacement for CNTs. To be more precise, the development of NCs as novel eco-friendly solids having varied possible applications has boosted their applications as innovative bioadsorbents. Additionally, the modification of their surface to adapt various surface properties, such as reactivity, surface charge density, processability, and functionality, further enlarges their applications for many pollutants. Compared to CNTs, the NC adsorbents exhibit many benefits, including environmental sustainability, cost-effective nature, adequate adsorption capacity, and adsorption selectivity by suitable functionalization of the surface.

Acknowledgment

We are thankful to Professor Aditya Shastri, Vice-Chancellor of Banasthali Vidyapith, to kindly extend the facilities of Banasthali Centre for Education and Research in Basic Sciences.

REFERENCES

[1] De las Casas, Charles, and Wenzhi Li. "A review of application of carbon nanotubes for lithium ion battery anode material." *Journal of Power Sources* 208 (2012): 74–85. doi.org/10.1016/j.jpowsour.2012.02.013.

[2] Rao, Rahul, Cary L. Pint, Ahmad E. Islam, Robert S. Weatherup, Stephan Hofmann, Eric R. Meshot, Fanqi Wu et al. "Carbon nanotubes and related nanomaterials: critical advances and challenges for synthesis toward mainstream commercial applications." *ACS Nano* 12 (2018): 11756–11784. http://doi.org/10.1021/acsnano.8b06511.

[3] Rahman, Gul, Zainab Najaf, Asad Mehmood, Salma Bilal, Shabeer Ahmad Mian, and Ghulam Ali. "An overview of the recent progress in the synthesis and applications of carbon nanotubes." *C—Journal of Carbon Research* 5 (2019): 3. doi.org/10.3390/c5010003.

[4] Green, Alexander A., Matthew C. Duch, and Mark C. Hersam. "Isolation of single-walled carbon nanotube enantiomers by density differentiation." *Nano Research* 2 (2009): 69–77. doi.org/10.1007/s12274-009-9006-y.

[5] Liu, Huaping, Daisuke Nishide, Takeshi Tanaka, and Hiromichi Kataura. "Large-scale single-chirality separation of single-wall carbon nanotubes by simple gel chromatography." *Nature Communications* 2 (2011): 1–8. doi.org/10.1038/ncomms1313.

[6] Xu, Zhaoyang, Xiangdong Jiang, Sicong Tan, Weibing Wu, Jiangtao Shi, Huan Zhou, and Peng Chen. "Preparation and characterisation of CNF/MWCNT carbon aerogel as efficient adsorbents." *IET Nanobiotechnology* 12 (2018): 500–504. doi:10.1049/iet–nbt.2017.0234.

[7] Ma, Peng-Cheng, Naveed A. Siddiqui, Gad Marom, and Jang-Kyo Kim. "Dispersion and functionalization of carbon nanotubes for polymer-based nanocomposites: a review." *Composites Part A: Applied Science and Manufacturing* 41 (2010): 1345–1367. doi.org/10.1016/j.compositesa.2010.07.003.

[8] Imai, Masanori, Kousuke Akiyama, Tomo Tanaka, and Eiichi Sano. "Highly strong and conductive carbon nanotube/cellulose composite paper." *Composites Science and Technology* 70 (2010): 1564–1570. doi.org/10.1016/j.compscitech.2010.05.023.

[9] Lu, Chungsying, Huantsung Chiu, and Hsunling Bai. "Comparisons of adsorbent cost for the removal of zinc (II) from aqueous solution by carbon nanotubes and activated carbon." *Journal of Nanoscience and Nanotechnology* 7 (2007): 1647–1652. doi.org/10.1166/jnn.2007.349.

[10] Cherian, Bibin Mathew, Alcides Lopes Leão, Sivoney Ferreira De Souza, Sabu Thomas, Laly A. Pothan, and M. Kottaisamy. "Isolation of nanocellulose from pineapple leaf fibres by steam explosion." *Carbohydrate Polymers* 81 (2010): 720–725. doi.org/10.1016/j.carbpol.2010.03.046.

[11] Lu, Zexiang, Liwei Fan, Huaiyu Zheng, Qilin Lu, Yiqiang Liao, and Biao Huang. "Preparation, characterization and optimization of nanocellulose whiskers by simultaneously ultrasonic wave and microwave assisted." *Bioresource Technology* 146 (2013): 82–88. doi.org/10.1016/j.biortech.2013.07.047.

[12] Deepa, B., Eldho Abraham, Nereida Cordeiro, Miran Mozetic, Aji P. Mathew, Kristiina Oksman, Marisa Faria, Sabu Thomas, and Laly A. Pothan. "Utilization of various lignocellulosic biomass for the production of nanocellulose: a comparative study." *Cellulose* 22 (2015): 1075–1090. doi.org/10.1007/s10570-015-0554-x.

[13] Le Bras, David, Maria Strømme, and Albert Mihranyan. "Characterization of dielectric properties of nanocellulose from wood and algae for electrical insulator applications." *Journal of Physical Chemistry B* 119 (2015): 5911–5917. doi.org/10.1021/acs.jpcb.5b00715.

[14] Abraham, E., B. Deepa, L. A. Pothan, M. Jacob, S. Thomas, U. Cvelbar, and R. Anandjiwala. "Extraction of nanocellulose fibrils from lignocellulosic fibres: a novel approach." *Carbohydrate Polymers* 86 (2011): 1468–1475. doi.org/10.1016/j.carbpol.2011.06.034.

[15] Iijima, Sumio, and Toshinari Ichihashi. "Single-shell carbon nanotubes of 1-nm diameter." *Nature* 363 (1993): 603–605. doi.org/10.1038/363603a0.

[16] Peng, Bei, Mark Locascio, Peter Zapol, Shuyou Li, Steven L. Mielke, George C. Schatz, and Horacio D. Espinosa. "Measurements of near-ultimate strength for multiwalled carbon nanotubes and irradiation-induced crosslinking improvements." *Nature Nanotechnology* 3 (2008): 626. doi.org/10.1038/nnano.2008.211.

[17] Esconjauregui, Santiago, Rongsie Xie, Martin Fouquet, Richard Cartwright, David Hardeman, Junwei Yang, and John Robertson. "Measurement of area density of vertically aligned carbon nanotube forests by the weight-gain method." *Journal of Applied Physics* 113 (2013): 144309. doi.org/10.1063/1.4799417

[18] Krishnan, A., E. Dujardin, T. W. Ebbesen, P. N. Yianilos, and M. M. J. Treacy. "Young's modulus of single-walled nanotubes." *Physical Review B* 58 (1998): 14013. doi.org/10.1103/PhysRevB.58.14013.

[19] Filleter, T., R. Bernal, S. Li, and Horacio Dante Espinosa. "Ultrahigh strength and stiffness in crosslinked hierarchical carbon nanotube bundles." *Advanced Materials* 23 (2011): 2855–2860. doi.org/10.1002/adma.201100547.

[20] Kim, Philip, Li Shi, Arun Majumdar, and Paul L. McEuen. "Thermal transport measurements of individual multiwalled nanotubes." *Physical Review Letters* 87 (2001): 215502. doi.org/10.1103/PhysRevLett.87.215502.

[21] Pop, Eric, David Mann, Qian Wang, Kenneth Goodson, and Hongjie Dai. "Thermal conductance of an individual single-wall carbon nanotube above room temperature." *Nano Letters* 6 (2006): 96–100. doi.org/10.1021/nl052145f.

[22] Deng, Libo, Robert J. Young, Ian A. Kinloch, Rong Sun, Guoping Zhang, Laure Noé, and Marc Monthioux. "Coefficient of thermal expansion of carbon nanotubes measured by Raman spectroscopy." *Applied Physics Letters* 104 (2014): 051907. doi.org/10.1063/1.4864056.

[23] An, K. H., Kwan Koo Jeon, Won Seok Kim, Young Soo Park, Seong Chu Lim, Dong Jae Bae, and Young Hee Lee. "Characterization of supercapacitors using single walled carbon nanotube electrodes." *Journal – Korean Physical Society* 39 (2001): S511–S517. doi:10.1.1.468.7119.

[24] Ye, Jian-Shan, Xiao Liu, Hui Fang Cui, Wei-De Zhang, Fwu-Shan Sheu, and Tit Meng Lim. "Electrochemical oxidation of multi-walled carbon nanotubes and its application to electrochemical double layer capacitors." *Electrochemistry Communications* 7 (2005): 249–255. doi.org/10.1016/j.elecom.2005.01.008.

[25] Hata, Kenji, Don N. Futaba, Kohei Mizuno, Tatsunori Namai, Motoo Yumura, and Sumio Iijima. "Water-assisted highly efficient synthesis of impurity-free single-walled carbon nanotubes." *Science* 306 (2004): 1362–1364. doi:10.1126/science.1104962.

[26] Yu, LePing, Cameron Shearer, and Joseph Shapter. "Recent development of carbon nanotube transparent conductive films." *Chemical Reviews* 116 (2016): 13413–13453. doi.org/10.1021/acs.chemrev.6b00179.

[27] Umemura, Kazuo. "Hybrids of nucleic acids and carbon nanotubes for nanobiotechnology." *Nanomaterials* 5 (2015): 321–350. doi.org/10.3390/nano5010321.

[28] Zheng, Ming, Anand Jagota, Ellen D. Semke, Bruce A. Diner, Robert S. McLean, Steve R. Lustig, Raymond E. Richardson, and Nancy G. Tassi. "DNA-assisted dispersion and separation of carbon nanotubes." *Nature Materials* 2 (2003): 338–342. doi.org/10.1038/nmat877.

[29] Nakashima, Naotoshi, Shingo Okuzono, Hiroto Murakami, Tonau Nakai, and Kenichi Yoshikawa. "DNA dissolves single-walled carbon nanotubes in water." *Chemistry Letters* 32 (2003): 456–457. doi. org/10.1246/cl.2003.456.

[30] Schroeder, Vera, Suchol Savagatrup, Maggie He, Sibo Lin, and Timothy M. Swager. "Carbon nanotube chemical sensors." *Chemical Reviews* 119 (2018): 599–663. doi.org/10.1021/acs.chemrev.8b00340.

[31] Zaporotskova, Irina V., Natalia P. Boroznina, Yuri N. Parkhomenko, and Lev V. Kozhitov. "Carbon nanotubes: sensor properties. A review." *Modern Electronic Materials* 2 (2016): 95–105. doi.org/10.1016/j. moem.2017.02.002.

[32] Lam, Chiu-wing, John T. James, Richard McCluskey, Sivaram Arepalli, and Robert L. Hunter. "A review of carbon nanotube toxicity and assessment of potential occupational and environmental health risks." *Critical Reviews in Toxicology* 36 (2006): 189–217. doi.org/10.1080/10408440600570233.

[33] Upadhyayula, Venkata K. K., Shuguang Deng, Martha C. Mitchell, and Geoffrey B. Smith. "Application of carbon nanotube technology for removal of contaminants in drinking water: a review." *Science of the Total Environment* 408 (2009): 1–13. doi.org/10.1016/j.scitotenv.2009.09.027.

[34] Wick, Peter, Pius Manser, Ludwig K. Limbach, Ursula Dettlaff-Weglikowska, Frank Krumeich, Siegmar Roth, Wendelin J. Stark, and Arie Bruinink. "The degree and kind of agglomeration affect carbon nanotube cytotoxicity." *Toxicology Letters* 168 (2007): 121–131. doi.org/10.1016/j.toxlet.2006.08.019.

[35] Magrez, Arnaud, Sandor Kasas, Valerie Salicio, Nathalie Pasquier, Jin Won Seo, Marco Celio, Stefan Catsicas, Beat Schwaller, and Laszlo Forro. "Cellular toxicity of carbon-based nanomaterials." *Nano Letters* 6 (2006): 1121–1125. doi.org/10.1021/nl060162e.

[36] Dumortier, Hélène, Stéphanie Lacotte, Giorgia Pastorin, Riccardo Marega, Wei Wu, Davide Bonifazi, Jean-Paul Briand, Maurizio Prato, Sylviane Muller, and Alberto Bianco. "Functionalized carbon nanotubes are non-cytotoxic and preserve the functionality of primary immune cells." *Nano Letters* 6 (2006): 1522–1528. doi.org/10.1021/nl061160x.

[37] Caneba, G. T., C. Dutta, V. Agrawal, and M. Rao. "Novel ultrasonic dispersion of carbon nanotubes." *Journal of Minerals and Materials Characterization and Engineering* 9 (2010): 165. https://www.scirp. org/html/20718.html.

[38] Olivito, R. S., O. A. Cevallos, and A. Carrozzini. "Development of durable cementitious composites using sisal and flax fabrics for reinforcement of masonry structures." *Materials & Design* 57 (2014): 258–268. doi.org/10.1016/j.matdes.2013.11.023.

[39] Chakraborty, Sumit, Sarada P. Kundu, Aparna Roy, Basudam Adhikari, and Subhasish B. Majumder. "Effect of jute as fiber reinforcement controlling the hydration characteristics of cement matrix." *Industrial & Engineering Chemistry Research* 52 (2013): 1252–1260. doi.org/10.1021/ie300607r.

[40] Boufi S., Kaddami H., and Dufresne A., "Mechanical performance and transparency of nanocellulose reinforced polymer nano composites." *Macromolecular Material Engineering* 299 (2014): 560–568. doi.org/10.1002/mame.201300232.

[41] Takagi H., and Asano A. "Characterization of green composites reinforced by cellulose nanofibers." *Key Engineering Materials* 334 (2007): 389–392. doi.org/10.4028/www.scientific.net/KEM.334–335.389.

[42] Vartiainen, Jari, Tiina Pöhler, Kristiina Sirola, Lea Pylkkänen, Harri Alenius, Jouni Hokkinen, Unto Tapper et al. "Health and environmental safety aspects of friction grinding and spray drying of microfibrillated cellulose." *Cellulose* 18 (2011): 775–786. doi.org/10.1007/s10570–011–9501–7.

[43] Bhatnagar, Amit, Mika Sillanpaa, and Anna Witek-Krowiak. "Agricultural waste peels as versatile biomass for water purification: a review." *Chemical Engineering Journal* 270 (2015): 244–271. doi. org/10.1016/j.cej.2015.01.135.

[44] Li, Qingqing, Sean McGinnis, Cutter Sydnor, Anthony Wong, and Scott Renneckar. "Nanocellulose life cycle assessment." *ACS Sustainable Chemistry & Engineering* 1, no. 8 (2013): 919–928. doi.org/10.1021/ sc4000225.

[45] Hokkanen, Sanna, Amit Bhatnagar, and Mika Sillanpää. "A review on modification methods to cellulose-based adsorbents to improve adsorption capacity." *Water Research* 91 (2016): 156–173. doi. org/10.1016/j.watres.2016.01.008.

[46] Liu, Peng, Pere Ferrer Borrell, Mojca Božič, Vanja Kokol, Kristiina Oksman, and Aji P. Mathew. "Nanocelluloses and their phosphorylated derivatives for selective adsorption of Ag⁺, Cu²⁺ and Fe³⁺ from industrial effluents." *Journal of Hazardous Materials* 294 (2015): 177–185. doi.org/10.1016/j.jhazmat.2015.04.001.

[47] Moon, Robert J., Gregory T. Schueneman, and John Simonsen. "Overview of cellulose nanomaterials, their capabilities and applications." *JOM* 68 (2016): 2383–2394. doi.org/10.1007/s11837–016–2018–7.

[48] Hu, Zhao-Hong, Ahmed Mohamed Omer, Xiao-kun Ouyang, and Di Yu. "Fabrication of carboxylated cellulose nanocrystal/sodium alginate hydrogel beads for adsorption of Pb (II) from aqueous solution." *International Journal of Biological Macromolecules* 108 (2018): 149–157. doi.org/10.1016/j.ijbiomac.2017.11.171.

[49] Liu, Huijuan, Fan Yang, Yuming Zheng, Jin Kang, Jiuhui Qu, and J. Paul Chen. "Improvement of metal adsorption onto chitosan/*Sargassum* sp. composite sorbent by an innovative ion-imprint technology." *Water Research* 45 (2011): 145–154. doi.org/10.1016/j.watres.2010.08.017.

[50] Mautner, Andreas, H. A. Maples, T. Kobkeatthawin, V. Kokol, Zoheb Karim, K. Li, and A. Bismarck. "Phosphorylated nanocellulose papers for copper adsorption from aqueous solutions." *International Journal of Environmental Science and Technology* 13 (2016): 1861–1872. doi:10.1007/s13762–016–1026–z.

[51] Dou, Jibo, Defu Gan, Qiang Huang, Meiying Liu, Junyu Chen, Fengjie Deng, Xiaoli Zhu, Yuanqing Wen, Xiaoyong Zhang, and Yen Wei. "Functionalization of carbon nanotubes with chitosan based on MALI multicomponent reaction for Cu²⁺ removal." *International Journal of Biological Macromolecules* 136 (2019): 476–485. doi.org/10.1016/j.ijbiomac.2019.06.112.

[52] Li, Cong, Hongyang Ma, Shyam Venkateswaran, and Benjamin S. Hsiao. "Highly efficient and sustainable carboxylated cellulose filters for removal of cationic dyes/heavy metals ions." *Chemical Engineering Journal* 389 (2020): 123458. doi.org/10.1016/j.cej.2019.123458.

[53] Rashid, Jamshaid, Rafia Azam, Rajeev Kumar, Mahtab Ahmad, Adeela Rehman, and M. A. Barakat. "Sulfonated polyether sulfone reinforced multi-wall carbon nanotubes composite for the removal of lead in wastewater." *Applied Nanoscience* 9 (2019): 1695–1705. doi.org/10.1007/s13204–019–00953–2.

[54] Yu, Haibiao, Shaobo Zhang, Yu Wang, Dingwen Yin, and Jintian Huang. "Covalent modification of nanocellulose (NCC) by functionalized graphene oxide (GO) and the study of adsorption mechanism." *Composite Interfaces* (2020): 1–14. doi.org/10.1080/09276440.2020.1731276.

[55] Huang, Xiangyu, Guilherme Dognani, Pejman Hadi, Mengying Yang, Aldo Eloizo Job, and Benjamin S. Hsiao. "Cationic dialdehyde nanocellulose from sugarcane bagasse for efficient chromium (VI) removal." *ACS Sustainable Chemistry & Engineering* 8 (2020): 4734–4744. doi.org/10.1021/acssuschemeng.9b06683.

[56] Kumar, Nitesh, Abhishek Kardam, V. K. Jain, and Suman Nagpal. "A rapid, reusable polyaniline-impregnated nanocellulose composite-based system for enhanced removal of chromium and cleaning of waste water." *Separation Science and Technology* 55 (2020): 1436–1448. doi.org/10.1080/01496395.2019.1600552.

[57] Sambaza, Shepherd S., Monaheng L. Masheane, Soraya P. Malinga, Edward N. Nxumalo, and Sabelo D. Mhlanga. "Polyethyleneimine-carbon nanotube polymeric nanocomposite adsorbents for the removal of Cr⁶⁺ from water." *Physics and Chemistry of the Earth, Parts A/B/C* 100 (2017): 236–246. doi.org/10.1016/j.pce.2016.08.002.

[58] Wei, Jie, Zhixing Yang, Yun Sun, Changkai Wang, Jilong Fan, Guoyin Kang, Rong Zhang, Xiaoying Dong, and Yongfeng Li. "Nanocellulose-based magnetic hybrid aerogel for adsorption of heavy metal ions from water." *Journal of Materials Science* 54 (2019): 6709–6718. doi.org/10.1007/s10853–019–03322–0.

[59] Hayati, Bagher, Afshin Maleki, Farhood Najafi, Hiua Daraei, Fardin Gharibi, and Gordon McKay. "Super high removal capacities of heavy metals (Pb²⁺ and Cu²⁺) using CNT dendrimer." *Journal of Hazardous Materials* 336 (2017): 146–157. doi.org/10.1016/j.jhazmat.2017.02.059.

[60] Mo, Liuting, Huiwen Pang, Yi Tan, Shifeng Zhang, and Jianzhang Li. "3D multi-wall perforated nanocellulose-based polyethylenimine aerogels for ultrahigh efficient and reversible removal of Cu (II) ions from water." *Chemical Engineering Journal* 378 (2019): 122157. doi.org/10.1016/j.cej.2019.122157.

[61] Guo, Xin, Dong Xu, Hanmeng Yuan, Qiuyan Luo, Shiyao Tang, Liu, and Yiqiang Wu. "A novel fluorescent nanocellulosic hydrogel based on carbon dots for efficient adsorption and sensitive sensing in heavy metals." *Journal of Materials Chemistry A* 7 (2019): 27081–27088. doi.org/10.1039/C9TA11502A.

[62] Rahimi, Esmaeil. "Toxic metal removal from aqueous solution by advanced carbon allotropes: a case study from the Sungun copper mine." *Advances in Environmental Technology* 3 (2017): 77–87. doi:10.22104/AET.2017.507.

[63] Varghese, Anitha George, Sherely Annie Paul, and M. S. Latha. "Remediation of heavy metals and dyes from wastewater using cellulose-based adsorbents." *Environmental Chemistry Letters* 17 (2019): 867–877. doi.org/10.1007/s10311–018–00843–z.

[64] Arabi, Seyed Masoud Seyed, Rahman Shakibaei Lalehloo, Mohamad Reza Talei Bavil Olyai, Gomaa A. M. Ali, and Hamidreza Sadegh. "Removal of Congo red azo dye from aqueous solution by ZnO nanoparticles loaded on multi-wall carbon nanotubes." *Physica E: Low-Dimensional Systems and Nanostructures* 106 (2019): 150–155. doi.org/10.1016/j.physe.2018.10.030.

[65] Jin, Liqiang, Weigong Li, Qinghua Xu, and Qiucun Sun. "Amino-functionalized nanocrystalline cellulose as an adsorbent for anionic dyes." *Cellulose* 22 (2015): 2443–2456. doi.org/10.1007/s10570–015–0649–4.

[66] Tavakolian, Mandana, Hannah Wiebe, Mohammad Amin Sadeghi, and Theo G. M. van de Ven. "Dye removal using hairy nanocellulose: experimental and theoretical investigations." *ACS Applied Materials & Interfaces* 12 (2019): 5040–5049. doi.org/10.1021/acsami.9b18679.

[67] Bhakta, Arvind K., Sunita Kumari, Sahid Hussain, Praveen Martis, Ronald J. Mascarenhas, Joseph Delhalle, and Zineb Mekhalif. "Synthesis and characterization of maghemite nanocrystals decorated multi-wall carbon nanotubes for methylene blue dye removal." *Journal of Materials Science* 54 (2019): 200–216. doi.org/10.1007/s10853–018–2818–y.

[68] Kokol, Vanja, Vera Vivod, Suzana Arnus, Urh Cernigoj, Betka Galicic, Kristina Obu Vazner, Branko Neral, and Ales Mihelic. "Zeolite integrated nanocellulose films for removal of loose anionic reactive dye by adsorption vs. filtration mode during textile laundering." *Fibers and Polymers* 19 (2018): 1556–1566. doi.org/10.1007/s12221–018–8036–z.

[69] Maleki, Afshin, Unes Hamesadeghi, Hiua Daraei, Bagher Hayati, Farhood Najafi, Gordon McKay, and Reza Rezaee. "Amine functionalized multi-walled carbon nanotubes: single and binary systems for high capacity dye removal." *Chemical Engineering Journal* 313 (2017): 826–835. doi.org/10.1016/j.cej.2016.10.058.

[70] Zhou, Chengjun, Sunyoung Lee, Kerry Dooley, and Qinglin Wu. "A facile approach to fabricate porous nanocomposite gels based on partially hydrolyzed polyacrylamide and cellulose nanocrystals for adsorbing methylene blue at low concentrations." *Journal of Hazardous Materials* 263 (2013): 334–341. doi.org/10.1016/j.jhazmat.2013.07.047.

[71] Song, Kaili, Helan Xu, Lan Xu, Kongliang Xie, and Yiqi Yang. "Cellulose nanocrystal-reinforced keratin bioadsorbent for effective removal of dyes from aqueous solution." *Bioresource Technology* 232 (2017): 254–262. doi.org/10.1016/j.biortech.2017.01.070.

[72] Bazrafshan, Edris, Ferdos Kord Mostafapour, Somaiieh Rahdar, and Amir Hossein Mahvi. "Equilibrium and thermodynamics studies for decolorization of Reactive Black 5 (RB5) by adsorption onto MWCNTs." *Desalination and Water Treatment* 54 (2015): 2241–2251. doi.org/10.1080/19443994.2014.895778.

[73] Sehaqui, Houssine, Andreas Mautner, Uxua Perez de Larraya, Numa Pfenninger, Philippe Tingaut, and Tanja Zimmermann. "Cationic cellulose nanofibers from waste pulp residues and their nitrate, fluoride, sulphate and phosphate adsorption properties." *Carbohydrate Polymers* 135 (2016): 334–340. doi.org/10.1016/j.carbpol.2015.08.091.

[74] Jodeh, Shehdeh, Inas Ibsharat, Bayan Khalaf, Othman Hamed, Diana Jodeh, and Dagdag Omar. "The use of magnetic multiwalled carbon nanotubes functionalized with chitosan for nitrate removal from wastewater." *Chemistry Africa* 2 (2019): 321–333. doi.org/10.1007/s42250–019–00056–z.

[75] Liu, Jie, Eden Morales-Narváez, Teresa Vicent, Arben Merkoçi, and Guo-Hua Zhong. "Microorganism-decorated nanocellulose for efficient diuron removal." *Chemical Engineering Journal* 354 (2018): 1083–1091. doi.org/10.1016/j.cej.2018.08.035.

[76] Al-Shaalan, Nora Hamad, Imran Ali, Zeid A. ALOthman, Lamya Hamad Al-Wahaibi, and Hadeel Alabdulmonem. "High performance removal and simulation studies of diuron pesticide in water on MWCNTs." *Journal of Molecular Liquids* 289 (2019): 111039. doi.org/10.1016/j.molliq.2019.111039.

[77] Jeddi, Mohammad Karzar, Ossi Laitinen, and Henrikki Liimatainen. "Magnetic superabsorbents based on nanocellulose aerobeads for selective removal of oils and organic solvents." *Materials & Design* 183 (2019): 108115. doi.org/10.1016/j.matdes.2019.108115.

[78] Gu, Hongbo, Xiaomin Zhou, Shangyun Lyu, Duo Pan, Mengyao Dong, Shide Wu, Tao Ding et al. "Magnetic nanocellulose-magnetite aerogel for easy oil adsorption." *Journal of Colloid and Interface Science* 560 (2020): 849–856. doi.org/10.1016/j.jcis.2019.10.084.

[79] Moustafa, Y. M., A. M. Al-Sabagh, S. A. Younis, Mostafa M. H. Khalil, and M. O. Abdel-Salam. "Preparation of magnetic carbon nanotube nanocomposite for enhancing the separation of dissolved hydrocarbon from petroleum wastewater." *Journal of Environmental Chemical Engineering* 5 (2017): 2240–2250. doi.org/10.1016/j.jece.2017.04.023.

8

Potential Differences of Plant Nanocellulose and Bacterial Nanocellulose in Water Purification

W. M. Dimuthu Nilmini Wijeyaratne
University of Kelaniya, Kelaniya, Sri Lanka

CONTENTS

8.1 Introduction

Cellulose is an abundant raw material used for a variety of biocompatible products. The chirality, biodegradability, hydrophilicity, high chemical-modifying capacity, and multipurpose semi-crystalline fiber morphologies are of greater importance for using cellulose in omnipresent environment-friendly applications. Cellulose is the main fundamental building block of plant, algal, and some bacterial cell walls. The microfibrillar arrangement of cellulose strengthens the cell wall and controls the cell growth [1]. Crystallite shape, nanometric size, and crystallinity are important physical features of cellulose microfibrils that help maintain the cell's physical stability. The chemical structure of these microfibrils is made up of linear chains of glucose units linked by β-1,4-glycosidic bonds. Two crystalline phases of cellulose with similar conformations are cellulose Iα and Iβ. The crystal structures of these two phases are different from each other. In cellulose Iα, triclinic unit cells are present, wherein cellulose Iβ monoclinic unit cells are present. The proportion between Iα and Iβ is different among different groups of organisms. The cell walls of higher plant cells contain a high percentage of Iβ monoclinic unit cells, and the cell walls of algal and bacteria cells have a high percentage of Iα. Cellulose is the starting material for nanocellulose [2].

The nanoscale cellulose fibers are attained by separating the microfibrils from a cellulosic material. These raw cellulosic materials are obtained from various substances, such as trees, algae, plant-based waste materials, and some bacteria, and are known as nanocellulose. The nanoscale cellulose fibers contain many hydroxyl groups as a structural component, resulting in their high hydrophilicity [3]. There are three main types of nanocellulose materials: Cellulose nanofibers (CNFs), cellulose nanocrystals (CNCs), and bacterial nanocellulose (BC). The nanofibrillated cellulose (NFCs) and CNCs are plant-based nanocelluloses. The preparation method of these three types of nanocellulose differs from each other. High-pressure homogenization, grinding, and chemical or enzymatic treatment are used for the isolation of NFCs. However, CNFs are produced using low-impact fiber separation techniques compared to those used in preparing NFCs. Physical separation of cellulose fibers, such as grinding, homogenization, and ultrasonication, is the major method for producing CNFs [4–6]. *Acetobacter, Rhizobium, Agrobacterium, Aerobacter, Achromobacter, Azotobacter, Salmonella, Escherichia,* and *Sarcina* are the major bacterial genera used in preparing BCs [7].

Wide applicability of nanocellulose-based raw materials in many fields is due to their biocompatibility, mechanical strength, renewability, and low cost of preparation. These nanocelluloses are applied as raw materials in bio-composites, pharmaceuticals, tissue engineering, bio-sensing, and pollution remediation. This chapter reviews the applicability of plant nanocellulose and BC in water purification. Further, it focuses on the prospects and potential improvements incorporated into nanocellulose-based water purification.

8.2 Plant Nanocellulose

Plant nanocellulose is the cellulosic material extracted from plant fibers. Natural plant cellulose microfibrils contain well-arranged crystalline portions and some disorganized amorphous portions. The proportions of crystalline and amorphous portions are variable among different groups of plants. When nanocellulose is extracted, mechanical, chemical, and enzyme treatments are applied to separate these cellulose microfibrils. The plant nanocellulose extraction involves several steps. Among them, the first step is the pretreatment process to remove lignin and hemicellulose. There are two pretreatment methods: Alkali treatment and acid chlorite treatment. In alkali treatment, alkali reagents such as sodium hydroxide or potassium hydroxide are used to remove the amorphous portion of hemicelluloses and lignin from the cellulosic fibers. The acid chlorite treatment is a delignification process used to remove lignin from the lignified cellulose. After the pretreatment process, the isolation of nanocellulose is done by either acid hydrolysis, enzymatic hydrolysis, or mechanical treatment processes. The detailed process of plant nanocellulose extraction is given in Fig. 8.1.

There are two types of plant nanocelluloses, i.e., CNF and CNC.

8.2.1 Cellulose Nanofibers (CNFs)

Synonyms for CNFs are micro-fibrillated cellulose (MFC), nanofibrils, micro-fibrils, and NFCs. These are extracted using mechanical treatment process, and pretreatment process can be present or absent depending on the source of cellulose. CNFs are three-dimensional (3D) structures made up of very long and ultrathin nanofibril structures. The arrangement of the fibrils provides a high aspect ratio for NFC. The high strength and high water-holding capacity of NFC compared to other types of nanocellulose are provided by this fibrils arrangement [8]. CNFs have a larger particle size compared to CNCs. The microscopic structure of NFCs consists of a complex, highly entangled, web-like structure with twisted/untwisted, curled/straight, and entangled/separate bundles of nanofibrils [9].

8.2.2 Cellulose Nanocrystals (CNCs)

CNCs are also known as crystallites, whiskers, rod-like cellulose, and microcrystals. The acid hydrolysis of cellulose microfibers digests the amorphous regions, and the crystalline regions are separated to form CNCs [10]. These are stiff rod-like structures and of lower viscosity, lower yield strength, and lower water-holding capacity than NFCs [11].

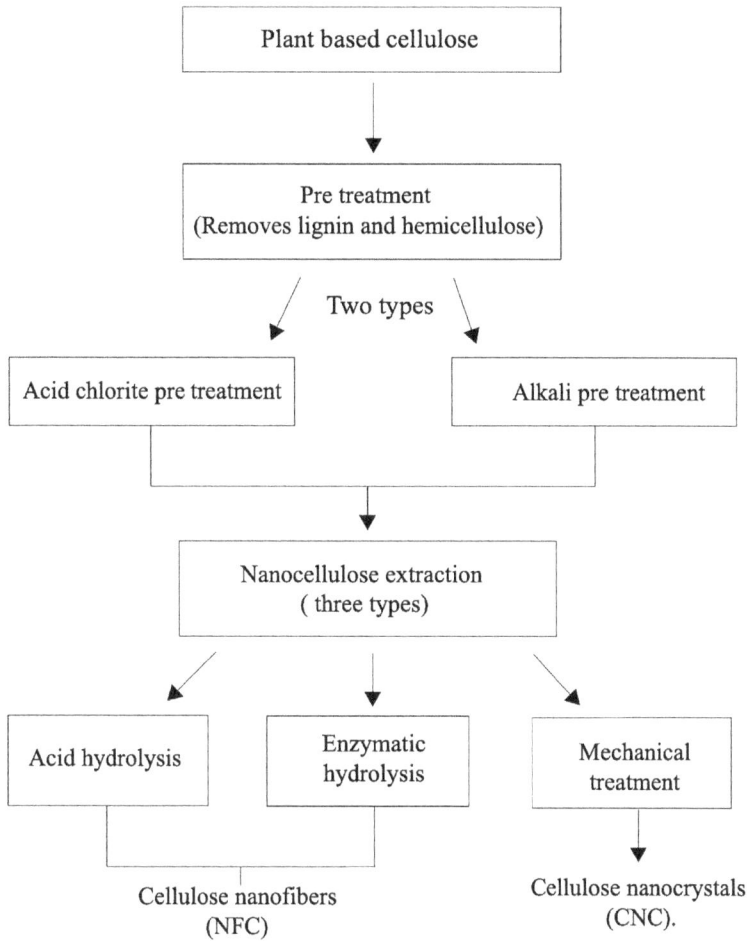

FIGURE 8.1 The plant nanocellulose extraction procedure.

The high surface-to-volume ratios and occurrence of many hydroxyl groups make CNCs proper for several surface functionalizations. The surface functionality of CNCs can be easily modified by esterification, oxidation, amidation, carbamation, nucleophilic substitution, silylation, and polymer grafting. These modifications can change the electrostatic charge on the CNC surface, improve its dispersion in any solvent/polymer, and change its surface energy characteristics [11]. CNCs are used in materials science, electronics, and medicine [12].

8.3 Bacterial Nanocellulose

BC/BNC is known as bio-cellulose, bacterial cellulose, and microbial cellulose. These are extracted from extracellular biosynthetic products from some species of gram-negative bacteria. *Gluconacetobacter, Acetobacter, Salmonella, Achromobacter, Pseudomonas, Sarcina, Azobacter, Rhizobium, Agrobacterium,* and *Alcaligenes* are the most common genera used in extraction of BC. The steps adopted in the extraction of BC are different from the extraction method of plant-based nanocellulose. The initial step in BC extraction is maintaining a bacterial culture using a suitable nutrient medium to increase bacterial cells. The pH, temperature, components, and concentrations of nutrients in the media and the time can influence the growth rate of bacterial cells for BC extraction [13]. When the bacterial culture achieves an optimum cell density, the extracellular enzymatic processes that produce cellulose fibers are stimulated using

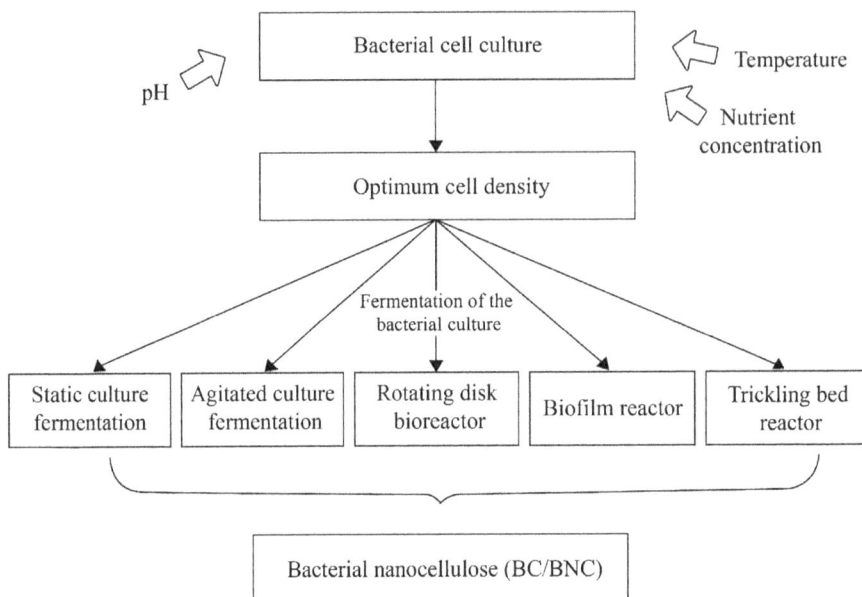

FIGURE 8.2 The bacterial nanocellulose extraction procedure.

different fermentation methods. Most common fermentation methods include static culture fermentation, agitated culture fermentation, rotating-disk bioreactor, biofilm reactor, and trickling-bed reactor. The resulting BC films from the fermentation are deposited as dense pellicles on the culture broth's outer surface [14]. In large-scale BC production, agitating the culture media and using different bioreactors are practiced to reduce the time needed for BC film formation [15]. Yeast or yeast extract in the culture medium or symbiotic co-cultivation with *Medusomyces gisevii* can improve BC yield [16]. Furthermore, the mechanical treatment of BC using a pressure homogenizer produces bacterial nanofiber suspension (BCNF), and acid hydrolysis forms a suspension of bacterial nanocrystals (BCNCs) [17, 18].

The purity of BC is higher compared to the plant-based nanocellulose, as BC is free of lignin, pectin, and hemicellulose. However, BC membranes can contain impurities such as bacterial cell residues, organic acids, salts, and residual sugars from the culture medium. Washing with NaOH solution, centrifugation, filtration, and chemical extraction are effective methods of removing these impurities [19]. However, BC extraction is comparatively less costly and less energy consuming compared to extraction of plant-based nanocellulose.

Further, mechanical strength, hydrophilicity, crystallinity, and BC porosity are higher than plant-based nanocellulose. BC's unique features make it widely applicable in medical applications such as wound-healing materials, artificial skin material in plastic surgeries, production of artificial blood vessels, and biomaterial in tissue engineering [20–22]. Besides, BC derivatives can be produced by carboxymethylation, acetylation, and phosphorylation reactions. These modified BCs also have wide applicability in many fields, including electronic, textile, paper, and cosmetics industries [11, 14, 23–25]. Fig. 8.2 shows BC extraction procedure.

8.4 Industrial Applications of Plant- and Bacterial-Based Nanocellulose

Both plant nanocellulose and BC have widespread industrial applications. Their stiffness, nanometric size, low density, non-abrasiveness, combustibility, non-toxicity, and biodegradability make them useable in the paper and pulp industry, electronics, construction, packaging, cosmetics, and pharmaceutical fields. The different sources of cellulose raw material are used in these industrial applications. Some examples of common industrial applications of nanocellulose are summarized in Table 8.1.

TABLE 8.1

Nanocellulose-Based Industrial Applications.

Industrial Sector	Application	References
Food	Fat replacer	[26, 27]
	Food stabilizer	[28–30]
	Food additive to improve thickening, gelling, and water-binding properties	[31, 32]
	Green food packaging	[33, 34]
Medicine	Tissue engineering	[35]
	Drug excipient and drug delivery	[36, 37]
	Antibacterial activity	[38, 39]
Textile	Improved absorbency and enhanced color strength	[40]
	Antibacterial and hydrophobic fabrics	[41, 42]
Paper	Improved tear strength, tensile strength, and folding endurance	[43, 44]
	Recycled paper production	[45]
Electronics	Thin film transistors, nonvolatile memory devices, and biodegradable computer chips	[46]

8.5 Plant and Bacterial Nanocellulose in Water Purification: Similarities and Differences of Application

Aquatic pollution is caused by point and non-point sources with addition of heavy metals, agricultural pollutants, organic chemicals, pathogens, and many other contaminants to the water bodies. This has drastically increased over the past decade because of the increase in population and intensification of agricultural and industrial activities. Access to clean water is a primary environmental concern, and pollution management and cleanup strategies to overcome aquatic pollution attract global attention.

Water purification removes undesirable substances from water to improve its portability and quality. Water purification can involve boiling, filtration, desalination, and ionization processes ranging from small-scale household applications to large-scale industrial applications. Different physical, chemical, and biological methodologies are involved in each water purification process. This section reviews the importance of nanocellulose in water purification to improve the quality of contaminated water.

Nanocellulose can be used as adsorbents, photocatalysts, flocculants, and membranes in water purification. Some examples of using nanocellulose in water purification are given in Table 8.2.

Nanocellulose is a natural polymer with high surface area, chemical resistance, surface tension, aspect ratio, mechanical strength, and stiffness. These features improve the applicability of nanocellulose as water purifiers. The availability of several hydroxyl groups on their surface facilitates the surface modification ability and further increases the performance of nanocellulose in water purification applications [47, 48].

8.5.1 Nanocellulose as Adsorbents in Water Purification

Pollution remediation by adsorption involves the attachment of pollutants, such as heavy metal ions, dyes, or organic compounds, onto a solid material's surface by electrostatic interaction or loose chemical bonds. The high surface area, crystalline structure, and occurrence of hydroxyl groups increase the adsorption capacity of nanocellulose. Both plant nanocellulose and BC are effective adsorbents that help in removing heavy metals [70–73]. Oils, pesticides, residuals, pharmaceutical waste products, persistent organic compounds, and their byproducts can also be effectively removed by adsorption onto nanocellulose surfaces—the negatively charged hydroxyl groups on the surface of nanocellulose increase the adsorption efficiency of positively charged contaminants. Surface modification reactions of NFC and

TABLE 8.2

Advantages and Disadvantages of Nanocellulose Applications in Water Purification.

Technique	Advantages	Disadvantages	Contaminant/(s) remediated	References
Adsorption	High surface area and availability of surface hydroxyl ions facilitates efficient adsorption High recyclability Non-toxic and does not add secondary pollutants to the environment	The presence of toxic compounds in the medium can be harmful to nanocellulose	Cationic heavy metals: Cd(II) ion, Pb(II) ion and Ni(II) ion	[47]
			Ag (I)	[48]
			Cu (II)	[49]
			V (V)	[50]
			Cationic dyes	[51]
			Anionic dyes	[52]
			Ionized drugs	[53, 54]
Photocatalysis	The organic matrix is environment-friendly and does not produce pollution-causing byproducts Low cost Complete degradation	The nanocellulose can degrade during the reactions and, therefore, may not be reusable and difficult to recover from a catalyst mixture. Nanocellulose can also agglomerate during the photocatalytic reactions and can reduce the efficiency	Organic dyes	[55–58]
			Heavy metals	[59]
Flocculation	Biodegradable Ability to flocculate particles in the nanoscale dimension High specific surface area	Flocculation capacity is affected by the pH of the medium. The small pore size of a nanocellulose may slow down the water flow The small pores can easily clog due to fine particles	Turbidity and COD	[60–62]
			Microalgae	[63]
Membrane filtration	Exchanger ions such as Na (I) and Ca(II) ions are not required Environment friendly Wide availability Low-cost approach Highly efficient	Fouling of nanocellulose after using several times	Heavy metals	[64–69]

CNC can convert the hydroxyl groups to phosphoric, sulfuric, or carboxyl functional groups and further enhance the adsorption capacity of these plant nanocellulose. These surface modifications can be performed by direct modification methods such as esterification, oxidation, and etherification or monomer grafting [74–76].

Compared to plant nanocellulose, the BC has less adsorption capacity as the number of surface hydroxyl groups of BCs are comparatively less than that of plant nanocellulose. However, the BC adsorbents' surface modification reactions can also improve their adsorbent capacity to remove metal ions and polar organic molecules present in industrial wastewater [77]. Recent research has found that rinsing with deionized water instead of using hot NaOH significantly improves BC's adsorption capacity. When deionized water is used in the washing step, the bacterial residues are not completely removed from BC. The embedded bacterial cell residues in the BC structure result in improved mechanical strength and modulus over bare BC. Further, functional groups, such as amide, in the cell residue increase the heavy metal adsorbent capacity [78].

8.5.2 Nanocellulose as Photocatalyst in Water Purification

A photocatalyst uses UV radiation to disintegrate various organic materials, organic acids, estrogens, pesticides, dyes, crude oil, microbes, inorganic molecules, and some metal ions in the contaminated environment.

Photocatalytic water treatment using nanocrystalline titanium dioxide (NTO) is a widely practiced advanced oxidation process for remediation of organically polluted environments. When nanocellulose are used as photocatalysts in water treatment, they have to be combined either with nano-titanium dioxide or with precious metal ion nanoparticles (i.e., Au, Ag, Pt nanoparticles). The titanium nanoparticles can form electrostatic bonds with the cellulose nanoparticles. The nano-TiO_2–cellulose complex is a very useful photocatalyst that can remove many organic materials, including pesticide residues and textile dyes, from the contaminated aquatic environment [79]. In addition to the photocatalytic activity of nano-TiO_2–cellulose composite, it can serve as an effective adsorbent in water purification. The hydrophilic nature of nano-TiO_2–cellulose composite is light-sensitive; at high UV illumination, the nano-TiO_2–cellulose composite is hydrophilic, and in the dark, it is hydrophobic. The lightweight nano-TiO_2–cellulose composite can float in the aquatic environment during the daytime or under controlled UV illumination. It can serve as an effective adsorbent to remove oil and other organic compounds from the contaminated water bodies [80]. Both plant-based nanocellulose and BC have been used in formation of nano-TiO_2–cellulose complex in photocatalytic water purification. However, BC is comparatively more effective in metal oxide nanoparticle support due to its high surface-area-to-volume ratio, high mechanical strength, and high hydrophilicity properties than plant nanocellulose [81]. In addition to removing organic compounds, the nano-TiO_2–nano-BC complex has been successfully used in mercury and cadmium removal.

However, both bacterial and plant-based nanocellulose are very sensitive to UV irradiation and are susceptible to disintegration during photocatalysis [82]. They can be protected from UV degradation by applying a covering made by inert UV-absorbing materials. Recent studies have identified that incorporation of magnetite to the nano-TiO_2–nanocellulose composite can prevent photodegradation of nanocellulose by UV radiation, and it enhances the catalytic activity of nanocellulose. Moreover, there is a high potential of recycling the catalyst by applying a magnetic field to the magnetite nano-TiO_2–nanocellulose composite [83].

8.5.3 Nanocellulose as Flocculants in Water Purification

In flocculation, fine particulates are stimulated to clump together to form a flock. These aggregations may float on the surface or sink to the bottom and can be easily separated from the liquid. The commonly used flocculants are aluminum, iron salts, or synthetic polyelectrolytes. However, these synthetic materials are not biodegradable and may cause unexpected environmental problems. Therefore, the applicability of natural flocculants in aquatic pollution remediation is considered of prompt importance. As a result of bioengineering technology advancements, chitosan, tannins, cellulose, and alginate are identified as sustainable biological alternatives for chemical flocculants.

These natural flocculants can be either used with chemical flocculants to function as semi-natural flocculants or can be used as pure natural flocculants. The flocculent ability of both plant nanocellulose and BC in water purification is tested in laboratory conditions. However, the presence of surface hydroxyl groups limits their applicability as flocculants. Therefore, the surface modification will enhance the functionality of both plant nanocellulose and BC when they are used as flocculants. The common surface modification methods such as carboxymethylation, periodate-chlorite oxidation, aminoguanidine reactions, 2,2,6,6-tetramethylpiperidin-1-oxyl (TEMPO) facilitation, and acid hydrolysis increase the anionic charge density or provide cationic charges to the nanocellulose surface [84, 85]. Surface modification by carboxymethylation, periodate-chlorite oxidation alters the hydroxyl groups of nanocellulose, thereby altering the carbon network, which increases the flocculation performance. The TEMPO-mediated oxidation converts surface hydroxymethyl groups to their respective carboxylic forms, thereby improving the electrostatic stability and increasing the flocculation ability [85].

Further, changing the physical properties of nanocellulose, such as the rod shape, size, and arrangement, have also helped increase their flocculation efficiency. Environmental factors, including pH and temperature of the medium, presence of other chelating agents, and the type of the contaminant, can also affect the flocculation ability of nanocellulose. However, many kinds of research are currently being conducted to improve the efficiency of both plant and bacterial nanocellulose to be used as effective and efficient biological flocculants to treat contaminated aquatic environments.

8.5.4 Nanocellulose Membranes in Water Purification

Membranes developed from CNFs or nanocrystals successfully removed heavy metals, microbes, dyes, pesticides, and other organic contaminants, including oil and cyclohexane. The nanocellulose-based membranes have become versatile due to their wide availability, sustainability, stability, and inert nature over a wide range of pH, environmental friendliness, and affordability. The performance of nanocellulose membranes used in water purification is based on selective adsorption of material from the medium. As described earlier in this chapter, the high surface area and surface hydroxyl groups availability facilitate their selective adsorption capacity.

The nanocellulose membranes can be either made up as pure nanocellulose membranes [86] or nanocellulose can be incorporated to other polymers to form composite nanofilter membranes [87]. The composite nanofilter membranes show more efficient water purification capacity and reduce bio-fouling and hydro-fouling of filtration membranes. The incorporation of nanocellulose to produce composite or pure nanofilter membranes can be done by three methods:

1. Impregnation
2. Vacuum filtration and coating
3. Freeze-drying

Impregnation of surface-modified CNCs of different aspect ratios onto the polymer mats produces very strong membrane filters that can be used for a considerably long duration of time in water purification. Further, these membranes have comparatively enhanced adsorption capacity, and they are capable of removing toxic chemicals and colors formed due to textile dyes [88].

The fabrication of CNCs on thin-film nanocomposite is a vacuum filtration process. In this process, both sides of the membrane are coated with CNCs using a vacuum-based incorporation process from an ultrathin membrane to remove heavy metal ions and fluorides, nitrates, phosphates, sulfates, and organic compounds from the contaminated aquatic environments [65]. The use of acetone-treated CNCs in these ultrathin membranes further enhances the filtration performance by controlling the pore structure, mechanical properties, and permeability [66].

Freeze-drying or lyophilization of nanocellulose is gradual freezing of CNF or nanocrystal suspension in steady-state laboratory conditions. During this process, gradual growth of ice crystals expels the cellulose nanoparticles to the space between the growing ice crystals. These aggregating nanocellulose are densely packed to form a membrane that could be used in water purification. The suspension concentration of nanocellulose used for freeze-drying can affect the properties of the membranes formed. High suspension concentrations produce a layered membrane with hydrogen bonding among neighboring microfibers. At dilute suspension concentrations, the weak inter-microfiber hydrogen bonding eases microfibers' self-assembling in different directions and sheet-like layered structure is not formed [89].

8.5.4.1 Pure Nanocellulose Filter Membranes

As the name implies, these membrane filters consist only of nanocellulose. Pure nanocellulose filter membranes can be made up of either plant nanocellulose or BC. These membranes can filter contaminants based on two strategies, i.e., pore size-based elimination of contaminants according to the size of the contaminant and adsorption of contaminants onto the membrane by electrostatic interaction.

The structure of the size-exclusion filter membranes has been developed over time by various research groups. However, the lower permeability of these membranes results in slowing down the rate of filtration. The increase in pore size or reduction of nanocellulose layers in the membrane negatively affects the filtration efficiency. Therefore, the researchers were challenged to investigate alternative mechanisms for improving the filtration efficiency without reducing the flow rate. Some research has shown that freeze-dried filters and filters prepared from organic dispersion successfully improved filtration efficiency without compromising the flow rate [90].

Besides, as a solution to overcome the flow rate reduction problem, the adsorption membranes were developed by improving nanocellulose surfaces' affinity to adsorb charged ions. The adsorption capacities of nanocellulose were mainly affected by the thickness of the structure and presence of functional groups, where thin membranes with more functional groups showed higher adsorption capacity. However, it is important to maintain a standard thickness and a pressure gradient for effective filtration and flow rate maintenance [91].

8.5.4.2 Composite Nanocellulose Filter Membranes

Composite nanocellulose filter membranes are produced by incorporating cellulose nanomaterials into a porous framework, the co-solvent casting of a resin formulation containing nanocellulose, or increasing the number of functional groups in interfacial polymerization. The composite nanocellulose membranes have significantly improved performance and flow rate compared to the pure nanocellulose filter membranes. These composite membranes can adsorb bacteria, positively charged dyes, and oil effectively from contaminated environments. Further, the composite membranes have greater hydrophilicity which increases the anti-fouling properties compared to pure nanocellulose membranes.

However, some synthetic polymers used in the composite nanocellulose filter membrane structures may not be biodegradable and may pose secondary environmental pollution problems and can be non-sustainable.

Plant-based nanocellulose is commonly used in membrane technologies for environmental remediation. Although BCs have many industrial applications, their applicability as water purification membranes is still at the research level. Compared to plant-based nanocellulose, BC has a highly porous network structure with small individual pore size, making them ideal for a filtration membrane. However, few published studies show that non-modified and modified BC as membranes for water treatment mainly affected the adsorption capacities of nanocellulose. In a study conducted by Periolatto et al. (2018), surface-modified BC using trimethyl silane successfully removed oil and organic compounds from the contaminated environment [92]. Further, silver-impregnated BC was used to produce environment-friendly antibacterial membranes which can be used to remove color and COD from textile wastewater [68].

Common drawbacks of water flow impairment due to small pores in both plant and BC membrane filters can be overcome by using electrical pumps to provide a sufficient flow through the filter. Another drawback of membrane filtration is that the membrane pores can be easily blocked by organic matter in water, and the clogged pores can provide habitats for unnecessary microorganisms. Therefore, as a solution, the incorporation of antibacterial metal nanoparticles into cellulose-based water filters has been successful in inactivating microorganisms. Both silver nanoparticles (AgNPs) and copper nanoparticles (CuNPs) can be used for this purpose. However, these nanoparticles can impose toxic effects on feral organisms in the aquatic ecosystems and may result in the development of silver- and copper-resistant microorganism strains. Therefore, it is important to take necessary steps to prevent unnecessary alterations to the aquatic ecosystems' natural performance during water purification using nanocellulose and their derivatives.

8.6 Future Prospects and Conclusions

Nanocellulose is a promising biological agent that can be used for purifying water in contaminated ecosystems. The functionality of nanocellulose is reliant on the physical and chemical features of the cellulose microfibrils. Also, environmental factors such as pH, temperature, and ionic composition of the medium can affect the purification efficiency of nanocellulose. Plant nanocellulose is used in water purification applications. Although BC has a wide range of industrial applications, their applicability in water purification is still at the research level. Appropriate modifications to the membranous structure, surface functional groups, and impregnation of other materials can significantly increase the performance of nanocellulose in water purification by increasing their affinity toward heavy metal ions, organic compounds, etc. However, the materials used in surface modification can compromise the non-toxic nature and biodegradability of cellulose nanoparticles. Therefore, the stability and the environmental effects of byproducts produced from the synthetic polymers and other materials used in surface modification have to be carefully analyzed, and remedial actions need to be proposed to minimize the unforeseen environmental effects. In addition, the production and operation costs in large-scale manufacturing and life cycle analysis of these nanocellulose-based water purification applications must be assessed to maintain the sustainability of these applications. Besides, in this regard, it would be preferable to use a supportive substrate consisting of a sustainable raw material rather than synthetic polymer material, which can cause secondary pollution effects in the natural environment.

Acknowledgment

The author wishes to acknowledge Mr. R.P.K.C. Rajapakse of the Department of Zoology and Environmental Management, University of Kelaniya, Sri Lanka, for his artwork assistance.

REFERENCES

[1] Jordan, B. M., and J. Dumais. *Biomechanics of plant cell growth. Encyclopedia of life sciences.* Hoboken, NJ: Wiley. 2010.

[2] Rongpipi, Sintu, Dan Ye, Enrique D. Gomez, and Esther W. Gomez. "Progress and opportunities in the characterization of cellulose—an important regulator of cell wall growth and mechanics." *Frontiers in Plant Science* 9 (2019): 1894. doi:10.3389/fpls.2018.01894.

[3] Uetani, Kojiro, and Hiroyuki Yano. "Nanofibrillation of wood pulp using a high-speed blender." *Biomacromolecules* 12 (2011): 348–353. doi:10.1021/bm101103p.

[4] Abe, Kentaro, Shinichiro Iwamoto, and Hiroyuki Yano. "Obtaining cellulose nanofibers with a uniform width of 15 nm from wood." *Biomacromolecules* 8 (2007): 3276–3278. doi:10.1021/bm700624p.

[5] Chen, Wenshuai, Qing Li, Youcheng Wang, Xin Yi, Jie Zeng, Haipeng Yu, Yixing Liu, and Jian Li. "Comparative study of aerogels obtained from differently prepared nanocellulose fibers." *ChemSusChem* 7 (2014): 154–161. doi:10.1002/cssc.201300950.

[6] Paakko, M., M. Ankerfors, H. Kosonen, A. Nykanen, S. Ahola, M. Osterberg, J. Ruokolainen, et al. "Enzymatic hydrolysis combined with mechanical shearing and high-pressure homogenization for nanoscale cellulose fibrils and strong gels." *Biomacromolecules* 8 (2007): 1934–1941. doi:10.1021/bm061215p.

[7] Ullah, Muhammad W., Mazhar Ul-Islam, Shaukat Khan, Yeji Kim, and Joong K. Park. "Structural and physico-mechanical characterization of bio-cellulose produced by a cell-free system." *Carbohydrate Polymers* 136 (2016): 908–916. doi:10.1016/j.carbpol.2015.10.,010.

[8] Sharma, Amita, Manisha Thakur, Munna Bhattacharya, Tamal Mandal, and Saswata Goswami. "Commercial application of cellulose nanocomposites—a review." *Biotechnology Reports* 21 (2019). doi:e00316. doi:10.1016/j.btre.2019.e00316.

[9] Xu, Xuezhu, Fei Liu, Long Jiang, J. Y. Zhu, Darrin Haagenson, and Dennis P. Wiesenborn. "Cellulose nanoculose nanofibrils: a comparative study on their microsmprucurend effects as polymer reinforcing agents." *ACS Applied Materials & Interfaces* 5 (2013): 2999–3009. doi:10.1021/am302624t.

[10] Domingues, Rui M., Manuela E. Gomes, and Rui L. Reis. "The potential of cellulose nanocrystals in tissue engineering strategies." *Biomacromolecules* 15 (2014):2327–2346. doi:10.1021/bm500524s.

[11] Habibi, Youssef, Lucian A. Lucia, and Orlando J. Rojas. "Cellulose nanocrystals: chemistry, self-assemhalppplications." *Chemical Reviews* 110 (2010):3479–3500. doi:10.1021/cr900339w.

[12] George, Johnsy, and S N. Sabapathi. "Cellulose nanocrystals: synthesis, functional properties, and applications." *Nanotechnology, Science and Applications* 8 (2015): 45. doi:10.2147/nsa.s64386.

[13] Trovatti, Eliane. "Bacterial cellulose." in *Biopolymer nanocomposites*, ed. Alain Dufresne, Sabu Thomas, Laly A. Pothen, 339–366. Wiley (2013). doi:10.1002/9781118609958.ch15.

[14] Dujardin, Erik, Matthew Blaseby, and Stephen Mann. "Synthesis of mesoporous silica by sol–gel mineralisation of cellulose nanorod nematic suspensions." *Journal of Materials Chemistry* 13 (2003): 696–699. doi:10.1039/b212689c.

[15] Campano, Cristina, Ana Balea, Angeles Blanco, and Carlos Negro. "Enhancement of the fermentation process and properties of bacterial cellulose: a review." *Cellulose* 23 (2015): 57–91. doi:10.1007/s10570-015-0802-0.

[16] Wiegand, Cornelia, Sebastian Moritz, Nadine Hessler, Dana Kralisch, Falko Wesarg, Frank A. Müller, Dagmar Fischer, and Uta-Christina Hipler. "Antimicrobial functionalization of bacterial nanocellulose by loading with polihexanide and povidone-iodine." *Journal of Materials Science: Materials in Medicine* 26 (2015). doi:10.1007/s10856-015-5571-7.

[17] Saska, Sybele, Lucas N. Teixeira, Larissa M. De Castro Raucci, Raquel M. Scarel-Caminaga, Leonardo P. Franchi, Raquel A. Dos Santos, Silvia H. Santagneli, et al. "Nanocellulose-collagen-apatite composite associated with osteogenic growth peptide for bone regeneration." *International Journal of Biological Macromolecules* 103 (2017): 467–476. doi:10.1016/j.ijbiomac.2017.05.086.

[18] Balea, Ana, Jose L. Sanchez-Salvador, M. C. Monte, Noemi Merayo, Carlos Negro, and Angeles Blanco. "*In situ* production and application of cellulose nanofibers to improve recycled paper production." *Molecules* 24 (2019): 1800. doi:10.3390/molecules24091800.

[19] Stumpf, Taisa R., Xiuying Yang, Jingchang Zhang, and Xudong Cao. "*In situ* and *ex situ* modifications of bacterial cellulose for applications in tissue engineering." *Materials Science and Engineering: C* 82 (2018): 372–383. doi:10.1016/j.msec.2016.11.121.

[20] Siqueira, Gilberto, Julien Bras, and Alain Dufresne. "Cellulosic bionanocomposites: a review of preparation, properties and applications." *Polymers* 2 (2010): 728–765. doi:10.3390/polym2040728.

[21] Chinga-Carrasco, Gary. "Cellulose fibres, nanofibrils and microfibrils: the morphological sequence of MFC components from a plant physiology and fibre technology point of view." *Nanoscale Research Letters* 6 (2011). doi:10.1186/1556-276x-6-417.

[22] Mondal, Subrata. "Preparation, properties and applications of nanocellulosic materials." *Carbohydrate Polymers* 163 (2017): 301–316. doi:10.1016/j.carbpol.2016.12.050.

[23] Garcia de Rodriguez, Nancy L., Wim Thielemans, and Alain Dufresne. "Sisal cellulose whiskers reinforced polyvinyl acetate nanocomposites." *Cellulose* 13 (2006): 261–270. doi:10.1007/s10570-005-9039-7.

[24] Habibi, Youssef, Henri Chanzy, and Michel R. Vignon. "TEMPO-mediated surface oxidation of cellulose whiskers." *Cellulose* 13 (2006): 679–687. doi:10.1007/s10570-006-9075-y.

[25] Svagan, Anna J., My A. Azizi Samir, and Lars A. Berglund. "Biomimetic polysaccharide nanocomposites of high cellulose content and high toughness." *Biomacromolecules* 8 (2007): 2556–2563. doi:10.1021/bm0703160.

[26] Marchetti, Lucas, Bianca Muzzio, Patricia Cerrutti, Silvina C. Andres, and Alicia N. Califano. "Bacterial nanocellulose as novel additive in low-lipid low-sodium meat sausages. Effect on quality and stability." *Food Structure* 14 (2017): 52–59. doi:10.1016/j.foostr.2017.06.004.

[27] Capron, Isabelle, Orlando J. Rojas, and Romain Bordes. "Behavior of nanocelluloses at interfaces." *Current Opinion in Colloid & Interface Science* 29 (2017): 83–95. doi:10.1016/j.cocis.2017.04.001.

[28] Golchoobi, Laleh, Mazdak Alimi, Shirin Shokoohi, and Hossein Yousefi. "Interaction between nanofibrillated cellulose with guar gum and carboxy methyl cellulose in low-fat mayonnaise." *Journal of Texture Studies* 47 (2016): 403–412. doi:10.1111/jtxs.12183.

[29] Winuprasith, Thunnalin, and Manop Suphantharika. "Microfibrillated cellulose from mangosteen (*Garcinia mangostana* L.) rind: preparation, characterization, and evaluation as an emulsion stabilizer." *Food Hydrocolloids* 32 (2013): 383–394. doi:10.1016/j.foodhyd.2013.01.023.

[30] Gomez H., C., A. Serpa, J. Velasquez-Cock, P. Ganan, C. Castro, L. Velez, and R. Zuluaga. "Vegetable nanocellulose in food science: a review." *Food Hydrocolloids* 57 (2016): 178–186. doi:10.1016/j.foodhyd.2016.01.023.

[31] Kuo, Chia-Hung, Jing-Hua Chen, Bo-Kang Liou, and Cheng-Kang Lee. "Utilization of acetate buffer to improve bacterial cellulose production by gluconacetobacter xylinus." *Food Hydrocolloids* 53 (2016): 98–103. doi:10.1016/j.foodhyd.2014.12.034.

[32] Corral, Mariela L., Patricia Cerrutti, Analia Vazquez, and Alicia Califano. "Bacterial nanocellulose as a potential additive for wheat bread." *Food Hydrocolloids* 67 (2017): 189–196. doi:10.1016/j.foodhyd.2016.11.037.

[33] Regubalan, Baburaj, Pintu Pandit, Saptarshi Maiti, Gayatri T. Nadathur, and Aranya Mallick. "Potential bio-based edible films, foams, and hydrogels for food packaging." in *Bio-based Materials for Food Packaging*, ed. Shakeel Ahmad, 105–123. Springer (2018). doi:10.1007/978-981-13-1909-9_5.

[34] Criado, Paula, Farah M. Hossain, Stéphane Salmieri, and Monique Lacroix. "Nanocellulose in food packaging." in *Composites Materials for Food Packaging* (2018): 297–329. doi:10.1002/9781119160243.ch10.

[35] Zmejkoski, Danica, Dragica Spasojevic, Irina Orlovska, Natalia Kozyrovska, Marina Soković, Jasmina Glamočlija, Svetlana Dmitrovic, et al. "Bacterial cellulose-lignin composite hydrogel as a promising agent in chronic wound healing." *International Journal of Biological Macromolecules* 118 (2018): 494–503. doi:10.1016/j.ijbiomac.2018.06.067.

[36] Kolakovic, Ruzica, Leena Peltonen, Timo Laaksonen, Kaisa Putkisto, Antti Laukkanen, and Jouni Hirvonen. "Spray-dried cellulose nanofibers as novel tablet excipient." *AAPS Pharm SciTech* 12 (2011): 1366–1373. doi:10.1208/s12249-011-9705-z.

[37] Lin, Ning, Jin Huang, Peter R. Chang, Liangdong Feng, and Jiahui Yu. "Effect of polysaccharide nano-crystals on structure, properties, and drug release kinetics of alginate-based microspheres." *Colloids and Surfaces B: Bio interfaces* 85 (2011): 270–279. doi:10.1016/j.colsurfb.2011.02.039.

[38] Maneerung, Thawatchai, Seiichi Tokura, and Ratana Rujiravanit. "Impregnation of silver nanoparticles into bacterial cellulose for antimicrobial wound dressing." *Carbohydrate Polymers* 72 (2008): 43–51. doi:10.1016/j.carbpol.2007.07.025.

[39] Luan, Jiabin, Jian Wu, Yudong Zheng, Wenhui Song, Guojie Wang, Jia Guo, and Xun Ding. "Impregnation of silver sulfadiazine into bacterial cellulose for antimicrobial and biocompatible wound dressing." *Biomedical Materials* 7 (2012): 065006. doi:10.1088/1748-6041/7/6/065006.

[40] Chattopadhyay, D. P., and B. H. Patel. "Synthesis, characterization and application of nano cel-lulose for enhanced performance of textiles." *Journal of Textile Science & Engineering* 6 (2016). doi:10.4172/2165-8064.1000248.

[41] Huang, Tianxiao, Chao Chen, Dongfang Li, and Monica Ek. "Hydrophobic and antibacterial textile fibres prepared by covalently attaching betulin to cellulose." *Cellulose* 26 (2019): 665–677. doi:10.1007/s10570-019-02265-8.

[42] Jafary, Razeah, Mohammad Khajeh Mehrizi, Seyed H. Hekmatimoghaddam, and Ali Jebali. "Antibacterial property of cellulose fabric finished by allicin-conjugated nanocellulose." *The Journal of The Textile Institute* 106 (2014): 683–689. doi:10.1080/00405000.2014.954780.

[43] Johnson, Donna, Mark Papadis, Michael Bilodeau, Bruce Crossley, Marc Foulger, and Pierre Gelinas. "Effects of cellulosic nanofibrils on papermaking properties of fine papers." *TAPPI Journal* 15 (2016): 395–402. doi:10.32964/tj15.6.395.

[44] Adnan, S., A. H. Azhar, L. Jasmani, and M. F. Samsudin. "Properties of paper incorporated with nano-cellulose extracted using microbial hydrolysis assisted shear process." *IOP Conference Series: Materials Science and Engineering* 368 (2018): 012022. doi:10.1088/1757-899x/368/1/012022.

[45] Balea, Ana, Elena Fuente, M. C. Monte, Noemi Merayo, Cristina Campano, Carlos Negro, and Angeles Blanco. "Industrial application of nanocelluloses in papermaking: a review of challenges, technical solutions, and market perspectives." *Molecules* 25 (2020): 526. doi:10.3390/molecules25030526.

[46] Nagashima, Kazuki, Hirotaka Koga, Umberto Celano, Fuwei Zhuge, Masaki Kanai, Sakon Rahong, Gang Meng, et al. "Cellulose nanofiber paper as an ultra flexible nonvolatile memory." *Scientific Reports* 4 (2014). doi:10.1038/srep05532.

[47] Kardam, Abhishek, Kumar R. Raj, Shalini Srivastava, and M. M. Srivastava. "Nanocellulose fibers for biosorption of cadmium, nickel, and lead ions from aqueous solution." *Clean Technologies and Environmental Policy* 16 (2013): 385–393. doi:10.1007/s10098-013-0634-2.

[48] Liu, Peng, Houssine Sehaqui, Philippe Tingaut, Adrian Wichser, Kristiina Oksman, and Aji P. Mathew. "Cellulose and chitin nanomaterials for capturing silver ions (Ag^+) from water via surface adsorption." *Cellulose* 21 (2013): 449–461. doi:10.1007/s10570-013-0139-5.

[49] Zhang, Xiaofang, Jiangqi Zhao, Long Cheng, Canhui Lu, Yaru Wang, Xu He, and Wei Zhang. "Acrylic acid grafted and acrylic acid/sodium humate grafted bamboo cellulose nanofibers for Cu^{2+} adsorption." *RSC Advances* 4 (2014): 55195–55201. doi:10.1039/c4ra08307e.

[50] Sirviö, Juho A., Tapani Hasa, Tiina Leiviskä, Henrikki Liimatainen, and Osmo Hormi. "Bisphosphonate nanocellulose in the removal of vanadium (V) from water." *Cellulose* 23 (2015): 689–697. doi:10.1007/s10570-015-0819-4.

[51] Qiao, Han, Yanmei Zhou, Fang Yu, Enze Wang, Yinghao Min, Qi Huang, Lanfang Pang, and Tongsen Ma. "Effective removal of cationic dyes using carboxylate-functionalized cellulose nanocrystals." *Chemosphere* 141 (2015): 297–303. doi:10.1016/j.chemosphere.2015.07.078.

[52] Jin, Liqiang, Weigong Li, Qinghua Xu, and Qiucun Sun. "Amino-functionalized nanocrystalline cellu-lose as an adsorbent for anionic dyes." *Cellulose* 22 (2015): 2443–2456. doi:10.1007/s10570-015-0649-4.

[53] Letchford, Jackson, Ben Wasserman, Ye, Wadood Hamad, and Helen Burt. "The use of nanocrystalline cellulose for the binding and controlled release of drugs." *International Journal of Nanomedicine* 6 (2011): 321. doi:10.2147/ijn.s16749.

[54] Espino-Pérez, Etzael, Julien Bras, Violette Ducruet, Alain Guinault, Alain Dufresne, and Sandra Domenek. "Influence of chemical surface modification of cellulose nanowhiskers on thermal, mechanical, and barrier properties of poly(lactide) based bionanocomposites." *European Polymer Journal* 49 (2013): 3144–3154. doi:10.1016/j.eurpolymj.2013.07.017.

[55] Anirudhan, T.S., and J.R. Deepa. "Nano-zinc oxide incorporated graphene oxide/nanocellulose composite for the adsorption and photo catalytic degradation of ciprofloxacin hydrochloride from aqueous solutions." *Journal of Colloid and Interface Science* 490 (2017): 343–356. doi:10.1016/j.jcis.2016.11.042.

[56] Lefatshe, Kebadiretse, Cosmas M. Muiva, and Lemme P. Kebaabetswe. "Extraction of nanocellulose and *in-situ* casting of ZnO/cellulose nanocomposite with enhanced photocatalytic and antibacterial activity." *Carbohydrate Polymers* 164 (2017): 301–308. doi:10.1016/j.carbpol.2017.02.020.

[57] Chen, Pengpeng, Xiaoyan Liu, Rundong Jin, Wangyan Nie, and Yifeng Zhou. "Dye adsorption and photo-induced recycling of hydroxypropyl cellulose/molybdenum disulfide composite hydrogels." *Carbohydrate Polymers* 167 (2017): 36–43. doi:10.1016/j.carbpol.2017.02.094.

[58] Jiang, Yifan, Ibrahim Lawan, Weiming Zhou, Mingxin Zhang, Gerard F. Fernando, Liwei Wang, and Zhanhui Yuan. "Synthesis, properties and photocatalytic activity of a semiconductor/cellulose composite for dye degradation—a review." *Cellulose* 27 (2019): 595–609. doi:10.1007/s10570-019-02851-w.

[59] Wittmar, Alexandra S., Qian Fu, and Mathias Ulbricht. "Photocatalytic and magnetic porous cellulose macrospheres for water purification." *Cellulose* 26 (2019): 4563–4578. doi:10.1007/s10570-019-02401-4.

[60] Wang, Ran, Sihui Guan, Anna Sato, Xiao Wang, Zhe Wang, Rui Yang, Benjamin S. Hsiao, and Benjamin Chu. "Nanofibrous microfiltration membranes capable of removing bacteria, viruses and heavy metal ions." *Journal of Membrane Science* 446 (2013): 376–382. doi:10.1016/j.memsci.2013.06.020.

[61] Kimura, S., and T. Itoh. "New cellulose synthesizing complexes (terminal complexes) involved in animal cellulose biosynthesis in the tunicate *Metandrocarpa uedai*." *Protoplasma* 194 (1996): 151–163. doi:10.1007/bf01882023.

[62] Sharma, Priyanka R., Aurnov Chattopadhyay, Sunil K. Sharma, Lihong Geng, Nasim Amiralian, Darren Martin, and Benjamin S. Hsiao. "Nanocellulose from *spinifex* as an effective adsorbent to remove cadmium (ii) from water." *ACS Sustainable Chemistry & Engineering* 6 (2018): 3279–3290. doi:10.1021/acssuschemeng.7b03473.

[63] Onyianta, Amaka J., Mark Dorris, and Rhodri L. Williams. "Aqueous morpholine pretreatment in cellulose nanofibril (CNF) production: comparison with carboxymethylation and TEMPO oxidisation pretreatment methods." *Cellulose* 25 (2017): 1047–1064. doi:10.1007/s10570-017-1631-0.

[64] Chitpong, Nithinart, and Scott M. Husson. "Polyacid functionalized cellulose nanofiber membranes for removal of heavy metals from impaired waters." *Journal of Membrane Science* 523 (2017): 418–429. doi:10.1016/j.memsci.2016.10.020.

[65] Karim, Zoheb, Minna Hakalahti, Tekla Tammelin, and Aji P. Mathew. "*In situ* TEMPO surface functionalization of nanocellulose membranes for enhanced adsorption of metal ions from aqueous medium." *RSC Advances* 7 (2017): 5232–5241. doi:10.1039/c6ra25707k.

[66] Karim, Zoheb, Aji P. Mathew, Vanja Kokol, Jiang Wei, and Mattias Grahn. "High-flux affinity membranes based on cellulose nanocomposites for removal of heavy metal ions from industrial effluents." *RSC Advances* 6 (2016): 20644–20653. doi:10.1039/c5ra27059f.

[67] Karim, Zoheb, Aji P. Mathew, Mattias Grahn, Johanne Mouzon, and Kristiina Oksman. "Nanoporous membranes with cellulose nanocrystals as functional entity in chitosan: Removal of dyes from water." *Carbohydrate Polymers* 112 (2014): 668–676. doi:10.1016/j.carbpol.2014.06.048.

[68] Isik, Zelal, Ali Unyayar, and Nadir Dizge. "Filtration and antibacterial properties of bacterial cellulose membranes for textile wastewater treatment." *Avicenna Journal of Environmental Health Engineering* 5 (2018): 106–114. doi:10.15171/ajehe.2018.14.

[69] Cruz-Tato, Perla, Edwin O. Ortiz-Quiles, Karlene Vega-Figueroa, Liz Santiago-Martoral, Michael Flynn, Liz M. Diaz-Vazquez, and Eduardo Nicolau. "Metalized nanocellulose composites as a feasible material for membrane supports: design and applications for water treatment." *Environmental Science & Technology* 51 (2017): 4585–4595. doi:10.1021/acs.est.6b05955.

[70] Phanthong, Patchiya, Prasert Reubroycharoen, Xiaogang Hao, Guangwen Xu, Abuliti Abudula, and Guoqing Guan. "Nanocellulose: extraction and application." *Carbon Resources Conversion* 1 (2018): 32–43. doi:10.1016/j.crcon.2018.05.004.

[71] Hokkanen, Sanna, Eveliina Repo, Amit Bhatnagar, Walter Z. Tang, and Mika Sillanpaa. "Adsorption of hydrogen sulphide from aqueous solutions using modified nano/micro fibrillated cellulose." *Environmental Technology* 35 (2014): 2334–2346. doi:10.1080/09593330.2014.903300.

[72] Sun, Xitong, Liangrong Yang, Qian Li, Junmei Zhao, Xiaopei Li, Xiaoqin Wang, and Huizhou Liu. "Amino-functionalized magnetic cellulose nanocomposite as adsorbent for removal of Cr(VI): synthesis and adsorption studies." *Chemical Engineering Journal* 241 (2014): 175–183. doi:10.1016/j.cej.2013.12.051.

[73] Hokkanen, Sanna, Amit Bhatnagar, Eveliina Repo, Song Lou, and Mika Sillanpaa. "Calcium hydroxyapatite microfibrillated cellulose composite as a potential adsorbent for the removal of Cr(VI) from aqueous solution." *Chemical Engineering Journal* 283 (2016): 445–452. doi:10.1016/j.cej.2015.07.035.

[74] Zhang, Kaitao, Peipei Sun, He Liu, Shibin Shang, Jie Song, and Dan Wang. "Extraction and comparison of carboxylated cellulose nanocrystals from bleached sugarcane bagasse pulp using two different oxidation methods." *Carbohydrate Polymers* 138 (2016): 237–243. doi:10.1016/j.carbpol.2015.11.038.

[75] Zhang, Nan, Guo-Long Zang, Chen Shi, Han-Qing Yu, and Guo-Ping Sheng. "A novel adsorbent TEMPO-mediated oxidized cellulose nanofibrils modified with PEI: preparation, characterization, and application for Cu(II) removal." *Journal of Hazardous Materials* 316 (2016): 11–18. doi:10.1016/j.jhazmat.2016.05.018.

[76] Putro, Jindrayani N., Alfin Kurniawan, Suryadi Ismadji, and Yi-Hsu Ju. "Nanocellulose based biosorbents for wastewater treatment: Study of isotherm, kinetic, thermodynamic and reusability." *Environmental Nanotechnology, Monitoring & Management* 8 (2017): 134–149. doi:10.1016/j.enmm.2017.07.002.

[77] Reeve, B., S. Petkiewicz, H. Hagemann, G. Santosa, M. Florea, and T. Ellis. "Modified bacterial nanocellulose as a bioadsorbent material." *IET/SynbiCITE Engineering Biology Conference* (2016): 2. doi:10.1049/cp.2016.1252.

[78] Wan, Yizao, Jie Wang, Miguel Gama, Ruisong Guo, Quanchao Zhang, Peibiao Zhang, Fanglian Yao, and Honglin Luo. "Biofabrication of a novel bacteria/bacterial cellulose composite for improved adsorption capacity." *Composites Part A: Applied Science and Manufacturing* 125 (2019): 105560. doi:10.1016/j.compositesa.2019.105560.

[79] Liebner, Falk, Nicole Pircher, and Thomas Rosenau. "Bacterial nanocellulose aerogels." *Bacterial Nanocellulose* (2016): 73–108. doi:10.1016/b978-0-444-63458-0.00005-6.

[80] Korhonen, J. T., M. Kettunen, R. H. A. Ras, and O. Ikkala. "Hydrophobic nanocellulose aerogels as floating, sustainable, reusable, and recyclable oil absorbents". *ACS Applied Materials and Interfaces* 3 (2011): 1813–1816. doi:https://doi.org/10.1021/am200475b.

[81] Li, Guohui, Avinav G. Nandgaonkar, Qingqing Wang, Jinning Zhang, Wendy E. Krause, Qufu Wei, and Lucian A. Lucia. "Laccase-immobilized bacterial cellulose/TiO₂ functionalized composite membranes: evaluation for photo- and bio-catalytic dye degradation." *Journal of Membrane Science* 525 (2017): 89–98. doi:10.1016/j.memsci.2016.10.033.

[82] Cheng, Fei, Mark Lorch, Seyed M. Sajedin, Stephen M. Kelly, and Andreas Kornherr. "Whiter, brighter, and more stable cellulose paper coated with TiO₂/SiO₂ core/shell nanoparticles using a layer-by-layer approach." *ChemSusChem* 6 (2013): 1392–1399. doi:10.1002/cssc.201300305.

[83] An, Xingye, Dong Cheng, Lei Dai, Baobin Wang, Helen J. Ocampo, Joseph Nasrallah, Xu Jia, Jijun Zou, Yunduo Long, and Yonghao Ni. "Synthesis of nano-fibrillated cellulose/magnetite/titanium dioxide (NFC@Fe₃O₄@TNP) nanocomposites and their application in the photocatalytic hydrogen generation." *Applied Catalysis B: Environmental* 206 (2017): 53–64. doi:10.1016/j.apcatb.2017.01.021.

[84] Liimatainen, Henrikki, Terhi Suopajärvi, Juho Sirvio, Osmo Hormi, and Jouko Niinimäki. "Fabrication of cationic cellulosic nanofibrils through aqueous quaternization pretreatment and their use in colloid aggregation." *Carbohydrate Polymers* 103 (2014): 187–192. doi:10.1016/j.carbpol.2013.12.042.

[85] Suopajarvi, Terhi, Juho A. Sirvio, and Henrikki Liimatainen. "Cationic nanocelluloses in dewatering of municipal activated sludge." *Journal of Environmental Chemical Engineering* 5 (2017): 86–92. doi:10.1016/j.jece.2016.11.021.

[86] Wang, Ran, Sihui Guan, Anna Sato, Xiao Wang, Zhe Wang, Rui Yang, Benjamin S. Hsiao, and Benjamin Chu. "Nanofibrous microfiltration membranes capable of removing bacteria, viruses and heavy metal ions." *Journal of Membrane Science* 446 (2013): 376–382. doi:10.1016/j.memsci.2013.06.020.

[87] Kong, Linlin, Dalun Zhang, Ziqiang Shao, Baixin Han, Yuxia Lv, Kezheng Gao, and Xiaoqing Peng. "Superior effect of TEMPO-oxidized cellulose nanofibrils (TOCNs) on the performance of cellulose triacetate (CTA) ultrafiltration membrane." *Desalination* 332 (2014): 117–125. doi:10.1016/j. desal.2013.11.005.

[88] Ma, Hongyang, Christian Burger, Benjamin S. Hsiao, and Benjamin Chu. "Nanofibrous microfiltration membrane based on cellulose nanowhiskers." *Biomacromolecules* 13 (2011): 180–186. doi:10.1021/ bm201421g.

[89] Han, Jingquan, Chengjun Zhou, Yiqiang Wu, Fangyang Liu, and Qinglin Wu. "Self-assembling behavior of cellulose nanoparticles during freeze-drying: effect of suspension concentration, particle size, crystal structure, and surface charge." *Biomacromolecules* 14 (2013): 1529–1540. doi:10.1021/bm4001734.

[90] Mautner, Andreas, Thawanrat Kobkeatthawin, Florian Mayer, Christof Plessl, Selestina Gorgieva, Vanja Kokol, and Alexander Bismarck. "Rapid water softening with TEMPO-oxidized/phosphorylated nanopapers." *Nanomaterials* 9 (2019): 136. doi:10.3390/nano9020136.

[91] Mautner, Andreas. "Nanocellulose water treatment membranes and filters: a review." *Polymer International* (2020). doi:10.1002/pi.5993.

[92] Periolatto, Monica, and Giuseppe Gozzelino. "Surface modification and characterization of cellulose-based filters for water-oil separation." *AIP Conference Proceedings 1981, 020017* (2018). doi:10.1063/1.5045879.

9

Nanocellulose-Based Membrane for Water Purification

Theivasanthi Thirugnanasambandan
Kalasalingam Academy of Research and Education (Deemed University), Krishnankoil, India

CONTENTS

9.1 Introduction

Contaminants present in water, such as dye and bacterial and fungal toxins, are incredibly harmful to humans. Methods like filtration, sedimentation, and distillation are used to purify water. Biological applications (like slow sand filters or bioactivated carbon), electromagnetic irradiation (UV light), and chemical practices (like chlorination) are used in the water purification process. New advanced technologies such as the reverse osmosis (RO) process use high-quality membranes for the filtration of water. Cellulose is created by biological sources like plants, algae, animals, and microorganisms (like bacteria and fungi). Membranes used in water purification methods like microfiltration and ultrafiltration need expensive synthetic materials that are non-sustainable. Nanocellulose membranes can be developed from agriculture residues and underutilized biomass waste at a low cost. Nanocellulose is available in various forms like bacterial cellulose nanofibrils (CNFs) and cellulose nanocrystals (CNCs). Nanocellulose possesses a large surface area, hydrophilic surface, chemical inertness, and more strength, suitable for making high-performance membranes. The chemical inertness of nanocellulose (NC) in aqueous media arises due to the high degree of crystallinity and hydrophilicity of nanocellulose. These properties reduce the biofouling in the membrane.

Nanocellulose-based membranes are applied to remove metal ions, dyes, and microbes. The selectivity of a membrane depends upon its structure and the chemical nature of the material. Adsorption of metallic ions (like Cu(II), Fe(III), and Ag(I)), sulfates, nitrates, phosphates, fluorides, humic acid, and other organic compounds can be performed with this membrane. The freeze-drying process produces low mechanical stability due to the lack of H-bond.

Membrane technology is considered as a process to remove the contaminants from water using less energy. The dimensions of CNF of approximately 5 nm in diameter and few micrometers in length make it suitable for making barrier layers in pressure-driven membranes. Nanocellulose can be crosslinked with the RO membrane, i.e., polyimide, by interfacial polymerization, which forms interlinked waterway. This increases the permeation of water without loss of selectivity [1].

Nanocellulose having hydrophobic functional groups (like silyl groups) can clean the organic pollutants (like oils and cyclohexanes). Nanocellulose is a natural polymer available in abundance and is also renewable. Nanocellulose possesses chiral nature because of repeating β-ᴅ-glucopyranose units covalently connected by acetal groups between the hydroxyl groups (related to C_4 and C_1 carbon atoms). Nanocellulose can be used to make valuable filtration material since it is steady at a wide range of pH and ionic concentration. Functional groups present on the nanocellulose surface are used to target various contaminants in water [2].

Developing new materials with properties like long durability, cost-effective, less pollution, and advanced structures with directed waterway create more permeation flux with less energy consumption. These materials maintain high selectivity or rejection rate. For microfiltration and ultrafiltration processes, ultrafine CNFs (approximate size 5 nm) are applied. These nanofibers are prepared by 2,2,6,6-tetramethylpiperidin-1-yl-oxidanyl (TEMPO)/NaBr/NaClO oxidation of natural cellulose like wood pulp. In addition to less diameter, large surface area, simple functionalization, and better mechanical/chemical resistances are also useful in filtration processes. The electrospun nanofibrous scaffolds with small pore sizes are superior to commercial MF membranes (Millipore GS9035). They can be utilized to remove waterborne bacteria to apply in the ultrafiltration process of oil and water emulsions and to purify the bilge water of ships or wastewater of industries [3].

Membrane technology involves using a semipermeable membrane, usually a finely porous polymeric or inorganic film, to separate the liquid into two distinct streams. This technique is applied in several fields, from dairy processing to wastewater treatment. Cellulose materials (collodion) and polymers (like polyamide nylon, polycarbonate, polypropylene, and polytetrafluoroethylene-Teflon) are utilized in preparing membranes. However, nanocellulose has properties better than cellulose and other materials. These properties are useful to construct the membranes with exceptional properties and overcome the filtration problems.

Membranes produced by using CNCs are low-cost membranes. Their cost is low compared to commercial membranes like Nafion membrane. Cellulose-nanocrystals-based membrane is naturally derived. Its cellular structure has many hydroxyl groups. These groups have hydrogen and oxygen bonds [4]. In this chapter, nanocellulose-based membranes and various methods involved in their fabrication are explained.

9.2 Fabrication of Nanocellulose Membranes

Nanocellulose membranes are produced by different methods such as electrospinning, vacuum filtration, and freeze-drying. Dai et al. have stated some methods to fabricate nanocellulose-based membranes. These fabrication methods are shown in Fig. 9.1 [5]. Nanocellulose-based membranes are also called nanopapers. Vacuum filtration technique followed by a hot pressing method is a rapid, easy, and scalable technique to fabricate the nanocellulose-based membranes.

Vacuum filtration is a technique where the solution is poured via a filter paper present in a Buchner funnel. During this process, the solid is retained by the filter. The liquid is drained via the funnel by vacuum. To prepare a multilayered nanocellulose membrane, the vacuum filtration process involving CNFs suspensions and dip coating processes using CNCs is useful. The pore size of a nanocellulose membrane and its surface area can be regulated by processing, drying nature, and treating the membrane with acetone. The mechanical properties of the membrane result from the processing method. The membrane's tensile strength (pore size 74 Å) is found to be 95, 3.7, and 2.7 MPa under dry conditions, wet conditions, and acetone treatment, respectively. The water flux of the acetone-treated membranes is 25 L/m²/h at 0.45 MPa. Metal ions are removed more effectively by adsorption on the NC membrane's surface because of more functional groups in NC. The membrane can clean Ag(I), Fe (III)/Fe(II), and Cu(II) ions present in the effluents of the mirror industry. Removal of ions from water by RO technique requires the membrane with 1–10 nm pore size, 15–30 bars pressure, and more energy.

In contrast, the adsorption process demands lower pressure, lower energy, and high selectivity. The membrane properties like the pore structure and thickness play a major role in water flux. The porosity is increased while using lower polarity solvents like acetone, methanol, and ethanol. These solvents decrease H-bonding in nanocellulose networks [6].

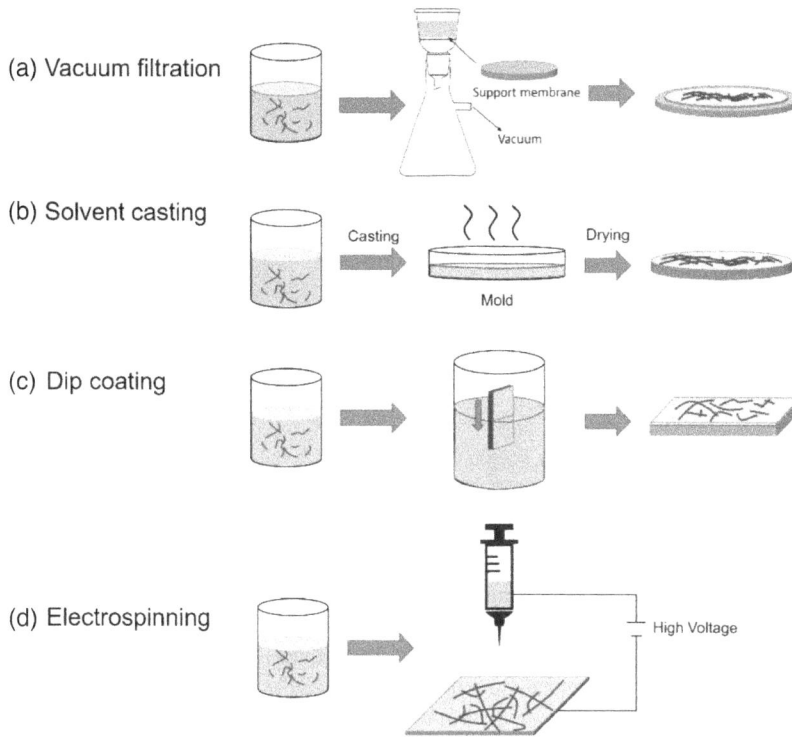

FIGURE 9.1   Schematic diagram shows the fabrication of nanocellulose membrane by various methods: (a) vacuum filtration, (b) solvent coating, (c) dip coating, and (d) electrospinning [5].

Nanocellulose papers can be made by filtering 0.2% microfibrillated cellulose (MFC) suspension using filter membrane (having 0.65 μm pore size) and dried at 363 K with 70 mbar. The papers thus prepared possess more mechanical strength. Since bacterial cellulose (BC) naturally forms membranes, there is no need for filtration. BC membrane can be created by pressing the BC hydrogel at 393 K with a pressure of 2 MPa. In another method, MFC's aqueous suspension is poured on a plastic plate and dried for 24 hours at room temperature. The created film has thickness of 3–100 μm and more tensile strength. Nanoparticles like calcium carbonate nanoparticles and montmorillonite nanoplatelets are added to nanocellulose membrane to improve the mechanical strength. Silver nanoparticles are attached to produce antimicrobial nanocellulose filters to remove viruses and bacteria from water. Ag NPs inhibit the surface biofouling and enhance the life of the membrane. Nanocellulose can hold more Ag NPs, and leakage of Ag NPs is avoided. Nanocellulose/Ag NPs nanocomposite is prepared by reducing silver ions to produce antimicrobial membranes [7].

Vacuum filtration of aqueous nanocellulose suspension through a filter paper substrate makes nanocellulose/filter paper (NC/FP) composite filtration membrane. The membrane made in this process can purify water. The filtration performance of the membranes is determined by the methods applied during the drying of the membrane. Excellent ultrafiltration properties with a retention rate of 97.14% and water flux (46,279 L/m²/h) is achieved for the membrane made using 0.1% CNFs with vacuum dried at 60 °C [8].

Cellulose nanopapers can be prepared similar to traditional papermaking. Several nanocelluloses such as BC, TEMPO-oxidized cellulose nanofibrils (TOCNFs), CNCs, and wood-derived nanocellulose are used instead of microsized cellulose fibers. Nanopapers with transmembrane permeance commercial UF membranes with molecular weight cutoffs in the range of 6–25 kDa can be prepared [9].

Cellulose processing is complicated because cellulose has low solubility in common solvents. Because of the many intermolecular and intramolecular hydrogen bonds, cellulose cannot melt. The novel process can make CNFs by electrospinning or electrostatic fiber spinning. In this process, a high-voltage

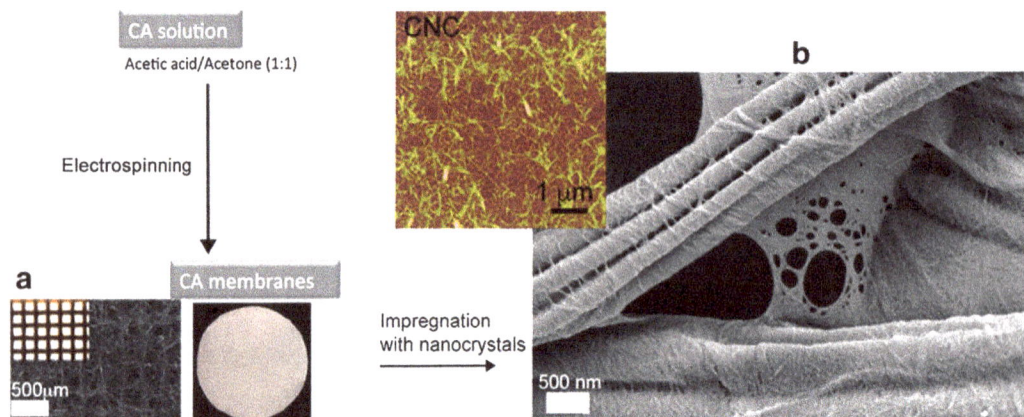

FIGURE 9.2 Schematic diagram shows electrospinning preparation of cellulose acetate (CA) membrane. (a) SEM image of prepared membrane exhibits mesh pattern (*Inset:* mesh collector used for collecting fibers) and photograph of CA membrane. (b) AFM analysis of the prepared cellulose nanocrystals (CNCs). SEM image exhibits the CNC-coated electrospun fibers of the membrane [11].

electric field is applied to the polymer solution. It flows out the of polymer solution through the needle tip. Finally, the fibers are collected on a grounded target plate [10].

All cellulose water purification membranes have been made by electrospinning cellulose acetate (CA) fibers collected on a mesh template and impregnating CNCs. The as-prepared membranes exhibit improved mechanical properties, and the hydrophilicity is enhanced from a hydrophobic contact angle range 102° to 0°. The membrane can reject microorganisms and dyes. Fig. 9.2a exhibits the prepared cellulose acetate membrane using electrospinning method. Fig. 9.2b demonstrates the AFM image of CNCs and SEM image of CNC-coated CA fibers [11].

Cellulose acetate nanofiber membrane is processed by electrospinning of cellulose acetate (7–19%) using high-boiling-point solvent mixture of methyl ethyl ketone and N,N-dimethylacetamide in three different ratios (2:1, 1:1, 1:2). The membrane possesses a more hydrophilic nature and more wettability with a water flux of 10,197.044 L/m²/h. It can be reusable up to four times without rupture [12].

TOCNF mixed in water/*tert*-butyl alcohol (TBA) suspension is freeze-dried to make TOCNF aerogels for designing the high-performance air filter. This suspension is combined with commercial high-efficiency particulate air (HEPA) filter to achieve superior filtration properties. A high specific surface area of more than 300 m²/g is obtained in this air filter. *tert*-Butyl alcohol concentration can be varied between 20 and 50% w/w [13].

9.3 Functionalization of Nanocellulose Membranes

TEMPO-mediated oxidized cellulose nanofiber is well known for various applications. It is demonstrated to adsorb multiple heavy metal ions present in water. The adsorption performance of nanocellulose is 72.8, 30.15, and 25.5 mg/g for the nanocellulose with phosphoryl groups, carboxyl groups, sulfonic acid groups, respectively [14].

Moisture sorption decreases the mechanical properties of plant-fibers-based biocomposites. Acetylation of nanofibrillated cellulose (NFC) based on wood and BC decreases the moisture sorption in nanopaper structures. Fig. 9.3 exhibits the FE-SEM images of three different NFC surfaces. After completing the preparation, these samples have been dried by a critical point drier (CPD). NFC samples without acetylation, acetylated at 50 °C, and over-acetylated NFC are shown in Fig. 9.3a–c, respectively. In Fig. 9.3c, a film-like form can be seen due to the increasing thickness of the nanofibrils [15]. It is observed from these images that the moisture sorption in nanopaper structures varies, depending upon the conditions of acetylation.

Cellulose nanopapers are phosphorylated by reacting CNF with phosphoric acid and analyzed for copper ion adsorption. Due to phosphorylation, phosphate groups are present on the surface of the

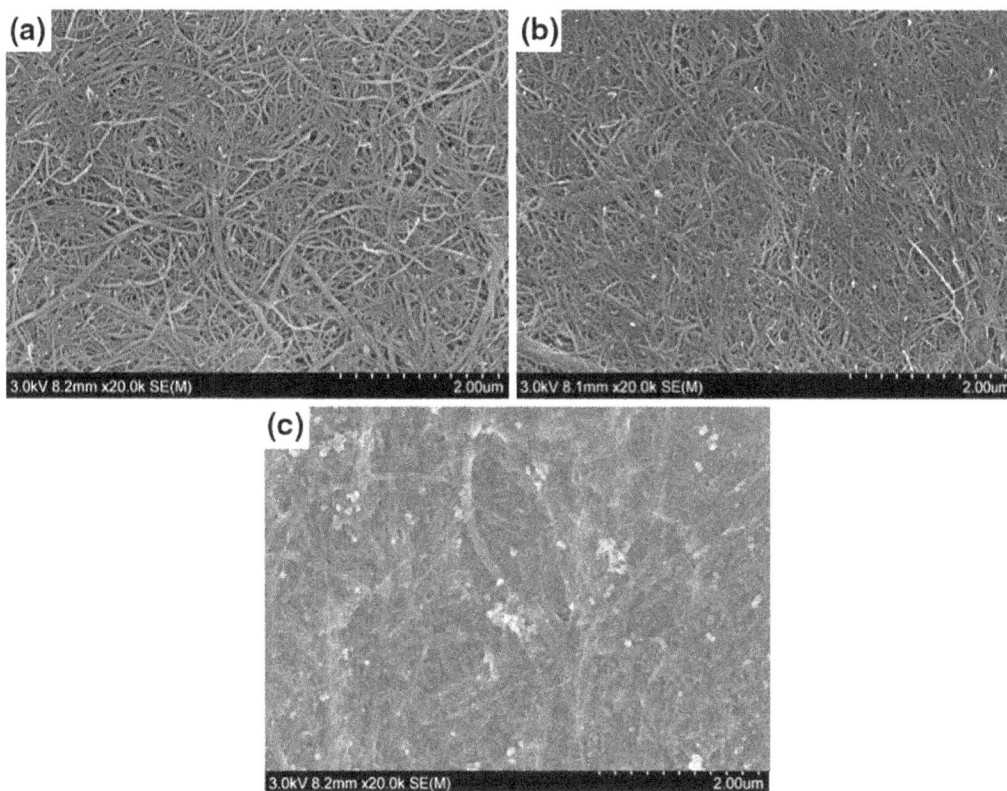

FIGURE 9.3 FE-SEM surface images of nanofibrillated cellulose (NFC) samples dried by critical point drier (CPD). (a) NFC samples CPD dried only, without acetylation. (b) NFC samples dried by CPD and acetylated at 50°C. (c) CPD dried and over-acetylated NFC nanopapers [15].

nanopaper. More functional groups are available in the nanopapers after phosphorylation than in the nanopapers before the phosphorylation. Fig. 9.4 exhibits the chemical structures of CNFs and phosphorylated CNFs. The functional groups can be seen from these structures. The phosphorylation contributes to better copper adsorption. Phosphorylated nanofibrils are useful membranes for removing copper ions present in the water via electrostatic interactions [16].

The electrospinning method is prepared using tree-like CNF membranes with exceptional hydrophilicity, minute pore size, better mechanical properties, and resisting properties to organic solvents. Electrospinning is controlled by the addition of tetrabutylammonium chloride (TBAC) in cellulose acetate solution to get this tree-like structure [17].

Novel paper filters for filtration of viruses are constructed by crosslinking of nanocellulose using citric acid. Crosslinking is performed by dipping *Cladophora* cellulose paper sheets in 16 wt% citric acid containing 1 wt% sodium hypophosphate for 12 hours. These sheets are dried at 160 °C for 10 minutes in a hot-press. The retention ability is confirmed using SEM analysis and ultraviolet absorbance intensity of the initial solution, filtrate. Further, this filter can be utilized as paper-based industrial filters for sterile removal [18].

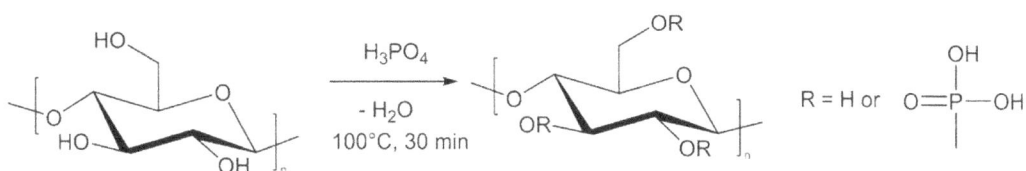

FIGURE 9.4 Schematic diagram illustrates the modification process of cellulose nanofibrils using phosphate groups [16].

Thiol functionalization of CNF membrane is useful in adsorbing heavy metal ions. The thiol functionalization is performed by esterification of a thiol precursor material. The adsorption capacity of the membrane for Cu(II), Cd(II), and Pb(II) ions are 49.0, 45.9, and 22.0 mg/g, respectively [19].

Nanocellulose is extracted from bagasse when functionalized with L-cysteine. It has both thiol and amine groups. The adsorption capacity of Cys-CNFs for Hg(II) ions is 116.822 mg/g [20]. Meldrum's acid (2,2-dimethyl-1,3-dioxane-4,6-dione)-modified CNF/PVDF membranes are analyzed for the filtration of Fe_2O_3 nanoparticles adsorption of crystal violet dye. The membrane's adsorption capacity for crystal violet dye is found as 3.984 mg/g [21].

9.4 Nanocellulose/Graphene Oxide Composite Membranes

Graphene oxide (GO) layer is deposited on the CNF membranes to improve wet strength, water flux, and water pollutants adsorbing capacity. Also, it reduces the swelling and deformations of CNFs produced during the process of water filtration. A membrane of GO:CNF (with 1:100 mass ratio) is obtained by vacuum filtration of the CNF suspension, followed by GO suspension (0.48 g CNF and 4.8 mg GO in 600 mL of distilled water). The thickness of the membrane is 19.3 ± 4.0 µm (measured from SEM) with the GO layer of 110 nm and pore size 7 nm (Barrett-Joyner-Halenda [BJH]). The structural development (dry and swollen conditions) during filtration and ion adsorption can be analyzed using *in situ* SAXS. The composite membrane can reject both negatively and positively charged dyes greater than 90% [22].

GO layer is coated on the CNF membrane to achieve synergistic water flux and separation performance. This hybrid membrane possesses enhanced mechanical stability (both in dry and wet conditions) and shows negative surface zeta potential. Its water flux is $18,123 \pm 574$ L/m²/h/bar. It is higher than the level of hydrophilic commercial reference membrane, i.e., Nylon 66, with pore size 0.2 µm. The high permeation of water is due to the GO's numerous water transport nanochannels. The adsorption of dyes is due to electrostatic/hydrophobic and molecular size [23].

Graphene oxide-CNC composite is functionalized using polymer (PVDF). This composite is used as MF membrane. It has more wettability, less adsorption (for materials like protein and polysaccharides), and more permeability. When applying in MBR, the nanocomposite membrane exhibits more antifouling performances with low EPS accumulation and lesser irreversible fouling. Also, it enhances the flux recovery ratio, cleaning cycle, hydrophilicity, porosity, and negative zeta potential [24].

Anti-biofouling ultrafiltration membrane based on the composite of reduced graphene oxide (RGO) and bacterial nanocellulose (BNC) is designed to face the membrane technology challenges related to the water flux and economic viability. The membrane shows excellent bactericidal activity because of its photothermal property under light illumination. This technology purifies water. Simultaneously, it prevents biofouling (due to harmful bacteria/microorganisms that decrease the water flow). The membranes can be made on large scale, which will help develop those countries where clean water is scarce. *Gluconacetobacter hansenii* form CNFs in water. Graphene oxide is incorporated into this BNC (during the growing stage). It makes the membrane stable and durable. Then, the membrane is treated with a base solution to remove the *Gluconacetobacter* and oxygen groups of GO. Hence, the GO is converted as RGO. While sunlight is falling on this membrane, heat is generated by the RGO that kills the bacteria. The membrane can be heated above 70 °C, which can kill harmful bacteria within 3 minutes. The membrane can work like commercial ultrafiltration membranes that operate under high pressure [25].

9.5 Polymer/Nanocellulose Composite Membranes

TEMPO-oxidized cellulose nanocrystals (TOCNs) are inserted in electrospun polyacrylonitrile (PAN) for the production of nanofiber membrane. This membrane is supported on a non-woven substrate (made with polyethylene terephthalate [PET] polymer material). Since CNCs possess a negative charge on their surface, positively charged materials like CV dye can be adsorbed readily. CNCs can also remove bacteria (such as *Escherichia coli* and *Brevundimonas diminuta*) and viruses (like bacteriophage MS2) under low pressure with high flux. Fig. 9.5 exhibits the functioning method of nanocellulose-based membrane in clean water production by adsorbing and filtering the contaminants from polluted water [26].

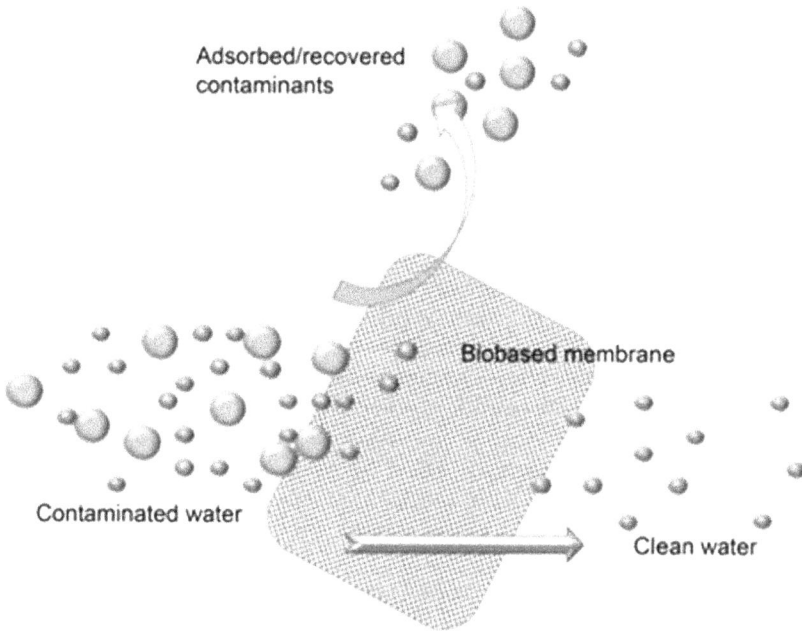

FIGURE 9.5 Schematic diagram shows nanocellulose membrane. This membrane adsorbs and filters the contaminants presented in the polluted water that leads to clean water production [26].

In electrospinning CNFs, poor mechanical performance is caused by some solvents. Hydroxyethyl cellulose (HEC) is coated over NFC to enhance the membrane's mechanical properties. The membrane is prepared by applying the vacuum filtration technique on NFC/HEC aqueous suspensions in the initial step. Then supercritical CO_2 drying is applied. The membrane has a high specific surface area. Also, it has more strength and more strain. The failure level is up to 55% only [27].

Membranes possessing more water flux property (i.e., six times more than the commercially available membranes) have been prepared from CNFs. These membranes are useful in the ultrafiltration of water. CNFs are coated layer by layer on an electrospun scaffold for preparing ultrafiltration membranes. PAN and PET are utilized as electrospun scaffolds and substrates. The membrane pore size is 55 nm, calculated using the molecular weight cutoff (MWCO) method. The membrane's permeate flux is compared with PAN10 (commercial ultrafiltration membrane). The membrane can also be used in the ultrafiltration process of oil/water emulsions. The membrane shows some remarkable properties: exceptional chemical resistance, more anti-biodegradation, and hypochlorite resistance. Also, it applies in a wide pH range [28].

The ion-exchange membranes of cellulose acetate nanofiber are made by grafting and electrospinning polymer material like polyacrylic acid (PAA). The permeability of these membranes strongly depends on the molecular weight of PAA. For removing Cd ions from water, the membrane's adsorption performance is decided by the solvents used for PAA grafting [29]. Polyurethane foam immobilized with carboxymethylated CNFs has been applied in the purification of industrial wastewater. Carboxymethylated CNFs are well dispersed in polyurethane foam. The hydrogen bonds of this nanocomposite enhance the strength and other mechanical properties. This nanocomposite has the ability for high adsorption of heavy metal ions [30].

Polyethylenimine is coated on bacterial CNF membranes by flushing wet BC membranes (present in the polyethyleneimine mixed with glycerol diglycidyl ether aqueous solution) under vacuum suction. The membrane is then heated up to 70 °C for crosslinking the coating on the CNFs surface. The membrane adsorbs Cu(II) and Pb(II) metal ions present in the aqueous solutions. These ions can be desorbed by treating with ethylene diamine tetraacetic acid. The membrane can be reused for several cycles [31].

Polydopamine/BNC hybrid membrane is prepared by *in situ* methods. PDA particles are incorporated into BNC matrix (during the growth of bacteria). This membrane can remove various metal ions (like lead and cadmium) and organic dyes (like rhodamine 6G, methylene blue, and methyl orange) from contaminated water. It is confirmed by a simple filtration test conducted at pH 4–7 [32].

The regenerated cellulose (RC)-based electrospun nanofiber membranes (ENMs) have micrometer/ sub-micrometer-level pores. More water flux and low-level fouling are the major advantages of ENMs. The small-sized pores of ENMs are useful in microfiltration applications, including water purification. The RC ENMs (to apply in ultrafiltration) are prepared in an atom transfer radical polymerization (ATRP) process using RC, water-insoluble material, and water-soluble material. In this process, water-insoluble monomers of 2-hydroxyethyl methacrylate (HEMA) and water-soluble sodium acrylate (AAS) are grafted RC nanofibers surface for preparing ENMs. The prepared RC ENMs can be applied for the ultrafiltration of nanoparticles (size ~40 nm) and bovine serum albumin (BSA) molecules having a size up to the level of ~10 nm from water [33].

TOCNF mixed with starch-based materials like hydroxypropyl starch (HPS), acetyl starch (AS), and acetyl oxidized starch (AOS) exhibit more stability and more mechanical properties in water. Hemiacetal bonding created in between the TCNF and starch decreases the swelling in water. Also, it improves mechanical properties and stability while in wet conditions. A tensile modulus of 7 MPa is exhibited by TCNF/HPS membrane with less swelling while in water [34].

The cellulose of agave bagasse (CNF) and polycaprolactone (PCL) are utilized in preparing PCL/CNF composite membranes (three different ratios of PCL/CNF) by electrospinning method. FTIR and XRD analyses of these composite membranes (CNF and PCL) are shown in Figs. 9.6 and 9.7, respectively. The prepared composite membranes exhibit excellent porosity, structure on a nanometric scale, and good mechanical properties [35].

Increasing the proportion of PCL (60% and 80%) increases the composite's vicious nature and the average diameter of the electrospun nanofibers. The membrane of PCL80:CNF20 composite is more

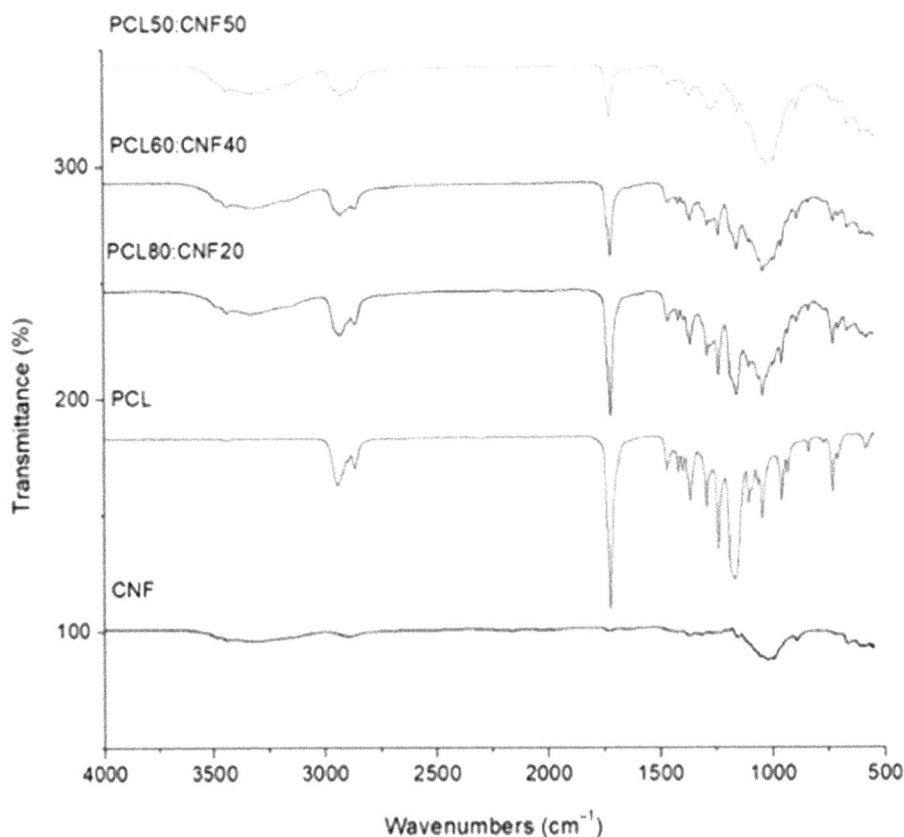

FIGURE 9.6 FTIR spectra related to the CNF, polycaprolactone (PCL), and their composites (utilized to produce electrospun composite membranes). The PCL/CNF composite has been prepared with three different ratios (v/v) of PCL/CNF like 50:50, 60:40, and 80:20 [35].

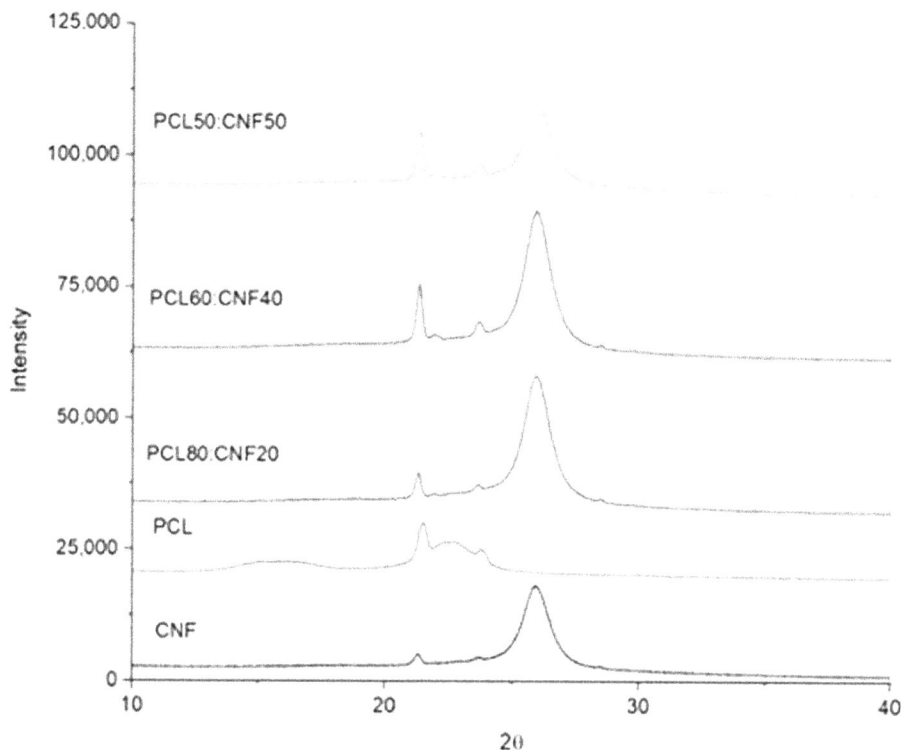

FIGURE 9.7 XRD spectra related to the CNF, polycaprolactone (PCL), and their composites (utilized to produce electrospun composite membranes). The PCL/CNF composite has been prepared with three different ratios (v/v) of PCL/CNF—50:50, 60:40, and 80:20 [35].

permeable than the membrane of both PCL50:NFC50 and PCL60:NFC40 composites. It is observed from the XRD analysis, increasing of CNF content in PCL/CNF composite decreases the intensity of the peaks. PCL50:CNF50 electrospun composite membrane keeps heavy metals (like iron and chromium—75% and 99%, respectively). Also, it removes the turbidity and conductivity of 100% level [35].

Nanocomposite membranes prepared by using crosslinking of polymer matrix, i.e., poly(ethylene glycol) and ultrafine CNFs, exhibit excellent antifouling properties. The antifouling nature of the prepared nanocomposite membrane has been tested by several short- and long-term fouling confirmation tests in the BSA solution with a concentration of 1 g/L [36].

9.6 Contaminants Removal by Nanocellulose Membranes

Water filters prepared using the composite material of natural fiber (like flax and agave fiber) and nanocellulose have high permeance and high metal ion adsorption capacity. The natural fiber and CNFs composite filter possess a more negative surface charge responsible for metal ions adsorption. The water flux measured from the membrane is found as 100,000 $L/m^2/h/MPa$. The composite membrane can adsorb 1200 mg/m of copper ions through its surface [37].

Cellulose-based foams with zeolitic imidazolate frameworks-8 (ZIF-8) material show exceptional properties such as heavy metal ions removal and gas adsorption activity. The ZIF-8@CNF@cellulose foam has a nitrogen adsorption property 30 times more than the pure cellulose foam. Also, it can adsorb heavy metal ions: Cr(VI) adsorption—35.6 mg/g, fluorescent dyes—24.6 mg/g adsorption of rhodamine B, and organic solvents 45.2 g/g adsorption of DMF. The ZIF-8@CNF@cellulose foam (having the quantity 40 wt% of CNF) exhibits compressive strength up to the level of 1.30 MPa [38].

The phase inversion process is employed to fabricate the nanocellulose membrane. In this process, a solution of cellulose (dissolved in acetone) and organic solvent like 1-ethyl-3-methylimidazolium acetate ([EMIM]OAc) with a proportion of 20% and 80%, respectively, is used. The cellulose solution mixed in [EMIM]OAc solution is dried after coagulation for membrane formation. The prepared membrane can reject 94% of bromothymol blue in ethanol (permeance is 0.3 L/h/m^2/bar). The permeance is increased to 8.4 L/h/m^2/bar for the solution of cellulose 12%, [EMIM]OAc 63%, and acetone 20% for pre-evaporation before coagulation. Electrostatic activities and hydrogen bonding interactions are responsible for this heavy dye adsorption [39].

To prepare a bilayered nanofiltration membrane, multiwalled carbon nanotubes (MWCNTS) are grafted on the surface of BC. MWCNT/BC nanofibers are electrospun as a membrane. Finally, chitosan hydrogel is coated. This membrane possesses more tensile strength and Young's modulus (respective values of tensile strength and Young's modulus are 11.75 and 244 MPa in dry state; 10.11 and 211 MPa in wet state). The water flux of the membrane increases from a level of 52.1 L/m^2/h to 140.7 L/m^2/h when the pressure of operation increases from 0.1 MPa to 0.6 MPa pressure level. The membrane can reject organic dyes like methylene blue, procion red mx-5B, direct orange, and stilbene yellow. It shows better antifouling properties in protein and oil materials [40].

Filter paper made with high crystalline CNFs is employed for virus removal. Size exclusion principle with LRV (log 10 reduction value) ≥6.3 is applied in this process. The prepared filter paper's performance is similar to the commercial virus removal filters (prepared with industrial synthetic polymer). *Cladophora* algae cellulose (with 300 mg) is sonicated and drained on a nylon filter (with 100 nm pore size distribution). The nylon support is removed by tweezers. Finally, the sample is dried using a heat-press at the temperature of 105 °C [41].

A microfiltration membrane is fabricated, having two layers. One layer is made with nanoscale PAN, and the other layer is made with microscale polyethylene terephthalate containing CNFs of diameter 5 nm. These membranes can remove *E. coli* bacteria by size extrusion. The LRV in MS2 virus removal is 4. It adsorbs 100 mg of Cr(VI) ions or 260 mg of Pb(II) ions per gram of CNF. The permeation rate is high i.e. 1300 L/m^2/h/psi [42].

The virus removing property of nanocellulose-based filter has been explored. Chemically defined Chinese hamster ovary (CHO) cells are utilized as a medium in this upstream bioprocess. The CHO cells medium has Pluronic F-68 material and insulin-transferrin-selenium (ITS) as supplement material. The nanocellulose-based filter has more virus retention capacity (LRV is above 4) and 180 L/m^2/h/bar of water flux [43].

The available technologies cannot separate the small-size oil droplets mixed in wastewater. Conventionally, oil/water emulsions separation is carried out by membranes with special wettability. However, this process has some challenges like high material cost, complex processing, and biofouling. New membrane produced by vacuum filtration method using tunicate CNCs and TiO$_2$ nanoparticles has a self-cleaning property. It can degrade oleic acid under the illumination of UV light. The produced nanocomposite membrane has high roughness, superhydrophilicity, and high underwater oleophobicity. Also, it has a high water flux that is useful in the separation of selective oil/water emulsions [44].

Never-dried BC collected after various incubation periods (2–10 days) is utilized to remove oil from the oil-in-water emulsions. It has a higher flux value. The photographs of never-dried BC harvested after 2, 6, and 10 days incubation periods are shown in Fig. 9.8. It has been observed that there is no

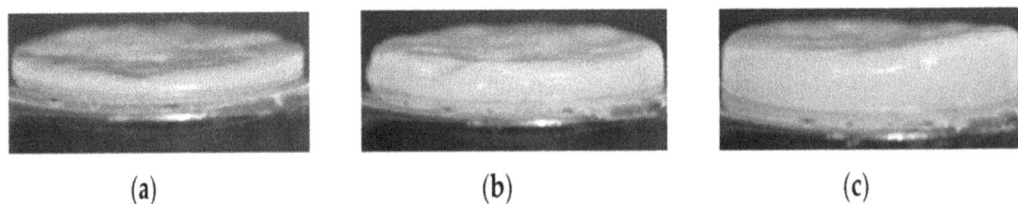

(a)   (b)   (c)

FIGURE 9.8 Photographs exhibit the thickness related to the never-dried BC collected after various incubation periods: (a) two days incubation, (b) six days incubation, and (c) ten days incubation [45].

TABLE 9.1

Weight (Wet and Dry), Thickness (Wet and Dry), and Water Content of the Never-Dried Bacterial Cellulose Collected After Various Incubation Periods.

Incubation Time (Days)	Wet Weight (g)	Dry Weight (g)	Water Content (%)	Wet Thickness (mm)	Dry Thickness (mm)
2	1.40	0.017	98.8	0.40	0.02
6	21.43	0.17	99.2	5.23	0.08
10	51.58	0.44	99.1	11.60	0.1

considerable variation in the width of BC nanofibers when the incubation period increases. However, the thickness increases because of the formation of a high-level mass of the BC. The weight (wet and dry), thickness (wet and dry), and the water content in BCs are enumerated in Table 9.1 [45].

9.7 Nanocellulose Nanofiltration Membranes

Ultrathin polymer membrane for nanofiltration is produced by a surface-modifying process. The surface of the ultrafine CNF membrane is modified through interfacial polymerization. The prepared membrane has thickness, pore size, and molecular weight cutoff of 77.4 nm, 0.45 nm and 824 g/mol, respectively. It has a smooth surface that allows speedy water permeation. The water flux value is 32.7 L/m^2/h/bar. Also, it can reject inorganic salts and organic dyes [46].

Nanofiltration membranes are fabricated through an interfacial polymerization process. In this process, ultrafine CNF material is polymerized using a polymer material, i.e., polyamide. Two interfacial polymerization methods (IP and IP-R) are tested related to the aqueous and organic phase arrangements. Interfacial polymerization with aqueous phase above organic phase (IP-R) enhances the filtration ability, i.e., rejection of MgCl$_2$ more than the IP-based membranes. When the IP-R based membrane is treated by 1% trimesoyl chloride (TMC) in hexane, the NaCl rejection rate is increased from 74% to 91%. This is due to the reduced pore size and crosslinking of TMC with secondary amino groups (present in the barrier layer of IP-R membranes). Permeability of the IP membranes is controlled by forming water channels at the CN and polymer matrix interface. This interface present in the IP membranes' barrier layer does not depend on the crosslinking reaction [47].

Nanofiltration membranes are created with a triple-layer using the process of interfacial polymerization. The triple-layer has CNC as an interlayer. Diamine and acyl chloride are present on each side of the interlayer. These layers are kept on a support of a microporous substrate. The CNC interlayer helps for the formation of polyamide skin layer and the nanofiltration. Aqueous diamine monomers storing and lessening the interfacial polymerization speed (for a reduced crosslinking degree of skin layer) are useful in nanofiltration and polyamide formation. This interlayer (with hydrophilic nature) enhances the water permeation. The water flux is 204 L/m^2/h under pressure 0.6 MPa, and the rejection rate of Na$_2$SO$_4$ is more than 97%. The reduced crosslinking degree of the skin layer separates the monovalent/divalent ions more [48].

Nanofiltration membranes are conventionally produced by using two polymers. By selective removal of one polymer by phase inversion process, pores with size 1—10 nm is created. CNCs are added with polysulfone (a commonly used ultrafiltration polymer membrane). But it is challenging to incorporate nanocellulose into water-insoluble polymers without aggregation. Membranes are made by using a solvent exchange process using an organic solvent *N*-methylpyrrolidone (NMP). Loading the filler more than 1% leads to the polymer's percolation, increasing the tensile modulus [49].

9.8 Polysulfone/Nanocellulose Composite Membranes

Water filtration membranes with varied content levels (0, 0.2, 0.5, and 1.0 wt%) of cellulose nanomaterials, polysulfone with different concentrations (20 and 30 wt%), and with varying thicknesses of casting (50, 100, and 200 μm) are produced for water purification applications. Membrane produced

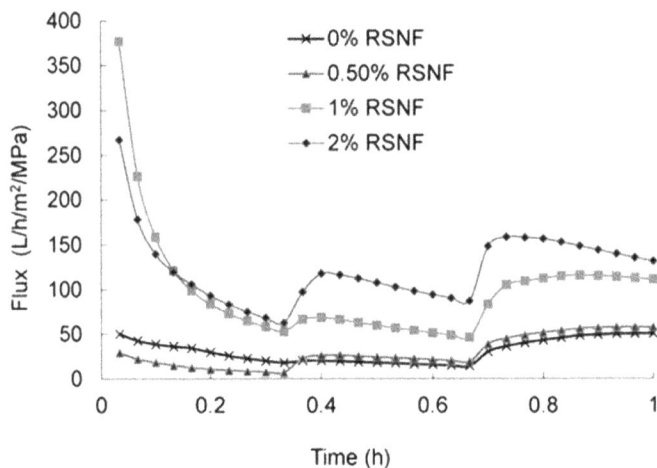

FIGURE 9.9 Shows the water flux versus time graph. Water flux of polysulfone (PSF) membrane and rice straw nanofibers (RSNF) loaded PSF/RSNF membranes. The loading quantity of RSNF on PSF membranes are 0.5%, 1%, and 2% [51].

with 30 wt% concentration of polysulfone, 0.5 wt% concentration of cellulose nanomaterials (CNC content), and 50 μm casting thickness has more water flux level, i.e., 22.9 L/m²/h/bar. This membrane has a 98.1% level of methylene blue rejection efficiency [50].

CNFs separated from the unbleached pulps possess more hydrophilicity and high mechanical strength. Polysulfone/rice straw nanofibers (PSF/RSNF) membrane for water purification is prepared by phase inversion of unbleached RSNF (separated from neutral sulfite pulp with 14% of lignin content) with PSF polymer. The PSF/RSNF membrane rejects lime nanoparticles from the aqueous suspension of lime up to the level of 98%. Tensile strength and Young's modulus of the membranes increase by ~29% and ~40%, respectively. Fig. 9.9 shows that increasing the loading of RSNF increases the water flux in PSF/RSNF membrane [51].

CNFs modify hollow fiber UF PSF membranes. In this modification process, CNF is introduced into the molding solution and internal coagulant. Both porosity and the permeability of hollow fiber PSF membranes increase. The permeability increases more than three times i.e. from 82 L/m²/h bar to 287 L/m²/h bar. The addition of CNF to the outside coagulant forms the inside, and the outer selective layer increases 1.5 times. Also, the rejection level of the dye Blue Dextran (MM = 69 kg/mol) increases—from 96% to 99% [52].

9.9 Nanocellulose Membranes in Reverse Osmosis

Thin film nanocomposite (TFN) membranes are formed with CNCs and polyamide active layer. These membranes are fabricated using the materials like CNCs, *m*-phenylenediamine (MPD), and TMC through *in situ* interfacial polymerization process. The fabrication process increases the hydrophilicity of the prepared membrane. It has been confirmed through contact angle measurements. Water flux of the membrane is in the range of 30 to 63 ± 10 L/m²·h. Also, the fabricated membrane has 97.8% salt rejection for 0.1% (w/v) CNCs loading [53].

TOCNFs are introduced into the polyamide layer present in the thin film composite (TFC) RO membrane. The interfacial polymerization process inserts CNFs into the polyamide layer. This membrane shows the water flux rate 29.8 L/m²/h at 1.5 MPa operating pressure and 96.2% of NaCl rejection rate. This CNFs nanocomposite membrane exhibits greater chlorine resistance than the bare membranes, as analyzed from the chlorine stability test [54].

Poly(acryloyl hydrazide) is grafted on CNC (CNC-PAH) to prepare a core-shell structure. The prepared core-shell is incorporated into a RO membrane using the LIP technique (layered interfacial polymerization). The dense-packed CNC-PAH offers reactive functional groups with more density. So, high water permeance and NaCl rejection are achieved with this new membrane. The abundantly available hydroxyl groups and amine groups in CNC-PAH provide excellent boron capturing ability more than the

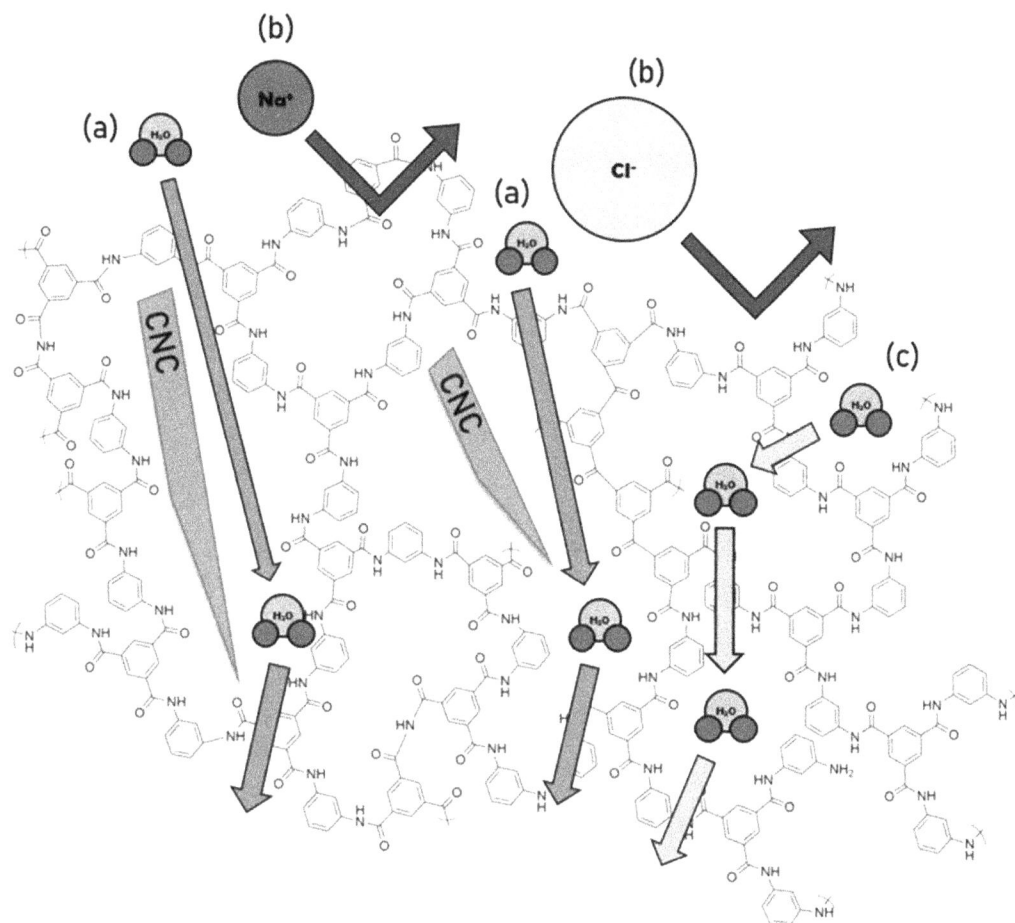

FIGURE 9.10 Exhibits the water transport mechanisms present inside the membrane of thin-film nanocomposite. (a) Enhancement in the water molecules transport through interfacial nanochannel development is due to the interactions between CNC and polymer (cellulose nanocrystal and polyamide). (b) Rejection of Na^+ and Cl^- ions by polyamide matrix. (c) Conventional method of water transport via polyamide matrix; no diffusion in the network. *Note:* Nanoparticles and ions shown here are not to scale [56].

commercial membrane. The surface smoothness, hydrophilicity, and organic fouling resistance are also high for CNC-PAH-assembled membrane than the commercial membrane [55].

Vacuum filtration and monomer dispersion techniques are performed to incorporate CNCs and TEMPO-oxidized CNCs in a polyamide matrix to form a water purification membrane. Low-level dispersion of nanocrystal in the polymer is obtained with vacuum filtration technique. But a higher consistent TFN is obtained by the monomer dispersion method. It improves salt rejection and water flux for 0.5 wt% of TOCNs loading. The development of nanochannels at the interface between the CNCs and the polyamide matrix produces the membrane's rapid transport pathways. The CNCs are high-aspect-ratio nanocrystals. Fig. 9.10 shows the water transport mechanisms of TFN membrane and conventional polyamide matrix [56].

9.10 Conclusion

Membranes made of nanocellulose can be produced by vacuum filtration, freeze-drying, and electrospinning methods. These membranes are used in water purification by eliminating dyes, bacteria, and viruses. They have properties like optimal pore size, high water flux, and low fouling, leading to competition with the commercial membranes. These membranes are biodegradable and can be produced at a low cost. Nanocellulose can also be incorporated with commercial RO membranes. Membranes made using polymer nanocomposites and nanocellulose are superior to microfiltration and ultrafiltration techniques.

Acknowledgments

The author expresses thanks to her husband, Mr. G. Sankar, for his help in this work. Also, she acknowledges the assistance of International Research Center, Kalasalingam Academy of Research and Education (Deemed University), Krishnankoil, India.

REFERENCES

[1] Sharma, Priyanka R., Sunil K. Sharma, Tom Lindstrom, and Benjamin S. Hsiao. "Nanocellulose-enabled membranes for water purification: perspectives." *Advanced Sustainable Systems* 4 (2020): 1900114. doi: https://doi.org/10.1002/adsu.201900114.

[2] Gopakumar, Deepu A., Vishnu Arumughan, Daniel Pasquini, Shao-Yuan Ben Leu, Abdul Khalil H. P. S., and Sabu Thomas. "Nanocellulose-based membranes for water purification." In *Nanoscale materials in water purification*, eds. S. Thomas, D. Pasquini, S. Y. Leu, D. A Gopakumar, 59–85. Oxfordshire, UK: Elsevier. 2019.

[3] Ma, Hongyang, Christian Burger, Benjamin S. Hsiao, and Benjamin Chu. "Ultra-fine cellulose nanofibers: new nanoscale materials for water purification." *Journal of Materials Chemistry* 21 (2011): 7507–7510. doi: https://doi.org/10.1039/C0JM04308G.

[4] Mukhopadhyay, Alolika, Zheng Cheng, Avi Natan, Yi Ma, Yang, Daxian Cao, Wei Wang, and Hongli Zhu. "Stable and highly ion-selective membrane made from cellulose nanocrystals for aqueous redox flow batteries." *Nano Letters* 19 (2019): 8979–8989. doi: https://doi.org/10.1021/acs.nanolett.

[5] Dai, Zhongde, Vegar Ottesen, Jing Deng, Ragne M. Lilleby Helberg, and Liyuan Deng. "A brief review of nanocellulose based hybrid membranes for CO_2 separation." *Fibers* 7 (2019): 40. doi: https://doi.org/10.3390/fib7050040.

[6] Karim, Zoheb, Simon Claudpierre, Mattias Grahn, Kristiina Oksman, and Aji P. Mathew. "Nanocellulose based functional membranes for water cleaning: tailoring of mechanical properties, porosity and metal ion capture." *Journal of Membrane Science* 514 (2016): 418–428. doi: https://doi.org/10.1016/j.memsci.2016.05.

[7] Wei, Haoran, Katia Rodriguez, Scott Renneckar, and Peter J. Vikesland. "Environmental science and engineering applications of nanocellulose-based nanocomposites." *Environmental Science: Nano* 1 (2014): 302–316. doi: 10.1039/C4EN00059E.

[8] Wang, Zhiguo, Wenwen Zhang, Juan Yu, Lijun Zhang, Liang Liu, Xiaofan Zhou, Chaobo Huang, and Yimin Fan. "Preparation of nanocellulose/filter paper (NC/FP) composite membranes for high-performance filtration." *Cellulose* 26 (2019): 1183–1194. doi: https://doi.org/10.1007/s10570-018-2121-8.

[9] Mautner, Andreas, Koon-Yang Lee, Tekla Tammelin, Aji P. Mathew, Alisyn J. Nedoma, Kang Li, and Alexander Bismarck. "Cellulose nanopapers as tight aqueous ultra-filtration membranes." *Reactive and Functional Polymers* 86 (2015): 209–214. doi: https://doi.org/10.1016/j.reactfunctpolym.2014.09.014.

[10] Prasanth, Raghavan, Shubha Nageswaran, Vijay Kumar Thakur, and Jou-Hyeon Ahn. "Electrospinning of cellulose: process and applications." In *Nanocellulose polymer nanocomposites: fundamentals and applications*, ed. V. K. Thakur, 311–340. Hoboken, NJ: John Wiley & Sons, Inc. 2014.

[11] Goetz, Lee A., Narges Naseri, Santhosh S. Nair, Zoheb Karim, and Aji P. Mathew. "All cellulose electrospun water purification membranes nanotextured using cellulose nanocrystals." *Cellulose* 25 (2018): 3011–3023. doi: https://doi.org/10.1007/s10570-018-1751-1.

[12] Naragund, Veereshgouda S., and P. K. Panda. "Electrospinning of cellulose acetate nanofiber membrane using methyl ethyl ketone and *N*,N-dimethylacetamide as solvents." *Materials Chemistry and Physics* 240 (2020): 122147. doi: https://doi.org/10.1016/j.matchemphys.2019.122147.

[13] Nemoto, Junji, Tsuguyuki Saito, and Akira Isogai. "Simple freeze-drying procedure for producing nanocellulose aerogel-containing, high-performance air filters." *ACS Applied Materials & Interfaces* 7 (2015): 19809–19815. doi: https://doi.org/10.1021/acsami.5b05841.

[14] Karim, Zoheb, Minna Hakalahti, Tekla Tammelin, and Aji P. Mathew. "*In situ* TEMPO surface functionalization of nanocellulose membranes for enhanced adsorption of metal ions from aqueous medium." *RSC Advances* 7 (2017): 5232–5241. doi: 10.1039/C6RA25707K.

[15] Cunha, Ana Gisela, Qi Zhou, Per Tomas Larsson, and Lars A. Berglund. "Topochemical acetylation of cellulose nanopaper structures for biocomposites: mechanisms for reduced water vapour sorption." *Cellulose* 21 (2014): 2773–2787. doi: http://dx.doi.org/10.1007/s10570-014-0334-z.

[16] Mautner, Andreas, H. A. Maples, T. Kobkeatthawin, V. Kokol, Zoheb Karim, K. Li, and A. Bismarck. "Phosphorylated nanocellulose papers for copper adsorption from aqueous solutions." *International Journal of Environmental Science and Technology* 13 (2016): 1861–1872. doi: https://doi.org/10.1007/s13762-016-1026-z.

[17] Zhang, Kai, Zongjie Li, Weimin Kang, Nanping Deng, Jing Yan, Jingge Ju, Yong Liu, and Bowen Cheng. "Preparation and characterization of tree-like cellulose nanofiber membranes via the electrospinning method." *Carbohydrate Polymers* 183 (2018): 62–69. doi: https://doi.org/10.1016/j.carbpol.2017.11.032.

[18] Quellmalz, Arne, and Albert Mihranyan. "Citric acid crosslinked nanocellulose-based paper for size-exclusion nanofiltration." *ACS Biomaterials Science & Engineering* 1 (2015): 271–276. doi: https://doi.org/10.1021/ab500161x.

[19] Choi, Hyeong Yeol, Jong Hyuk Bae, Yohei Hasegawa, Sol An, Ick Soo Kim, Hoik Lee, and Myungwoong Kim. "Thiol-functionalized cellulose nanofiber membranes for the effective adsorption of heavy metal ions in water." *Carbohydrate Polymers* 234 (2020): 115881. doi: https://doi.org/10.1016/j.carbpol.2020.115881.

[20] Bansal, Monica, Bhagat Ram, Ghanshyam S. Chauhan, and Anupama Kaushik. "L-Cysteine functionalized bagasse cellulose nanofibers for mercury (II) ions adsorption." *International Journal of Biological Macromolecules* 112 (2018): 728–736. doi: https://doi.org/10.1016/j.ijbiomac.2018.01.206.

[21] Gopakumar, Deepu A., Daniel Pasquini, Mariana Alves Henrique, Luis Carlos de Morais, Yves Grohens, and Sabu Thomas. "Meldrum's acid modified cellulose nanofiber-based polyvinylidene fluoride microfiltration membrane for dye water treatment and nanoparticle removal." *ACS Sustainable Chemistry & Engineering* 5 (2017): 2026–2033. doi: https://doi.org/10.1021/acssuschemeng.6b02952.

[22] Valencia, Luis, Susanna Monti, Sugam Kumar, Chuantao Zhu, Peng Liu, Shun Yu, and Aji P. Mathew. "Nanocellulose/graphene oxide layered membranes: elucidating their behaviour during filtration of water and metal ions in real time." *Nanoscale* 11 (2019): 22413–22422. doi: 10.1039/C9NR07116D.

[23] Liu, Peng, Chuantao Zhu, and Aji P. Mathew. "Mechanically robust high flux graphene oxide-nanocellulose membranes for dye removal from water." *Journal of Hazardous Materials* 371 (2019): 484–493. doi: https://doi.org/10.1016/j.jhazmat.2019.03.009.

[24] Lv, Jinling, Guoquan Zhang, Hanmin Zhang, and Fenglin Yang. "Graphene oxide-cellulose nanocrystal (GO-CNC) composite functionalized PVDF membrane with improved antifouling performance in MBR: behavior and mechanism." *Chemical Engineering Journal* 352 (2018): 765–773. doi: https://doi.org/10.1016/j.cej.2018.07.088.

[25] Jiang, Qisheng, Deoukchen Ghim, Sisi Cao, Sirimuvva Tadepalli, Keng-Ku Liu, Hyuna Kwon, Jingyi Luan, Yujia Min, Young-Shin Jun, and Srikanth Singamaneni. "Photothermally active reduced graphene oxide/bacterial nanocellulose composites as biofouling-resistant ultrafiltration membranes." *Environmental Science & Technology* 53 (2018): 412–421. doi: https://doi.org/10.1021/acs.est.8b02772.

[26] Voisin, Hugo, Lennart Bergström, Peng Liu, and Aji P. Mathew. "Nanocellulose-based materials for water purification." *Nanomaterials* 7 (2017): 57. doi: https://doi.org/10.3390/nano7030057.

[27] Sehaqui, Houssine, Seira Morimune, Takashi Nishino, and Lars A. Berglund. "Stretchable and strong cellulose nanopaper structures based on polymer-coated nanofiber networks: an alternative to nonwoven porous membranes from electrospinning." *Biomacromolecules* 13 (2012): 3661–3667. doi: https://doi.org/10.1021/bm301105s.

[28] Ma, Hongyang, Christian Burger, Benjamin S. Hsiao, and Benjamin Chu. "Fabrication and characterization of cellulose nanofiber based thin-film nanofibrous composite membranes." *Journal of Membrane Science* 454 (2014): 272–282. doi: https://doi.org/10.1016/j.memsci.2013.11.055.

[29] Chitpong, Nithinart, and Scott M. Husson. "Polyacid functionalized cellulose nanofiber membranes for removal of heavy metals from impaired waters." *Journal of Membrane Science* 523 (2017): 418–429. doi: https://doi.org/10.1016/j.memsci.2016.10.020.

[30] Hong, Hye-Jin, Jin Seong Lim, Jun Yeon Hwang, Mikyung Kim, Hyeon Su Jeong, and Min Sang Park. "Carboxymethlyated cellulose nanofibrils (CMCNFs) embedded in polyurethane foam as a

modular adsorbent of heavy metal ions." *Carbohydrate Polymers* 195 (2018): 136–142. doi: https://doi.org/10.1016/j.carbpol.2018.04.081.

[31] Wang, Jianqiang, Xinkun Lu, Pui Fai Ng, Ka I. Lee, Bin Fei, John H. Xin, and Jian-yong Wu. "Polyethylenimine coated bacterial cellulose nanofiber membrane and application as adsorbent and catalyst." *Journal of Colloid and Interface Science* 440 (2015): 32–38. doi: https://doi.org/10.1016/j.jcis.2014.10.035.

[32] Derami, Hamed Gholami, Qisheng Jiang, Deoukchen Ghim, Sisi Cao, Yatin J. Chandar, Jeremiah J. Morrissey, Young Shin Jun, and Srikanth Singamaneni. "A robust and scalable polydopamine/bacterial nanocellulose hybrid membrane for efficient wastewater treatment." *ACS Applied Nano Materials* 2 (2019): 1092–1101. doi: 10.1021/acsanm.9b00022.

[33] Wang, Zhao, Caitlin Crandall, Vicki L. Prautzsch, Rajesh Sahadevan, Todd J. Menkhaus, and Hao Fong. "Electrospun regenerated cellulose nanofiber membranes surface-grafted with water-insoluble poly (HEMA) or water-soluble poly (AAS) chains via the ATRP method for ultrafiltration of water." *ACS Applied Materials & Interfaces* 9 (2017): 4272–4278. doi: https://doi.org/10.1021/acsami.6b16116.

[34] Soni, Raghav, Taka-Aki Asoh, and Hiroshi Uyama. "Cellulose nanofiber reinforced starch membrane with high mechanical strength and durability in water." *Carbohydrate Polymers* 238 (2020): 116203. doi: https://doi.org/10.1016/j.carbpol.2020.116203.

[35] Palacios Hinestroza, Hasbleidy, Hilary Urena-Saborio, Florentina Zurita, Aida Alejandra Guerrero de Leon, Gunasekaran Sundaram, and Belkis Sulbaran-Rangel. "Nanocellulose and polycaprolactone nanospun composite membranes and their potential for the removal of pollutants from water." *Molecules* 25 (2020): 683. doi: https://doi.org/10.3390/molecules25030683.

[36] Wang, Zhe, Hongyang Ma, Benjamin S. Hsiao, and Benjamin Chu. "Nanofibrous ultrafiltration membranes containing crosslinked poly(ethylene glycol) and cellulose nanofiber composite barrier layer." *Polymer* 55 (2014): 366–372. doi: https://doi.org/10.1016/j.polymer.2013.10.049.

[37] Mautner, Andreas, Yosi Kwaw, Kathrin Weiland, Mlando Mvubu, Anton Botha, Maya Jacob John, Asanda Mtibe, Gilberto Siqueira, and Alexander Bismarck. "Natural fibre-nanocellulose composite filters for the removal of heavy metal ions from water." *Industrial Crops and Products* 133 (2019): 325–332. doi: https://doi.org/10.1016/j.indcrop.2019.03.032.

[38] Ma, Shanshan, Meiyun Zhang, Jingyi Nie, Jiaojun Tan, Shunxi Song, and Yanwei Luo. "Lightweight and porous cellulose-based foams with high loadings of zeolitic imidazolate frameworks-8 for adsorption applications." *Carbohydrate Polymers* 208 (2019): 328–335. doi: https://doi.org/10.1016/j.carbpol.2018.12.081.

[39] Sukma, Faqih Muhamad, and P. Z. Culfaz-Emecen. "Cellulose membranes for organic solvent nanofiltration." *Journal of Membrane Science* 545 (2018): 329–336. doi: https://doi.org/10.1016/j.memsci.2017.09.080.

[40] Zhijiang, Cai, Xiong Ping, Zhu Cong, Zhai Tingting, Guo Jie, and Zhao Kongyin. "Preparation and characterization of a bi-layered nanofiltration membrane from a chitosan hydrogel and bacterial cellulose nanofiber for dye removal." *Cellulose* 25 (2018): 5123–5137. doi: https://doi.org/10.1007/s10570-018-1914-0.

[41] Metreveli, Giorgi, Linus Wågberg, Eva Emmoth, Sandor Belak, Maria Stromme, and Albert Mihranyan. "A size-exclusion nanocellulose filter paper for virus removal." *Advanced Healthcare Materials* 3 (2014): 1546–1550. doi: https://doi.org/10.1002/adhm.201300641.

[42] Wang, Ran, Sihui Guan, Anna Sato, Xiao Wang, Zhe Wang, Rui Yang, Benjamin S. Hsiao, and Benjamin Chu. "Nanofibrous microfiltration membranes capable of removing bacteria, viruses and heavy metal ions." *Journal of Membrane Science* 446 (2013): 376–382. doi: https://doi.org/10.1016/j.memsci.2013.06.020.

[43] Manukyan, Levon, Justine Padova, and Albert Mihranyan. "Virus removal filtration of chemically defined Chinese hamster ovary cells medium with nanocellulose-based size exclusion filter." *Biologicals* 59 (2019): 62–67. doi: https://doi.org/10.1016/j.biologicals.2019.03.001.

[44] Zhan, Hui, Na Peng, Xiaojuan Lei, Yanan Huang, Dong Li, Rongjun Tao, and Chunyu Chang. "UV-induced self-cleanable TiO$_2$/nanocellulose membrane for selective separation of oil/water emulsion." *Carbohydrate Polymers* 201 (2018): 464–470. doi: https://doi.org/10.1016/j.carbpol.2018.08.093.

[45] Hassan, Enas, Mohammad Hassan, Ragab Abou-zeid, Linn Berglund, and Kristiina Oksman. "Use of bacterial cellulose and crosslinked cellulose nanofibers membranes for removal of oil from oil-in-water emulsions." *Polymers* 9 (2017): 388. doi: https://doi.org/10.3390/polym9090388.

[46] Soyekwo, Faizal, Qiugen Zhang, Runsheng Gao, Yan Qu, Chenxiao Lin, Xiaoling Huang, Aimei Zhu, and Qinglin Liu. "Cellulose nanofiber intermediary to fabricate highly-permeable ultrathin nanofiltration membranes for fast water purification." *Journal of Membrane Science* 524 (2017): 174–185. doi: https://doi.org/10.1016/j.memsci.2016.11.019.

[47] Wang, Xiao, Tsung-Ming Yeh, Zhe Wang, Rui Yang, Ran Wang, Hongyang Ma, Benjamin S. Hsiao, and Benjamin Chu. "Nanofiltration membranes prepared by interfacial polymerization on thin-film nanofibrous composite scaffold." *Polymer* 55 (2014): 1358–1366. doi: https://doi.org/10.1016/j. polymer.2013.12.007.

[48] Wang, Jing-Jing, Hao-Cheng Yang, Ming-Bang Wu, Xi Zhang, and Zhi-Kang Xu. "Nanofiltration membranes with cellulose nanocrystals as an interlayer for unprecedented performance." *Journal of Materials Chemistry A* 5 (2017): 16289–16295. doi: https://doi.org/10.1039/C7TA00501F.

[49] Noorani, Sweda, John Simonsen, and Sundar Atre. "Nano-enabled microtechnology: polysulfone nanocomposites incorporating cellulose nanocrystals." *Cellulose* 14 (2007): 577–584. doi: https://doi. org/10.1007/s10570-007-9119-y.

[50] Daria, Maral, Hossein Fashandi, Mohammad Zarrebini, and Zahra Mohamadi. "Contribution of polysulfone membrane preparation parameters on performance of cellulose nanomaterials." *Materials Research Express* 6 (2018): 015306. doi: https://doi.org/10.1088/2053-1591/aae600.

[51] Hassan, Mohammad, Ragab E. Abou Zeid, Wafaa S. Abou-Elseoud, Enas Hassan, Linn Berglund, and Kristiina Oksman. "Effect of unbleached rice straw cellulose nanofibers on the properties of polysulfone membranes." *Polymers* 11 (2019): 938. doi: https://doi.org/10.3390/polym11060938.

[52] Anokhina, Tatyana S., S. D. Bazhenov, Ilya L. Borisov, V. P. Vasilevsky, V. A. Vinokurov, and A. V. Volkov. "Nanocellulose as modifier for hollow fiber ultrafiltration PSF membranes." *Key Engineering Materials* 816 (2019): 238–43. doi: https://doi.org/10.4028/www.scientific.net/kem.816.238.

[53] Asempour, Farhad, Daryoush Emadzadeh, Takeshi Matsuura, and Boguslaw Kruczek. "Synthesis and characterization of novel cellulose nanocrystals-based thin film nanocomposite membranes for reverse osmosis applications." *Desalination* 439 (2018): 179–187. doi: https://doi.org/10.1016/j.desal.2018.04.009.

[54] Liu, Shasha, Ze-Xian Low, Hanaa M. Hegab, Zongli Xie, Ranwen Ou, Guang Yang, George P. Simon, Xiwang Zhang, Lian Zhang, and Huanting Wang. "Enhancement of desalination performance of thin-film nanocomposite membrane by cellulose nanofibers." *Journal of Membrane Science* 592 (2019): 117363. doi: https://doi.org/10.1016/j.memsci.2019.117363.

[55] Park, Chan Hyung, SungKwon Jeon, Sang-Hee Park, Min Gyu Shin, Min Sang Park, Sun-Young Lee, and Jung-Hyun Lee. "Cellulose nanocrystal-assembled reverse osmosis membranes with high rejection performance and excellent antifouling." *Journal of Materials Chemistry A* 7 (2019): 3992–4001. doi: https://doi.org/10.1039/C8TA10932J.

[56] Smith, Ethan D., Keith D. Hendren, James V. Haag, E. Johan Foster, and Stephen M. Martin. "Functionalized cellulose nanocrystal nanocomposite membranes with controlled interfacial transport for improved reverse osmosis performance." *Nanomaterials* 9 (2019): 125. doi: https://doi.org/10.3390/nano9010125.

10

Nanocellulose-Based Absorbent for Dye Adsorption

Pallavi Jain
SRM Institute of Science and Technology, Delhi-NCR Campus, Modinagar, India

A. Geetha Bhavani
Noida International University, Greater Noida, India

Sapna Raghav
Banasthali Vidyapith, Banasthali, India

CONTENTS

10.1 Introduction

The word "cellulose" is derived from the French word "cellule", meaning a small living cell composed of glucose. It was discovered in 1838 by Anselme Payen [1]. The cellulose-based materials are natural polymers available in abundance [2]. For example, cereal straws, hemp, cotton, and other plant-based materials are its sources [3, 4]. This polysaccharide having 3000 glucose units is the elementary constituent of plants. The reports assessed 10^{10}–10^{12} tons of cellulose production each year, of which 6×10^9 tons are used in paper and textile industries [5, 6]. The cellulose is semicrystalline (in both amorphous and crystalline regions), depending on its source [7, 8]. The semicrystalline property provides the substantial strength and flexibility to materials. Cellulose is a stable polymer compound having glucose-based monomers, which depend upon their sources. Cellulose is naturally abundant, derived from many renewable resources such as wood, plant, bacteria, algae, etc. [9]. The cellulose is extracted and isolated as nanoscale material known as nanocellulose (NC), with 100 nm or smaller in size. The NC materials offer promising physical and chemical characteristics like large surface area, exceptional tensile and mechanical strength, and flexibility that enables them to be easily modified for specific applications [10].

TABLE 10.1

Various Types of Cellulose Forms [11].

Cellulose Type	Production
I	Naturally available in bacteria, plant, etc.
II	Modification of cellulose I by dissolving or swelling in a solvent or alkaline or acid solutions
III	Cellulose I or II is treated with ammonia solution
IV	Cellulose III is treated thermally with glycerol up to 260 °C

10.2 Categories of Nanocellulose

The cellulose is usually considered in four forms—cellulose I, II, III, and IV—as provided in Table 10.1 [11]. These celluloses can be prepared in nanosize with a large surface area. These materials have versatile physiochemical properties: high mechanical strength, greater stiffness, and large hydroxyl groups with stable performance in dye removal [12, 13]. The new nanomaterials like nanoparticles (NPs) are often used to attain better performance. These rod-shaped NPs contain a crystalline phase interconnected with the amorphous phase of cellulose to form micro/macroscopic fibers through inter-fiber hydrogen bonding. NC reveals significant characteristics like enlarged surface area, greater stiffness with light weight, mechanical and thermal stability, highly reactive surface, optical transparency, extraordinary porosity, biodegradability, biocompatibility, and flexibility for modification [14].

NC materials are made up of both amorphous and crystalline regions to deliver significant flexibility and strength. The cellulose is modified with specific methods to prepare cellulose nanocrystals (CNCs) and cellulose nanofibrils (CNFs) for industrial applications.

   a. CNC is also known as nanocrystalline cellulose (NCC) and cellulose nanowhiskers.
   b. Nanofibrillar cellulose (NFC) is also known as NC, CNF, CNF, nanofibrillated cellulose, or microfibrillated cellulose (MFC).
   c. Bacterial nanocellulose (BNC) is also called biocellulose/bacterial cellulose (BC)/microbial cellulose (MC). BNC is isolated from bacterial genera. Gram-positive *Acetobacter xylinum* was found to be highly productive compared with aerobic, non-pathogenic bacteria.

The specific type of NC is possible to prepare by using various synthesis methods to alter as per the desired structure, morphology, size, and properties [15].

10.3 Adsorption Mechanism of NC-Based Adsorbents

The adsorption efficiency process depends on the interactions of adsorbent, adsorbate, adsorption models, and isotherms. The interactions are possible via various types of mechanisms through hydrogen bonding, ion exchange, complexation formation on pores and surface, chelation, van der Waals forces, dipolar interactions, and hydrophobicity [4].

10.4 Dye Removal

The removal of dyes from industrial waste through renewable sources is necessary as water is a prime living component [16]. Many researchers have come up with an innovative process using modified materials for specific dye compound removal. Xie et al. [17] explored the high NC hybrids adsorption for reactive cationic dyes like blue B-RN and yellow B-4RFN dyes. They assessed that the high adsorption is due to the organic functional groups (–C–N–) and nano-ranged cubic crystals present in NC hybrids.

10.4.1 Congo Red Dye

Beyki et al. [18] studied the removal of Congo red dye using Fe_3O_4/NC having the core-shell structured polymeric ionic liquid. They elevated the different factors that affected the adsorption of Congo red: (1) adsorbent surface (methylene groups) hydrophobic interaction with the Congo red, (2) π-π interactions between the Congo red molecules and bulk π phase present on imidazolium ring, (3) ion exchange between the Congo red and the adsorbent. Zheng et al. [19] tried to remove the Congo red from the composites material composed of chitosan (Cts)/dialdehyde microfribillated cellulose (DAMFC). The Cts/DAMFC has shown promising adsorption of Congo red because of the electrostatic attraction and H-bonding. The electrostatic attractions existed between protonated amino groups (of Cts) and sulfonate groups of Congo red. The H-bonding was involved between amino groups (Congo red) and hydroxyl groups (Cts/DAMFC).

A unique polymeric ionic liquid was prepared in a single-step synthesis process of Fe_3O_4/cellulose nanohybrid using epichlorohydrin and 1-methylimidazole to achieve sorption of Congo red efficiently. The adsorbent surface interactions played a vital role in an effective dye removal process. Congo red dye adsorption was faster at equilibrium duration of 11 min with 131 mg/g absorption capacity. The Congo red dye adsorption level is influenced by surface properties like adsorbent dosage, contact time, and ionic strength. The isotherm study showed that the Langmuir adsorption model showed promising adsorption behavior of dye. The composite material was regenerated into the mixture of CH_3COCH_3/CH_3OH/NaOH (3.0 mol/L). The nanohybrid composites regeneration proved to have effective reusability and environmental remediation resolution [18].

NC was also prepared through acid hydrolysis with cellulose. Further, polypyrrole was coupled to form NC polypyrrole composite (NCPPY). The NCPPY showed a higher conversion percentage to nano-ranged crystal structure with enhanced thermal stability and surface area. The batch-wise preparation was optimized with parameters like temperature, pH, contact time, adsorbent level, and initial Cr(VI) concentration. The optimization tool effectively removed Cr(VI) and Congo red dye up to 80% and 85%, respectively. The thermodynamic data through biosorption of Cr(VI) and Congo red over NCPPY showed a spontaneous, endothermic, and entropy-determined process. NCPPY showed outstanding adsorption for both metal and dye [20].

Nanoadsorbent is well known for its high adsorption capabilities and has limitations as a recovery after wastewater treatment. This drawback was rectified by Cts and NC entrapment with poly(hydroxyalkanoate) (PHA) through the electrospinning technique. Before electrospinning, the pickering emulsion was stabilized with Tween 80, which led to the formation of a homogeneous NC-Cts-PHA mixture. The prepared electrospun biocomposites encompassed enhanced porosity to react with Cts and NC for adsorption of dye. The incorporation of NC and Cts significantly increased 57.6–70.5% crystallinity through the electrospun biocomposites. The adsorption of Congo red dye was fitted within the Langmuir isotherm model, which suggested the chemisorption nature of the composite material. PHA/NC was found to be more efficient in dye removal than PHA/Cts. The pickering emulsion showed the highest dye removal percentage with PHA/NC/Cts (75.8%) [21].

The microgel-based NC and amphoteric polyvinylamine (PVAm) were developed in a two-step process. First, sodium periodate was used to oxidize CNC to yield dialdehyde NC (DANC). The crosslinker DANC was reacted with PVAm to form a microgel with particles sizes from 200 nm to 300 nm. The formed microgel was very useful for the removal of anionic dye, even in an acidic medium for the protonation of amino groups. The microgel nanocomposite materials were employed for the adsorption of various anionic dyes and Congo red. The anionic dyes adsorption followed pseudo-second-order kinetics and showed dye adsorption by chemisorption of the surface [16].

The synthesis of nanosized cellulose with high crystallinity from a renewable source, like used printed papers as adsorbent using a different organic solvent with a yield of 30%, was also reported. The hydroxy naphthol blue and Congo red adsorption capacity at pH 2 and pH 3 was found a maximum of 0.170 mmol/g and 0.156 mmol/g, respectively. The high adsorption of hydroxy naphthol blue was accounted for having an extra −OH group (in sorbate molecule), while the Congo red transition form ammonium and azonium ions. Both the dyes' adsorption was constructive and endothermic, and showed electrostatic forces favorable to the adsorption mechanism process. The Congo red and hydroxy naphthol blue were used as sorbate to test nanosized cellulose's adsorption capacity through Freundlich and Langmuir isotherm models from low to high temperatures. The kinetic models were assessed using pseudo-first-order and pseudo-second-order kinetics. The adsorption at elevated temperature resulted in higher adsorption capacity of dyes, suggesting the endothermic nature of adsorption through solution interface [22].

10.4.2 Methylene Blue Dye

Recently, hairy NC is technologically advanced to enhance the functional properties and to prepare biorenewable cellulose NPs. Electrostatically stabilized nanocrystalline cellulose (ENCC) was prepared from natural resources like wood pulp by two-step oxidation methods called periodate and chlorite. The ENCC was evaluated with the adsorption of cationic dye-methylene blue. The maximum adsorption capacity was due to the high degree of an ion-exchange mechanism, which influenced other ions. The sodium form of alginate and ENCC (ALG-ENCC beads) developed composite hydrogel with removal capacity (1250 mg/g) of methylene blue. ALG-ENCC beads made various NPs available to access the contaminants, and were found to be potential for dye removal in larger capacities from wastewaters [23].

Nanosized GO particles were explored through the conventional sol-gel process. The GO-silica loaded over CNF matrix through freeze-drying technique demonstrated the promising methylene blue degradation. It was observed that silica particles stuck to the GO surface, which has a lower crystallinity. At pH 7, the GO-silica showed a 608.4 mg/g adsorption value of methylene blue, which was outstanding compared with CNF aerogels (adsorption capacity of 328.3 mg/g). The better methylene blue degradation ability was attributed to high surface area and functional groups of GO. The adsorption degree of methylene blue over CNF aerogels was monitored with the Langmuir isotherm and pseudo-second-order kinetics. The above considerations over CNF aerogels bioadsorbents were promising alternatives for the degradation of organic dyes from polluted wastewater [24].

The Ag_3PO_4 particles with excellent crystallinity were prepared through the ion-exchange method using NC sheets *in situ* with particles in the range of 70–280 nm. Ag_3PO_4/NC composite was developed using a simple *in situ* method and tested in visible light to follow the photocatalytic reactions to degradation of methylene blue and methyl orange in wastewater matrix under solar irradiation. The photocatalytic activity over methyl orange led to the degradation rate by up to 90%. The excellent activity was due to strong NC support for Ag_3PO_4 particles dispersion and also prevented from aggregation. The Ag_3PO_4/NC showed excellent compatibility between cellulose support and Ag_3PO_4 and developed a perfect photocatalytic design for wastewater treatment (Fig. 10.1) [25].

Another attempt was made to isolate celery lignocellulosic to prepare NC films and biocomposite films for a potential refabrication into nanocomposite films. The structural properties relations of these hybrid films were analyzed systematically. Using alkali soakings of 4% NaOH at 30 °C for 12 hours, the celery was extracted as cellulose-rich solid and hemicellulose. The hemicellulose yield was 16.1%, whereas the CNF yield was 43.1% through hydrolysis of celery with oxalic acid. The microfluidization was employed after the hydrolysis process to separate CNF with various sizes from 5 nm to 6 nm and 1.3 μm to 1.7 μm. The hemicellulose fabricated with CNF potentially upgraded the properties like flexibility and mechanical strength. The results confirmed that enhanced properties improved methylene blue degradation over hybrid films (from 192.1 mg/g to 429.9 mg/g). The 30% montmorillonite clay was also mixed with hybrid films for further improvement in dye degradation ability. The hybrid film fabrication of composite enabled the crosslinking ability to increase tensile strength and flexibility. However, the blue ethylene degradation lowered with CNF content increased due to the transparencies and adsorption [26].

10.4.3 Rhodamine Dye

The CNFs are quite useful as a template in preparing magnetic nanoparticles (MNPs) by identical particle size with uniform distribution. The nanofibers surface was loaded with MNPs with metal precursors through *in situ* hydrolysis at room temperature. The synthesis parameters like technique, crystal size, morphology, metal precursors, and nanofibers ratio affected the properties like magnetic, thermal stability, and degradation ability. The optimized parameters resulted in the MNPs <20 nm size and the crystallite in the range of 96–130 Å. Due to its low-cost and eco-friendly nature, a magnetic membrane was used to catalyze the oxidation process in a sulfate radical. The prepared membranes were activated

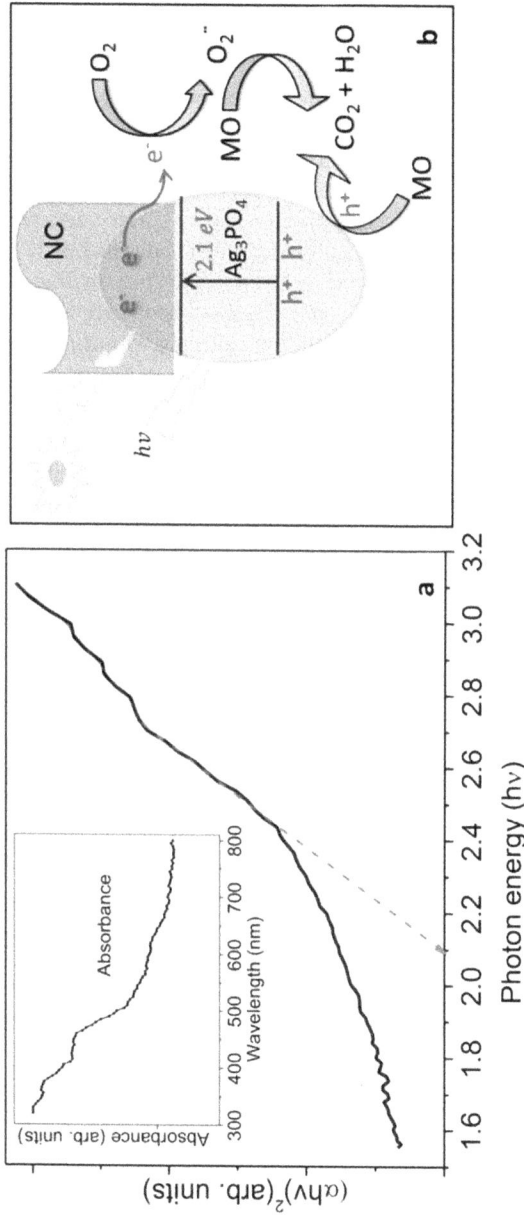

FIGURE 10.1 Ag$_3$PO$_4$/NC nanocomposite (Tauc's plot). (a) Absorbance spectra and (b) photocatalytic degradation mechanism of the azo dyes by Ag$_3$PO$_4$/NC catalyst under visible light (Reproduced with permission from Ref. [25]. © Elsevier 2019).

using peroxymonosulfate (PMS) to remove rhodamine B (RhB) by 94.9% within 300 minutes of duration at room temperature. This result confirmed the catalyzed PMS with MNC membrane degraded rhodamine B with effectivity.

The co-precipitation methods were also used for uniformity in a composite. The MNPs initial nucleation and further growth occur on the nanofiber's surface, which is densely dispersed and contributes to the high magnetic properties. The NP's size decreased with increasing NC content due to magnetic particle formation over nucleation sites. The MNPs possessed superparamagnetic performance with S-shaped hysteresis loops. The Fe_3O_4 NPs were loaded over the CNFs' surface to increase the magnetization, which ultimately improved the dye removal. The synthesis methods using metal oxides and membranes to form composites are applicable for various dye species removal with greater efficiency [27].

The CNF was modified with GO ultrathin film to achieve a layered membrane without crosslinkers, which showed remarkable separation performance and synergy with water flux. The simple GO or cellulosic membranes showed less separation performance than a modified composite like GO-CNF hybrid membranes. The membranes showed negative surface zeta potential. The GO-CNF membrane with a ratio of 1:100 demonstrated a high degree of water flux of $18,123 \pm 574$ L/m^2/h/bar, which was higher than the mere CNF membrane or Nylon 66, 0.2μm (hydrophilic commercial membrane). The GO nanosheets created abundant water transport nanochannels responsible for increasing water permeability. The layered membrane with anisotropic nature provided >90% elimination of both negative and positive charged dyes compounds because of the proper blend of hydrophobic, electrostatic interactions and compound size exclusion. The GO-CNF also provided unique properties like the massive scale of functional groups and scalable purification membranes with good flux benefits, thermal and mechanical stability, and eliminated dyes. The hydride composite membranes showed improvement pointedly in mechanical and thermal stability throughout wet and dry separation states [28–30].

10.4.4 Sudan Dyes

The Sudan dyes species (Sudan I, Sudan II, Sudan III, and Sudan IV) were identified through two techniques: (i) solid-phase extraction and (ii) capillary liquid chromatography presented in food samples. Both techniques were used over NC isolated from the agricultural waste material of bleached argan press cake (APC) through bleaching, refining, and hydrolysis with sulfuric acid. The first method included the separation of NC dispersion in waste effluents through centrifugation. In the second method, NC was fabricated with iron NPs for effective removal of Sudan dyes. The iron NPs component allowed dye removal through the centrifugation process efficiently and conveniently. Both processes effectively removed the Sudan dye up to 0.25–2.00 μg/L range. When NC is used as a nanoadsorbent, the limits of detections, quantifications, and standard deviations were lesser than that of 0.1 μg/L, 0.20 μg/L, and 3.46%, respectively. Conversely, using MNC, these values were found to be lesser than 0.07 μg/L, 0.23 μg/L, and 2.62%, respectively. The fabricated NC composite was applied to Sudan dyes present in ketchup sauce and barbeque sample and estimated up to 93.4% and 109.6%, respectively. The composite resulted in immense stability and reusability for more than three cycles [31].

The hydrogels were modified with an interpenetrating polymer network (IPN) by crosslinking with neutral polyacrylamide (PAM) and polyelectrolyte sodium alginate (SA) for enhancing dye removal. The three types of NC—BCFs, CNCs, 2,2,6,6-tetramethylpiperidine-1-oxy radical (TEMPO), and oxidized 2,2,6,6-tetramethylpiperidine-1-oxy cellulose nanofibers (TOCNs)—were used to disperse in the matrix of SA-PAM gel. The formed hydrogels were macroporous structures with a moisture content in the range of ~83%, a pore size of 60±51 μm, and excellent membrane transparency. The bacterial cellulose fibers are widely used for improving the hydrogel's functions because of their constrictive strength of SA/PAM fabricated with NC hydrogel, which shows 6.59 times higher performance on dye removal compared to simple SA/PAM. On the other hand, TOCNs are found to be a better candidate as the fillers for enhancing the adsorption level with the help of the carboxyl functional group present over the member surfaces. Finally, every fabrication's preparation to form hydride composition membranes aims to enhance the desired applications like dye removal in wastewater treatment tanks [32].

10.5 Sustainability and Challenges

The utilization of green materials from natural adsorbents for water pollution is currently being implemented internationally because of worldwide environmental concerns and continuous resource depletion. These materials are essential and promote the growth of human societies in terms of sustainable development. One of the green materials recognized for the adsorption of contaminants from water sources is cellulose and it is abundant in nature. Compared to other materials, the relatively high availability of cellulose and the versatility of its conversion to other diverse derivatives like NC and composites enable its manufacturing comparatively at less cost. It is also environment-friendly since microbial decay procedures can quickly recycle it back into the ecological system, i.e., carbon cycle [33]. Even then, amid these excellent cellulose properties, many drawbacks and problems remain unanswered. Cellulose is accountable for plants' rigid cell wall structure is a proven fact and it occurs naturally in the environment. However, one of its restrictions was the cultivation of natural forests that could lead to environmental hazards resulting in a loss of native vegetation [34]. Cellulose isolation is often a slow process, demands high energy, and involves harmful chemicals that damage the environment and human health [35, 36]. In fact, over the years, the limited synthesis of nanocomposites based on cellulose has become a resounding achievement. The advancement of technology to accomplish massive commercial production significantly seems to have been a challenge [37]. To counter this, attempts are being made to develop novel methods with an aim of ultimately enhancing the current process or facilitating artificial production on a large scale.

10.6 Conclusion

Owing of the significant contingent, NC absorbents have shown tremendous perspective to reduce various water pollutants. In this chapter, we focused on different NC composites and studied their absorption capacity to achieve dye removal. It also aimed to investigate the absorbate and absorbent interaction and affect the absorption process efficiency. It is known that, through various mechanisms like hydrogen bonding, van der Waals forces, electrostatic, and $\pi-\pi$ interactions, much of the adsorption technique involves materials and contaminants interaction. The interactions are possible via various types of mechanisms: hydrogen bonding, ion exchange, complexation formation on pores and surface, chelation, van der Waals forces, dipolar interactions, and hydrophobicity. The nanohybrid composites regeneration is significant for reusability and environmental remediation resolution.

Acknowledgment

Pallavi Jain is thankful to SRM Institute of Science and Technology, Delhi NCR Campus, for their support.

REFERENCES

[1] Shron Olivera, Handanahally Basavarajaiah Muralidhara, Krishna Venkatesh, Vijay Kumar Guna, Keshavanarayana Gopalakrishna, and Yogesh Kumar K. "Potential applications of cellulose and chitosan nanoparticles/composites in wastewater treatment: a review." *Carbohydrate Polymer* 153 (2016): 600–618. https://doi.org/10.1016/j.carbpol.2016.08.017.

[2] Mohamed E. Mahmoud, Azza E.H. Abdou, Mostafa E. Sobhy, and Nesma A. Fekry. "Solid-solid crosslinking of carboxymethyl cellulose nanolayer on titanium oxide nanoparticles as a novel biocomposite for efficient removal of toxic heavy metals from water." *International Journal of Biological Macromolecules* 105 (2017): 562–567. https://doi.org/10.1016/j.ijbiomac.2017.07.156.

[3] J. Jayaramudu Sonakshi Maiti, Kunal Das, Siva Mohan Reddy, Rotimi Sadiku, Suprakas Sinha Ray, and Dagang Liu. "Preparation and characterization of nano-cellulose with new shape from different precursor." *Carbohydrate Polymer* 98 (2013): 562–567. https://doi.org/10.1016/j.carbpol.2013.06.029.

[4] Junhui Si, Zhixiang Cui, Qianting Wang, Qiong Liu, and Chuntai Liu. "Biomimetic composite scaffolds based on mineralization of hydroxyapatite on electrospun poly(ε-caprolactone)/nanocellulose fibers." *Carbohydrate Polymer* 43 (2016): 270–278. https://doi.org/10.1016/j.carbpol.2016.02.015.

[5] Catalina Gómez Hoy, Aangelica Maria Serpa, Jorge Velasquez-Cock, Piedad Ganan Rojo, Cristina I. Castro, Lina Maria Velez and Robin Zuluaga. "Vegetable nanocellulose in food science: a review." *Food Hydrocolloids* 57 (2016): 178–186. https://doi.org/10.1016/j.foodhyd.2016.01.023.

[6] Nathalie Lavoine, Isabelle Desloges, Alain Dufresne, and Julien Bras. "Microfibrillated cellulose—its barrier properties and applications in cellulosic materials: a review." *Carbohydrate Polymer* 90 (2012): 735–764. https://doi.org/10.1016/j.carbpol.2012.05.026.

[7] Pareetha Balakrishnan, M.S. Sreekala, Matjaz Kunaver, Miroslav Huski, and Sabu Thomas. "Morphology transport characteristics and viscoelastic polymer chain confinement in nanocomposites based on thermoplastic potato starch and cellulose nanofibers from pineapple leaf." *Carbohydrate Polymer* 169 (2017): 176–188. https://doi.org/10.1016/j.carbpol.2017.04.017.

[8] Malladi Rajinipriya, Malladi Nagalakshmaiah, Mathieu Robert, and Saiid Elkoun. "Importance of agricultural and industrial waste in the field of nanocellulose and recent industrial developments of wood based nanocellulose: a review." *ACS Sustainable Chemical Engineering* 6 (2018) 2807–2828. https://doi.org/10.1021/acssuschemeng.7b03437.

[9] Prachiben Panchal, Emmanuel Ogunsona, and Tizazu Mekonnen. "Trends in advanced functional material applications of nanocellulose." *Processes* 7 (2018): 10. https://doi.org/10.3390/pr7010010.

[10] A. Tshikovhi, Shivani B. Mishra, and Ajay K. Mishra. "Nanocellulose-based composites for the removal of contaminants from wastewater." *International Journal of Biological Macromolecules* 152 (2020): 616–632. https://doi.org/10.1016/j.ijbiomac.2020.02.221.

[11] Ragvendra Kumar Mishra, Arjun Sabu, and Santosh Kumar Tiwari. "Materials chemistry and the futurist eco-friendly applications of nanocellulose: status and prospect." *Journal of Saudi Chemical Society* 22 (2018): 949–978. https://doi.org/10.1016/j.jscs.2018.02.005.

[12] Qinzhi Li, Bing Wei, Laiming Lu, Yibo Li, Yangbing Wen, Wanfen Pu, Hao Li, and Chongyang Wang. "Investigation of physical properties and displacement mechanisms of surface-grafted nano-cellulose fluids for enhanced oil recovery." *Fuel* 207 (2017): 352–364. https://doi.org/10.1016/j.fuel.2017.06.103.

[13] Mohammand Asim. "Nanocellulose: preparation method and applications." In *Cellulose-reinforced nanofibre composites: production properties and application*, eds. M. Jawaid, S. Boufi, and A.K. HPS, 261–276. Sawston, Cambridge: Woodhead Publishing. 2017.

[14] Dieter Klemm, Friederike Kramer, Sebastian Moritz, Tom Lindström, Mikael Ankerfors, Derek Gray, and Annie Dorris. "Nanocelluloses: a new family of nature-based materials." *Angewandte Chemie: International Edition* 50 (2011): 5438–5466. https://doi.org/10.1002/anie.201001273

[15] M. Skočaj. "Bacterial nanocellulose in papermaking." *Cellulose* 26 (2019): 6477–6488. https://doi.org/10.1007/s10570-019-02566-y.

[16] Liqiang Jin, Qiucun Sun, Qinghua Xu, and Yongjian Xu. "Adsorptive removal of anionic dyes from aqueous solutions using microgel based on nanocellulose and polyvinylamine." *Bioresource Technology* 197 (2015): 348–355. https://doi.org/10.1016/j.biortech.2015.08.093.

[17] Kongliang Xie, Weiguo Zhao, and Xuemei He. "Adsorption properties of nano-cellulose hybrid containing polyhedral oligomeric silsesquioxane and removal of reactive dyes from aqueous solution." *Carbohydrate Polymer* 83 (2011): 1516–1520. https://doi.org/10.1016/j.carbpol.2010.09.064.

[18] Mostafa Hossein Beyki, Mehrnoosh Bayat, and Farzaneh Shemirani. "Fabrication of core-shell structured magnetic nanocellulose base polymeric ionic liquid for effective biosorption of Congo red dye." *Bioresource Technology* 218 (2016): 326–334. https://doi.org/10.1016/j.biortech.2016.06.069.

[19] Xuejing Zheng, Xiaoxiao Li, Jinyang Li, Liwei Wang, Wenjing Jin, Jie Liu, Ying Pei, and Keyong Tang. "Efficient removal of anionic dye (Congo red) by dialdehyde microfibrillated cellulose/chitosan composite film with significantly improved stability in dye solution." *International Journal of Biological Macromolecules* 107 (2018): 283–289. https://doi.org/10.1016/j.ijbiomac.2017.08.169.

[20] Tasrin Shahnaz, Madhar Fazil S. Mohamed, V.C. Padmanaban, and Selvaraju Narayanasamy. "Surface modification of nanocellulose using polypyrrole for the adsorptive removal of Congo red dye and chromium in binary mixture." *International Journal of Biological Macromolecules* 151 (2020): 322–332. https://doi: 10.1016/j.ijbiomac.2020.02.181.

[21] Chu Yong Soona, Norizah Abdul Rahmanb, Yee Bond Teec, Rosnita A. Talibd, Choon Hui Tana, Khalina Abdane, and Eric Wei Chiang Chana. "Electrospun biocomposite: nanocellulose and chitosan

entrapped within a poly(hydroxyalkanoate) matrix for Congo red removal." *Journal of Materials Research and Technology* 8 (2019): 5019–5102. https://doi.org/10.1016/j.jmrt.2019.08.030

[22] Jindrayani Nyoo Putro, Shella Permatasari Santoso, Felycia Edi Soetaredjo, Suryadi Ismadji, and Yi-Hsu Ju. "Nanocrystalline cellulose from waste paper: adsorbent for azo dyes removal." *Environmental Nanotechnology, Monitoring & Management* 12 (2019): 5091–5102. https://doi.org/10.1016/j.enmm.2019.100260.

[23] Mandana Tavakolian, Hannah Wiebe, Mohammad Amin Sadeghi, and Theo G.M. van de Ven. "Removal using hairy nanocellulose: experimental and theoretical investigations." *ACS Applied Material Interfaces* 12 (2020): 5040–5049. https://doi: 10.1021/acsami.9b18679.

[24] Sijie Wang, Qinqin Zhang, Zhenxing Wang, and Junwen Pu. "Facile fabrication of an effective nanocellulose-based aerogel and removal of methylene blue from aqueous system." *Journal of Water Process Engineering* 37 (2020): 101511. https://doi.org/10.1016/j.jwpe.2020.101511.

[25] Lesedi Lebogang, Romang Bosigo, Kebadiretse Lefatshe, Cosmas Muiva, "Ag$_3$PO$_4$/nanocellulose composite for effective sunlight driven photodegradation of organic dyes in wastewater." *Materials Chemistry and Physics* 236 (2019): 121756. doi: https://doi.org/10.1016/j.matchemphys.2019.121756.

[26] Jing Luo, Kaixuan Huang, Xin Zhou, and Yong Xu, "Hybrid films based on holistic celery nanocellulose and lignin/hemicellulose with enhanced mechanical properties and dye removal." *International Journal of Biological Macromolecules* 147 (2020): 699–705. https://doi.org/10.1016/j.ijbiomac.2020.01.102.

[27] Amiraliana Nasim, Mislav Mustapicb, Md. Shahriar A. Hossaina, Chaohai Wanga, Muxina Konarova, Jing Tanga, Jongbeom Na, Aslam Khand, and Alan Rowana. "Magnetic nanocellulose: a potential material for removal of dye from water." *Journal of Hazardous Materials* 394 (2020): 122571. https://doi.org/10.1016/j.jhazmat.2020.122571.

[28] Peng Liu, Chuantao Zhu, and Aji P. Mathew. "Mechanically robust high flux graphene oxide: nanocellulose membranes for dye removal from water." *Journal of Hazardous Materials* 371 (2019): 484–493. https://doi.org/10.1016/j.jhazmat.2019.03.009.

[29] Qile Fang, Xufeng Zhou, Wei Deng, Zhi Zheng, and Zhaoping Liu. "Freestanding bacterial cellulose-graphene oxide composite membranes with high mechanical strength for selective ion permeation." *Scientific Reports* 6 (2016): 33185. https://doi: 10.1038/srep33185.

[30] Baoxia Mi. "Graphene oxide membranes for ionic and molecular sieving." *Science* 343 (2014): 740–742. https://doi.org/10.1126/science.1250247.

[31] Yassine Benmassaouda, María J. Villaseñora, Rachid Salghi, Shehdeh Jodeh, Manuel Algarra, Mohammed Zougagh, and Angel Ríosa. "Magnetic/non-magnetic argan press cake nanocellulose for the selective extraction of Sudan dyes in food samples prior to the determination by capillary liquid chromatograpy." *Talanta* 166 (2017): 63–69. https://doi.org/10.1016/j.talanta.2017.01.041.

[32] Yiying Yue, Xianhui Wang, Jingquan Han, and Lei Yu, Jianqiang Chen, Qinglin Wu, and Jianchun Jiang. "Effects of nanocellulose on sodium alginate/polyacrylamide hydrogel: mechanical properties and adsorption-desorption capacities." *Carbohydrate Polymer* 206 (2019): 289–301. https://doi.org/10.1016/j.carbpol.2018.10.105.

[33] Priyank L. Bhutiya, Mayur S. Mahajan, M. Abdul Rasheed, Manoj Pandey, S. Zaheer Hasan, and Nirendra Misra, "Zinc oxide nanorod clusters deposited seaweed cellulose sheet for antimicrobial activity." *International Journal of Biological Macromolecule* 112 (2018): 1264–1271. https://doi.org/10.1016/j.ijbiomac.2018.02.108.

[34] You Wei Chen, and HweiVoon Lee. "Revalorization of selected municipal solid wastes as new precursors of "green" nanocellulose via a novel one-pot isolation system: a source perspective." *International Journal of Biological Macromolecule* 107 (2018): 78–92. https://doi.org/10.1016/j.ijbiomac.2017.08.143.

[35] Patchiya Phanthong, Prasert Reubroycharoen, Xiaogang Hao, Guangwen Xu, Abuliti Abudula, and Guoqing Guan. "Nanocellulose: extraction and application." *Carbon Resources Conversion* 1 (2018): 32–43. https://doi.org/10.1016/j.crcon.2018.05.004.

[36] Ragab E. Abouzeid, Ramzi Khiari, Nahla El-wakil, and Alain Dufresne. "Current state and new trends in the use of cellulose nanomaterials for wastewater treatment." *Biomacromolecules* 20 (2019): 573–597. https://doi.org/10.1021/acs.biomac.8b00839.

[37] Istvan Siro, and David Plackett. "Microfibrillated cellulose and new nanocomposite materials: a review." *Cellulose* 17 (2010): 459–494. https://doi.org/10.1007/s10570-010-9405-y.

11

Nanocellulose-Based Materials for Desalination of Saline Water

Pallavi Jain
SRM Institute of Science and Technology, Delhi-NCR Campus, Modinagar, India

A. Geetha Bhavani
Noida International University, Greater Noida, India

Makarand Upadhyaya
College of Business Administration, Zallaq, Kingdom of Bahrain

CONTENTS

11.1 Introduction

Clean water is an indispensable commodity for any living being on Earth with a right to live. The statistics reveal that scarcity of clean water is increasing year by year, and by 2025, most of the population will face water scarcity [1]. The considerable option to obtain clean water is via desalination and freshwater and wastewater treatment for sustainable development. The desalination is proceeding with using natural and synthetic materials like membranes made with polymers, various types of filters, and different resins [2–4] with high efficiency techniques, which are economically viable and reusable. Ceramic membranes are displaying extensive applications in the treatment of water. The ceramic membranes with different fabrications are used to eliminate the various pollutants and waste with higher efficiency, while at bulk processes it consumes more energy and temperature [5]. Every treatment process is expected to be workable at low cost, with easy resource availability [6]. Cellulose is used in the water treatment process; it is a readily available biopolymer with various advantages: low cost, energy-efficient, biodegradability, and inexhaustible. The cellulose is copious in lignocellulosic biomass. The resultant from agricultural waste and forestry residues offers abundant renewable resources to prepare novel materials at low cost [7] and with non-toxicity [8–10]. Many researchers and commercial units have confirmed its better performance in many practical applications like water treatment [11, 12], paper manufacturing [13], energy sector [14], and biomedical industry [15]. Apart from many uses, still, cellulose materials are adequate [16, 17].

APPLICATIONS OF NANOCELLULOSE

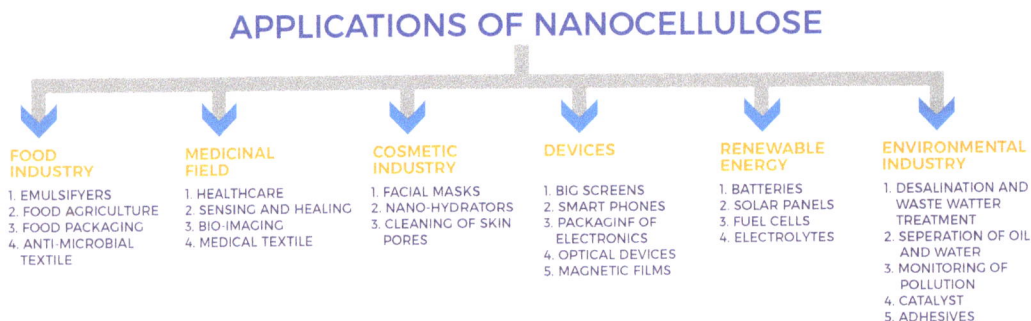

FOOD INDUSTRY
1. EMULSIFYERS
2. FOOD AGRICULTURE
3. FOOD PACKAGING
4. ANTI-MICROBIAL TEXTILE

MEDICINAL FIELD
1. HEALTHCARE
2. SENSING AND HEALING
3. BIO-IMAGING
4. MEDICAL TEXTILE

COSMETIC INDUSTRY
1. FACIAL MASKS
2. NANO-HYDRATORS
3. CLEANING OF SKIN PORES

DEVICES
1. BIG SCREENS
2. SMART PHONES
3. PACKAGINF OF ELECTRONICS
4. OPTICAL DEVICES
5. MAGNETIC FILMS

RENEWABLE ENERGY
1. BATTERIES
2. SOLAR PANELS
3. FUEL CELLS
4. ELECTROLYTES

ENVIRONMENTAL INDUSTRY
1. DESALINATION AND WASTE WATTER TREATMENT
2. SEPERATION OF OIL AND WATER
3. MONITORING OF POLLUTION
4. CATALYST
5. ADHESIVES

FIGURE 11.1 Applications of NCs.

In the quest to find new alternative materials for water treatment, nanocellulose (NC) is also known for better performance. Some applications of NCs are shown in Fig. 11.1. The chapter elaborates on the role of NC and the assessment of different methods of filters and membranes.

11.2 The Role of Filters and Membranes

The filters and membranes play a crucial role in water treatment by not allowing them to pass or stop the undesired chemical compounds and their derivatives [18]. The treatment capacity and selectivity are wholly based on filters and membranes structural arrangements and chemistry of the materials used. The membranes are tailored to achieve desired micropore structures and surface properties to influence the solutes adsorption [19, 20]. The filters are used to eliminate the pollutants by their particle sizes, particularly the micrometer-sized contaminants are eliminated through the pore flow model. Simultaneously, the membranes are efficiently refusing the pollutants of microsizes (1.5 nm) through the size exclusion model. The filters and membranes prepared with NC are functioned either (i) by size exclusion model or (ii) by affinity model. The size exclusion model is known for separation through filter or membrane based on size, which holds on the species through the pore sizes. Affinity membranes eliminate the pollutants through the electrostatic interface of surface functional groups with that of contaminants. The composite model works on both size exclusion and affinity for targeted pollutants. The pollutant sizes merely select the type of membrane to be used for specific rejection. The various membranes used for different processes to refuse with a size of the particle are reverse osmosis (RO) (particle size of 0.1–1 nm), nanofiltration (NF) (particle size of 1–2 nm), ultrafiltration (UF) (particle size of 2–100 nm), and microfiltration (MF) (particle size of 100 nm–10 μm). The small-sized pore membranes need the pressure to force so that water quickly passes through and interacts with the membrane surface [21]. The filters and membranes used to repel the pollutant rely on particle sizes and also remove the poisonous, heavy metals and ion species from the water stream in any treatment phase. The membranes are mixed with specific nano-objects and nanostructures with organic additives that are environment-friendly and non-toxic to increase filtration performance. The organic additives like poly(methylhydrosiloxane) (PMHS), polypropylene (PPy), glutardialdehyde (GA), polyacrylonitrile (PAN), *m*-phenylenediamine (MPDA), polyamide (PA), trimesoyl chloride (TMC), poly(acryloyl hydrazide) (PAH), polytetrafluoroethylene (PTFE), and polyethersulfone (PES) are found to be highly effective, depending upon preparation methods. This chapter aims to brief the desalination of water by the NC-based materials processes through biosorption and adsorption mechanisms to reject the salt.

11.3 Classification of NC

Cellulose materials with nanosized NC are classified into two major categories based on their size (nano-objects) and structure (nanostructures) [22]. These two major categories are further divided into subgroups: nano-object cellulose includes (i) cellulose microcrystal (CMC) and (ii) cellulose microfibers

TABLE 11.1

Comparison of Properties of CNCs and CNFs.

Properties	Cellulose Nanocrystals	Cellulose Nanofibers
Density	1.6 g/cm^3	1.57 g/cm^3
Diameter	5–70 nm	5–100 nm
Young's modulus	50–143 GPa	15–150 GPa
Degree of polymerization	500–15,000	≥500
Tensile strength	2–6 GPa	2–4 GPa
Crystallinity degree	55–88%	59–64%
Optical properties	Iridescent and transparent	Transparent

(CMFs); nanostructured cellulose includes (i) cellulose nanocrystal (CNC) and (ii) cellulose nanofiber (CNF). Both subgroups are differentiated by their size: nano-object cellulose materials are 10–100 μm in size, and nanostructured celluloses are 1–50 nm in size [23–26]. Further, the NCs are grouped into two main categories: CNC and CNF, with recommendations of governing bodies and Technical Association of the Pulp and Paper Industry (TAPPI) [19]. The CNCs and CNFs are nanosized particles with rod-like structures with different diameters and lengths derived by various synthesis methods with varying parameters. The synthesis methods mainly comprise hydrolysis via H_2SO_4 or HCl. The cellulose precursor reacts with H_2SO_4 to form cellulose surface hydroxyl groups to produce half sulfate esters. The half sulfate esters are negatively charged CNC NPs and electrostatically stabilized. The cellulose precursor is reacted with HCl to form unreacted surface hydroxyl groups of CNC particles with nominal charge leads inclined to flocculation. Table 11.1 shows physical properties of CNCs and CNFs [27–31].

Habibi et al. [32, 33] described the CNC synthesis methods, various applications, and properties. Comprehensive chemical modification methods for NCs are also illustrated [34]. The bacterial cellulose (BC) is also one of NC's types synthesized through a bioprocess with high crystallinity. The NCs are promising candidates because the properties with combinations and alterations lead to greater strength, chemically inert, higher surface area, and surface hydrophilicity. Fig. 11.2 shows some properties of NC. These properties of filters and membranes provide higher efficiency in being selective toward rejection of pollutants and chemical species from drinking and industrial waters. The NC membranes enhanced mechanical strength and rigidity to withstand the high pressures and pH ranges [35] at water flux [36]. The chemical stability [37] and essential hydrophilic character of NC can enhance antifouling from both the bioprocess and organic process.

The cellulose fiber treated through the defibrillation process to produce NC leads to an extreme surface area rise. The NC preparation route to treat cellulose fiber precursor is crucial to achieving a significant hike in the surface area of 500 m^2/g [38, 39]. In contrast, the pure soft cellulose pulp surface area is 1–4 m^2/g [40]. The increase in NC surface area is directly connected with the availability of surface hydroxyl groups. These functional groups are embedded through various reactions like sulfonation, carboxylation, 2,2,6,6-tetramethylpiperidin-1-yl)oxyl (TEMPO)-mediated oxidation, esterification, amidation, phosphorylation, and silylation [41–44]. The linkage of functional groups increases the efficiency

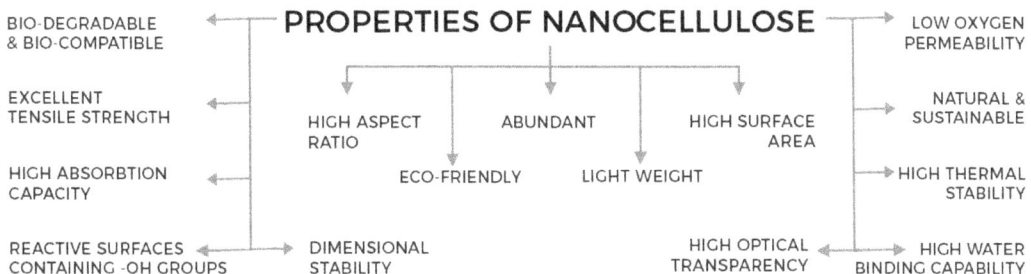

FIGURE 11.2 Properties of NC.

of rejection of contaminants and pollutants [45], water restoration [46–48], and elimination of metal ions [49], dyes species [50–53], metals compounds [54], and types of microbes [55]. Specifically, organic compounds like cyclohexenes and oils are also refused by fabrication of NC matrix with additives and grafting with functional groups [56, 57].

11.4 Key Remarks to Improve Adsorption Capacity of NC

The researcher confirmed that fabrication is effective in improving the filtration capacity. This filtration through adsorption is processed by perfect combination with physical and chemical synergy, thus leading to high performance. The fabrication and activation parameters influence the membrane to increase the surface area, which allows a rise in the pores volume. These properties enhance the adsorption by holding surface (membrane) and absorbents (salts, dye, metals, toxic chemical compounds) through strong interactions. The selective absorbents get fixed through specific surface modifications, which is the objective for particular substrates species. The locked absorbents will slip into the porous structure's inner core due to progressive concentrations [58].

The NC-based membranes have many hydroxyl and carboxyl groups over the surface, which leads to an increase in the affinity toward absorbents, depending upon the surface area. Inappropriately, this property also shows accumulation that provides less access to adsorption, which is a drawback. The accumulation was rectified using the ultrasonication process; still, these methods are varied, depending upon the membrane. The ultrasonication process over CN shows the irreversible accumulation leads to wet the membrane for the entire cycle. Desorption is the reverse process that depends upon the pore structure of the membrane. This process, used for the regeneration of the membrane, also increases the cost and pollution. The physical property was possible to change with a modification of bulk adsorbent with hydrogel casting, freeze-drying, etc. To achieve high performance, a specific polymer with chemical blending with additive and filters is necessary to be tailored to the adsorbent. For example, Hokkanen et al. [59] found that the chemical modifications are strategic interactions to achieve the desired changes in the bulk CNC surfaces, which are directly related to the total adsorption capacity toward the specifically targeted adsorbate. Peng et al. [60] studied the CNC drying effect concerning surface activity and found drying methods significantly improve the surface properties. Qu et al. [61] concluded that hydroxyl, carboxyl, alkyl, and amino groups are promising functional groups obtained by oxidation, esterification, polymer grafting, or etherification to improve the adsorption capacity. The finally obtained nanomaterial achieves new stable reactivity (for example, negatively charged surface react with targeted heavy metal cations) and high dispersion. The functional groups lead to a significant improvement in the surface area because of less accumulation of CN by modification in sustainable methods [45].

11.5 Functioning of Different NC Membrane in Desalination

Cellulose acetate (CA) membranes are covered with polytetrafluoroethylene (PTFE) membranes that are used to treat desalination of brackish water by varying the treatment parameters to achieve higher performance. The performance was found over a composite membrane that was prepared with 20% of PTFE at 90 °C. This study also observed that covering with PTFE improves the water permeability, flux, and salt rejection by more than 20% compared to conventional CA membranes [62]. CA was synthesized using both novel homogeneous and traditional heterogeneous acetylation methods. The homogeneous acetylation shows low salt rejection compared to heterogeneous acetylation methods [63]. The CA modifications improve the salt rejection with the drip in water flux because of a rise in size. The substitution of hydrophobic character delays the flow of water with solute by membranes. CA membranes are modified with various monomers or polymers to alter the properties for effective separation actively. For example, the accumulation of PMHS in less quantity for molding CA membranes leads significantly to improved salt removal and lowers the permeation. This activity is due to high surface hydrophobic property, high porosity, and microporous with thick membrane layers [64]. The CA membrane made with sugarcane bagasse was mixed with hydrolyzed polymethylmethacrylate (PMMA) in alkaline media, which leads to a highly effective desalination of brackish water through RO process with high thermal stability [65]. The CA membrane modified with polymer (e.g., PPy) leads to undesired separation phase with a decline in desalination treatment [66]. The recent report reveals only a few contributions of performing cellulosic-derived membranes with the addition of PA or derivatives and their properties [67].

11.5.1 NC-Based Nanofibrous Membranes via Electrospun Technique

The desalination and purification of water treatment are carried out over the traditional phase inversion technique to prepare polymeric membranes. These polymeric membranes have showed few setbacks like less distribution of pore size with porosity, inadequate water flux, and aggressive fouling during treatment [68]. Many reports trigger these limitations with many fabrication techniques and blending with additives to resolve. Among many methods, the electrospun approach for preparing the polymeric nanofibrous membranes is well appreciated due to the higher performance side. The electrospun technique confirms highly distributed pores with porosity inside and outside of the membranes of nanometer to micrometer range, creating a too high surface-area-to-volume ratio and internally connected pores' structures, which leads to great permeability [69, 70]. The cellulose and its derivatives as a precursor using the electrospun technique are used to prepare polysaccharide nanofibers to improve the characteristics like hydrophilicity, biocompatibility, potential functional ability, and non-toxicity. All the advantages mentioned above are achieved through the electrospun technique, despite that this technique shows less mechanical strength and less possibility of regeneration of nanofibers membranes [71].

11.5.2 NC TFC Membranes

The sugar-NaCl solution with cellulose composite membrane (low molecular weight) was effective for the highest retention of sugar (sucrose) and low NaCl rejection. Finally, the cellulose-based nanofibrous membranes are effectively used for desalination and water treatments with tailored membranes' characteristics for highly active and effective performance. The cellulose-based thin-film composites (TFC) are also promising membranes for effective desalination and water treatments. To prepare cellulose TFC membranes, initially, trimethylsilyl cellulose (TMSC) was coated over porous PAN supported through UF, but later cellulose was regenerated by mixing the GA and HCl vapors as a crosslinker. The prepared cellulose TFC membrane shows very high filtration efficiency by NaCl rejection is around zero while maintaining 80% of sucrose levels, which is relatively high compared to traditional membranes. The commercial units operated with NF membranes, however, give 30–60% of NaCl rejection [72]. Wu et al. [73] prepared the TFC membrane blended with cellulose triacetate (CTA) as support through thermally induced phase separation technique. It was observed that the essential hydrophilic character was enhanced with the addition of the CTA support due to the availability of highly porous structures with a selective layer of PA that was perfect to proceed with the desalination process. The high filtration ability of the cellulose TFC membrane is also accounted for, and massive ridges like structures over the surface provided exceptional thermal stability and superior crystallinity. The preparation methods surely increase the crosslinking capacity in cellulose-based membranes to provide the high mechanical ability for stable performance for long-run treatments over five days at 1.2 LMH/bar.

11.5.3 Modified NC Membranes for Desalination and Wastewater Treatments

From the above discussion, it is clear that cellulose membranes are modified with various additive types by blending to increase filtrations' performance purely depending upon applications. The biobased nanomaterials like cellulose NPs are used for eco-friendly, non-toxic membrane with the blending of organic additives. The cellulose nanomaterials are the main concern for alternative potential membrane as an adsorbent with high-performance efficiency with enhanced properties. NC membranes hold high functionalized groups over the surface, which can grab the chemical compounds by binding efficiency. The efficient properties of NC membranes are encouraged to the commercial unit with the features like high surface area, considerably low density, and greater hydrophilicity that influence the antifouling in a stream. Asempour et al. [74] integrated NPs in CNCs to prepare an active thin-film PA layer.

The thin-film nanocomposite (TFN) membrane was prepared through the *in situ* interfacial polymerization (IP) method by mixing the TMC with MPDA with varying CNCs. This preparation method was proved to increase the hydroxyl functional groups over CNCs surface, enrich the hydrophilic character of the membrane, and enhance antifouling. The PA was modified by inserting CNCs into active layer that provides a water flux of 100%, without adjustment in SR. In another case, the RO process was artificial brackish water treatment using the membranes prepared by controlled modification of CNC-TFN membranes with CNC loading of 0.05% (w/v) and monomer (TMC). This CNC-TFN membrane performance was not up to the mark due to low selectivity attributed to defects formed by the aggregation of CNCs.

CNCs were blended with PAH particles through the layered interfacial polymerization (LIP) method that results in CNC-PAH particles. The CNC-PAH particles provide polymer arms, with active and dense amine groups crosslinked in the membrane layers. The LIP method also features the thin, uniform, and closed installation of PAH over CNCs layers. This leads to high rejection of NaCl with higher permeability compared to commercial units of RO membrane. Another application of CNC-TFC was to eliminate the boron from the stream by surface hydroxyl and amine groups. CNC-TFC was used for removing fouling in treatment units. The core-shell structured materials modification by different functional coatings, films, and membranes is used for various self-cleaning, separation, and antifouling [75].

The unique vacuum filtration method was used to prepare TFN membranes with CNCs by nanocrystals dispersion in monomer solution to enhance the performance. This study of RO process was compared to the effect of polyamide amount over TFC membranes, which influence the performance of SR and water flux. The TFNs are also prepared through the vacuum filtration method to lower water flux with higher SR. The (TEMPO)-oxidized cellulose nanofibers (TOCNs) membranes prepared through vacuum filtration method shows improved water flux but low SR. The decrease in performance was further analyzed by AFM imaging of CNCs, and TOCNs shows nanocrystals are compact over the layer with aggregation in the membrane of PES support. The poor degree of polymerization disrupts the structure, causing decrease in performance because of low dispersion of nanocrystals, low water flux, and SR, which also accounted for aggregation of distraction polyamide layer. The addition of monomer was found to be reliable for CNCs, TOCNs, and TFN formation. The nanocrystals rate of dispersion during preparation caused the defects, which allows water transportation through the interface. The CNCs show that water transportation through the interface is too high, leading to high flux, and avoiding salt transformation leads to lower salt rejection. The performance of TOCNs membranes includes increase in SR, and high-water flux values are provided by better polymer/nanocrystal interactions, which are accredited with carboxyl groups and surface hydrogen bonding. The functional group interactions with PA matrix leads to interface in interconnected nanopores that quickly allow the water transport by refusing the ion transport. The high efficiency of quick water transport is also due to the newly formed nanoporous ratio and the polymer membrane's surface volume. This also implies that particles/polymer interactions are also controlled while preparations of nanocomposite membranes, promising to increase the molecular transport with high selectivity [76].

The modification of polymer membranes using NPs is an established technique for improving membrane efficiency in water treatment. The NPs that flush out of the membrane is difficult because the treated water is adulterating. The TOCNs are prepared by blending the mixture of PA and TFC layer using RO process with IP to progressive performance. Eco-friendly and safe CNFs are achieved by utilizing sustainable and renewable resources. The controlled addition of CNFs with TFC membrane shows better transport by 50% and SR of 96.2% at the parameters of 29.8 LMH at 1.5 MPa. The nanocomposite membranes show better chlorine stability compared to pure TFC membranes. The CNFs are significant for high performance with reasonable cost, non-toxic and antifouling properties may be used in the RO process of desalination. Finally, CNFs are prepared with sustainable resources through eco-friendly methods to improve the performance and stability of the membrane (CNF-TFC membranes) for desalination, water purification, and wastewater treatment. The modification with the addition of NPs to the top layer of the substrate was still not successful every time as it refuses at the pore opening in an aqueous medium. Many reports have shown improved performance, but the limitations of using nanocomposite membranes to commercialize on pilot scale still persist. The concerns related to using nanocomposite membranes are environmental risks, high process cost, and toxicity of NPs. The toxicity of NPs is increasing as NPs are infusing through a membrane and keeping in the water stream [77].

11.6 Conclusion

Clean water is the prime commodity for the survival of life. Clean water scarcity has made humans find alternatives through the processes like desalination and wastewater treatments using membranes and filters. The NCs membranes are the most promising for effective desalination and water treatments. The NCs membranes are blended with various polymers to increase the efficiency of treatment in the pilot scale. The preparation methods increase the crosslinking capacity of cellulose-based membranes to provide high mechanical ability for stable performance for long-run treatments with regeneration ability. The highly crystalline membranes provide modifications to improve the SR with a drip in water flux due to the rise in size. The substitution of hydrophobic character delays the flow of water with solute through membranes. The surface functional groups increase the efficiency of rejection of contaminants and pollutants, water restoration, and elimination of metal ions, dyes species, metals compounds, and microbes types. Finally, the CNF membranes are effectively used for desalination and water treatments with membranes' tailored characteristics for highly active and effective performance in desalination.

Acknowledgment

Pallavi Jain is grateful to SRM Institute of Science and Technology, Modinagar, for support and encouragement.

REFERENCES

[1] Mautner, Andreas. "Nanocellulose water treatment membranes and filters: a review." *Polymer International* (2020): 1–11. doi: 10.1002/pi.5993.

[2] Nur Durmaz, Elif, and P. Z. Culfaz-Emecen. "Cellulose-based membranes via phase inversion using [EMIM] OAC-DMSO mixtures as solvent." *Chemical Engineering Science* 178 (2018): 93–103. doi: 10.1016/J.CES.2017.12.020.

[3] Paul, Mou, and Steven D. Jons. "Chemistry and fabrication of polymeric nanofiltration membranes: a review." *Polymer* 103 (2016): 417–456. doi: 10.1016/j.polymer.2016.07.085.

[4] Mohammad, A. W, Y. H. Teow, W. L. Ang, Y. T. Chung, D. L. Oatley-Radcliffe, and N. Hilal. "Nanofiltration membranes review: recent advances and future prospects." *Desalination* 356 (2015): 226–254. doi: 10.1016/j.desal.2014.10.043.

[5] Li, Kang. *Ceramic membranes for separation and reaction.* Hoboken, NJ: John Wiley & Sons, Inc., 2007.

[6] Lu, Yuan, and Soydan Ozcan. "Green nanomaterials: on track for a sustainable future." *Nano Today* 10 (2015): 417–420. doi: 10.1016/j.nantod.2015.04.010.

[7] Loow, Yu-Loong, Eng K. New, Ge H. Yang, Lin Y. Ang, Luther Y. W. Foo, and Ta Y. Wu. "Potential use of deep eutectic solvents to facilitate lignocellulosic biomass utilization and conversion." *Cellulose* 24 (2017): 3591–3618. doi: 10.1007/s10570-017-1358-y.

[8] Mahfoudhi, Norhene, and Sami Boufi. "Nanocellulose as a novel nanostructured adsorbent for environmental remediation: a review." *Cellulose* 24 (2017): 1171–1197. doi: 10.1007/s10570-017-1194-0.

[9] Bhatnagar, Amit, Mika Sillanpaa, and Anna Witek-Krowiak. "Agricultural waste peels as versatile biomass for water purification: a review." *Chemical Engineering Journal* 270 (2017): 244–271. doi: 10.1016/j.cej.2015.01.135.

[10] Vartiainen, Jari, Tiina Pohler, Kristiina Sirola, Lea Pylkkanen, Harri Alenius, Jouni Hokkinen, Unto Tapper et al. "Health and environmental safety aspects of friction grinding and spray drying of microfibrillated cellulose." *Cellulose* 18 (2011): 775–786. doi: 10.1007/s10570-011-9501-7.

[11] Putro, J. Nyoo, Alfin Kurniawan, Suryadi Ismadji, and Yi-Hsu Ju. "Nanocellulose based biosorbents for wastewater treatment: study of isotherm, kinetic, thermodynamic and reusability." *Environmental Nanotechnology Monitoring & Management* 8 (2017): 134–149. doi: 10.1016/j.enmm.2017.07.002.

[12] Karim, Zoheb, Simon Claudpierre, Mattias Grahn, Kristiina Oksman, and Aji P. Mathew. "Nanocellulose based functional membranes for water cleaning: tailoring of mechanical properties, porosity and metal ion capture." *Journal of Membrane Science* 514 (2016): 418–428. doi: 10.1016/j.memsci.2016.05.018.

[13] Park, Nae-Man, Jae B. Koo, Ji-Young Oh, Hye J. Kim, Chan W. Park, Seong-Deok Ahn, and Soon W. Jung. "Electroluminescent nanocellulose paper." *Materials Letters* 196 (2017): 12–15. doi: 10.1016/j. matlet.2017.03.003.

[14] Du, Xu, Zhe Zhang, Wei Liu, and Yulin Deng. "Nanocellulose-based conductive materials and their emerging applications in energy devices: a review." *Nano Energy* 35 (2017): 299–320. doi: 10.1016/j. nanoen.2017.04.001.

[15] Liu, Jun, Stefan Willför, and Albert Mihranyan. "On importance of impurities, potential leachables and extractables in algal nanocellulose for biomedical use." *Carbohydrate Polymers* 172 (2017): 11–19. doi: 10.1016/j.carbpol.2017.05.002.

[16] Ngah, W. S. Wan, and M. A. K. M. Hanafiah. "Removal of heavy metal ions from wastewater by chemically modified plant wastes as adsorbents: a review." *Bioresource Technology* 99 (2008): 3935–3948. doi: 10.1016/j.biortech.2007.06.011.

[17] O'Connell, D. William, Colin Birkinshaw, and Thomas F. O'Dwyer. "Heavy metal adsorbents prepared from the modification of cellulose: a review." *Bioresource Technology* 99 (2008): 6709–6724. doi: 10.1016/j.biortech.2008.01.036.

[18] Richard, W. Baker. *Membrane technology and applications*, 3rd ed. Hoboken, NJ: John Wiley & Sons, Inc., 2012.

[19] Wells Carpenter, Alexis, Charles-François de Lannoy, and Mark R. Wiesner "Cellulose nanomaterials in water treatment technologies." *Environmental Science Technology* 49 (2015): 5277–5287. doi: 10.1021/es506351r.

[20] Ulbricht, Mathias. "Advanced functional polymer membranes." *Polymer* 47 (2006): 2217–2262. doi: 10.1016/j.polymer.2006.01.084.

[21] Yang, Zi, Yi Zhou, Zhiyuan Feng, Xiaobo Rui, Tong Zhang, and Zhien Zhang. "A review on reverse osmosis and nanofiltration membranes for water purification." *Polymer* 11 (2019): 1252. doi: 10.3390/ polym11081252.

[22] Kargarzadeh, Hanieh, Michael Ioelovich, Ishak Ahmad, Sabu Thomas, and Alain Dufresne. "Methods for extraction of nanocellulose from various sources." In *Handbook of nanocellulose and cellulose nanocomposites*, eds. K. Hanieh, M. Ioelovich, I. Ahmad, S. Thomas, and A. Dufresne, 1–49. Weinheim, Germany: Wiley-VCH Verlag GmbH & Co. 2017.

[23] Dat Nguyen, Huu, Thi T. T. Mai, Ngoc B. Nguyen, Thanh D. Dang, My L. P. Le, Tan T. Dang and Van M. Tran. "A novel method for preparing microfibrillated cellulose from bamboo fibers." *Advances in Natural Sciences: Nanoscience and Nanotechnology* 4 (2013): 015016. doi: 10.1088/2043-6262/4/1/015016.

[24] Kunusa, Wiwin Rewini, Ishak Isa, Lukman A. R. Laliyo, and Hendrik Iyabu. "FTIR, XRD and SEM analysis of microcrystalline cellulose (MCC) fibers from corncorbs in alkaline treatment." *Journal of Physics: Conference Series* 1028 (2018): 012199. doi: 10.1088/1742-6596/1028/1/012199.

[25] Huang, Lijie, Xiaoxiao Zhang, Mingzi Xu, Jie Chen, Yinghan Shi, Chongxing Huang, Shuangfei Wang, Shuxiang An, and Chunying Li. "Preparation and mechanical properties of modified nanocellulose/PLA composites from cassava residue." *AIP Advances* 8 (2018): 025116. doi: 10.1063/ 1.5023278.

[26] Razalli, Rawaida Liyana, Mahnaz M. Abdi, Paridah M. Tahir, Amin Moradbak, Yusran Sulaiman, and Lee Y. Heng. "Polyaniline-modified nanocellulose prepared from Semantan bamboo by chemical polymerization: preparation and characterization." *RSC Advances* 7 (2017): 25191–25198. doi: 10.1039/ C7RA03379F.

[27] Saito, Tsuguyuki, Ryota Kuramae, Jakob Wohlert, Lars A. Berglund, and Akira Isogai. "An ultra strong nanofibrillar biomaterial: the strength of single cellulose nanofibrils revealed via sonication-induced fragmentation." *Biomacromolecules* 14 (2013): 248–253. doi: 10.1021/bm301674e.

[28] Chul Kim, Hyung, and Vasilis Fthenakis. "Life cycle energy and climate change implications of nanotechnologies." *Journal of Industrial Ecology* 17 (2012): 528–541. doi: 10.1111/j.1530-9290.2012. 00538.x.

[29] Kumar Mishra, Raghvendra, and Arjun Sabu. "Materials chemistry and the futurist eco-friendly applications of nanocellulose: status and prospect." *Journal of Saudi Chemical Society* 22 (2018): 949–978. doi: 10.1016/j.jscs.2018.02.005.

[30] Klemm, Dieter, Friederike Kramer, Sebastian Moritz, Tom Lindstrom, Mikael Ankerfors, Derek Gra, and Annie Dorris. "Nanocelluloses: a new family of nature-based materials." *Angewandte Chemie* 50 (2011): 5438–5466. doi: 10.1002/anie.201001273.

[31] Klemm, Dieter, Brigitte Heublein, Hans-Peter Fink Habil, and Andreas Bohn. "Cellulose: fascinating biopolymer and sustainable raw material." *Angewandte Chemie* 44 (2005): 3358–3393. doi: 10.1002/anie.20046 0587.

[32] Habibi, Youssef, Lucian A. Lucia, and Orlando J. Rojas. "Cellulose nanocrystals: chemistry, self-assembly, and applications." *Chemical Reviews* 110 (2010): 3479–3500. doi: 10.1021/cr900339w.

[33] Habibi, Youssef. "Key advances in the chemical modification of nanocelluloses." *Chemical Society Reviews* 43 (2014): 1519–1542. doi: 10.1039/C3CS60204D.

[34] Salas, Carlos, Tiina Nypelo, Carlos Rodriguez-Abreu, Carlos Carrillo, and Orlando J. Rojas. "Nanocellulose properties and applications in colloids and interfaces." *Current Opinion in Colloid & Interface Science* 19 (2014): 383–396. doi: 10.1016/j.cocis.2014.10.003.

[35] Sturcova, Adriana, Geoffrey R. Davies, and Stephen J. Eichhorn. "Elastic modulus and stress-transfer properties of tunicate cellulose whiskers." *Biomacromolecules* 6 (2005): 1055–1061. doi: 10.1021/bm049291k.

[36] Mohmood, Iram, Cláudia B. Lopes, Isabel Lopes, Iqbal Ahmad, Armando C. Duarte, and Eduarda Pereira. "Nanoscale materials and their use in water contaminants removal: a review." *Environmental Science and Pollution Research International* 20 (2013): 1239–1260. doi: 10.1007/s11356-012-1415-x.

[37] Mansouri, Jaleh Simon Harrisson, and Vicki Chen. "Strategies for controlling biofouling in membrane filtration systems: challenges and opportunities." *Journal of Materials Chemistry* 20 (2010): 4567–4586. doi: 10.1039/B926440J.

[38] Sehaqui, Houssine, Uxua P. de Larraya, Peng Liu, Numa Pfenninger, Aji P. Mathew, Tanja Zimmermann, and Philippe Tingaut. "Enhancing adsorption of heavy metal ions onto biobased nanofibers from waste pulp residues for application in wastewater treatment." *Cellulose* 21 (2014): 2831–2844. doi: 10.1007/s10570-014-0310-7.

[39] Sehaqui, Houssine, Qi Zhou, Olli Ikkala, and Lars A. Berglund. "Strong and tough cellulose nano paper with high specific surface area and porosity." *Biomacromolecules* 12 (2011): 3638–3644. doi: 10.1021/bm2008907.

[40] Banavath, HussenNaik, Nishi K. Bhardwaj, and A. K. Ray. "A comparative study of the effect of refining on charge of various pulps." *Bioresource Technology* 102 (2011): 4544–4551. doi: 10.1016/j.biortech.2010.12.109.

[41] Singh, Kiran, Jyoti K. Arora, T. Jai M. Sinha, and Shalini Srivastava. "Functionalization of nanocrystalline cellulose for decontamination of Cr(III) and Cr(VI) from aqueous system: computational modeling approach." *Clean Technologies and Environmental Policy* 16 (2014): 1179–1191. doi: 10.1007/s10098-014-0717-8.

[42] Bozic, Mojca, Peng Liu, Aji P. Mathew, and Vanja Kokol. "Enzymatic phosphorylation of cellulose nanofibers to new highly-ions adsorbing, flame-retardant and hydroxyapatite-growth induced natural nanoparticles." *Cellulose* 21 (2014): 2713–2726. doi: 10.1007/s10570-014-0281-8.

[43] Volesky, Bohumil. "Biosorption and me." *Water Research* 41 (2007): 4017–4029. doi: 10.1016/j.watres.2007.05.062.

[44] Isogai, Akira, Tsuguyuki Saito, and Hayaka Fukuzumi. "TEMPO-oxidized cellulose nanofibers." *Nanoscale* 3 (2011): 71–85. doi: 10.1039/C0NR00583E.

[45] Klein, Elias. "Affinity membranes: a 10-year review." *Journal of Membrane Science* 179 (2000) 1–27. doi: 10.1016/S0376-7388(00)00514-7.

[46] Sehaqui, Houssine, Andreas Mautner, Uxua P. de Larraya, Numa Pfenninger, Philippe Tingaut, and Tanja Zimmermann. "Cationic cellulose nanofibers from waste pulp residues and their nitrate, fluoride, sulphate and phosphate adsorption properties." *Carbohydrate Polymers* 135 (2016): 334–340. doi: 10.1016/j.carbpol.2015.08.091.

[47] Liu, Peng, Houssine Sehaqui, Philippe Tingaut, Adrian Wichser, Kristiina Oksman, and Aji P. Mathew. "Cellulose and chitin nanomaterials for capturing silver ions (Ag$^+$) from water via surface adsorption." *Cellulose* 21 (2014): 449–461. doi: 10.1007/s10570-013-0139-5.

[48] Liu, Peng, Kristiina Oksman, and Aji P. Mathew. "Surface adsorption and self-assembly of Cu(II) ions on TEMPO-oxidized cellulose nanofibers in aqueous media." *Journal of Colloid and Interface Science* 464 (2016): 175–182. doi: 10.1016/j.jcis.2015.11.033.

[49] Wang, Ran, Sihui Guan, Anna Sato, Xiao Wang, Zhe Wang, Rui Yang, Benjamin S. Hsiao, and Benjamin Chu. "Nanofibrous microfiltration membranes capable of removing bacteria, viruses and heavy metal ions." *Journal of Membrane Science* 446 (2013): 376–382. doi: 10.1016/j.memsci.2013.06.020.

[50] Jin, Liqiang, Weigong Li, Qinghua Xu, and Qiucun Sun "Amino-functionalized nanocrystalline cellulose as an adsorbent for anionic dyes." *Cellulose* 22 (2015): 2443–2456. doi: 10.1007/s10570-015-0649-4.

[51] Suman, Abhishek Kardam, Meeta Gera, and V. K. Jain. "A novel reusable nanocomposite for complete removal of dyes, heavy metals and microbial load from water based on nanocellulose and silver nano-embedded pebbles." *Environmental Technology* 36 (2016): 706–714. doi: 10.1080/09593330.2014.959066.

[52] Salama, Ahmed, Nadia Shukry, and Mohamed El-Sakhawy. "Carboxymethyl cellulose-G-poly(2-(dimethylamino)ethylmethacrylate) hydrogel as adsorbent for dye removal." *International Journal of Biological Macromolecules* 73 (2015): 72–75. doi: 10.1016/j.ijbiomac.2014.11.002.

[53] Samadder, Rajib, Nahida Akter, Abinash C. Roy, Md. Mosfeq Uddin, Md. Jahangir Hossen, and Md. Shafiul Azam. "Magnetic nanocomposite based on polyacrylic acid and carboxylated cellulose nanocrystal for the removal of cationic dye." *RSC Advances* 10 (2020): 11945–11956. doi: 10.1039/D0RA00604A.

[54] Zhu, Chuantao, Illia Dobryden, Jens Ryden, Sven Oberg, Allan Holmgren, and Aji P. Mathew. "Adsorption behavior of cellulose and its derivatives toward Ag(I) in aqueous medium: an AFM, spectroscopic, and DFT study." *Langmuir* 31 (2015): 12390–12400. doi: 10.1021/acs.langmuir.5b03228.

[55] KumarThakur, Vijay, and Stefan I. Voicu. "Recent advances in cellulose and chitosan based membranes for water purification: a concise review." *Carbohydrate Polymers* 146 (2016): 148–165. doi: 10.1016/j.carbpol.2016.03.030.

[56] Wang, Xiao, Tsung-Ming Yeh, Zhe Wang, Rui Yang, Ran Wang, Hongyang Ma, Benjamin S. Hsiao, and Benjamin Chu. "Nanofiltration membranes prepared by interfacial polymerization on thin-film nanofibrous composite scaffold." *Polymer* 55 (2014): 1358–1366. doi: 10.1016/j.polymer.2013.12.007.

[57] Zhang, Zheng, Gilles Sèbe, Daniel Rentsch, Tanja Zimmermann, and Philippe Tingaut. "Ultralight weight and flexible silylated nanocellulose sponges for the selective removal of oil from water." *Chemistry of Materials* 26 (2014) 2659–2668. doi: 10.1021/cm5004164.

[58] Köse, Kazim, Miran Mavlan, and Jeffrey P. Youngblood. "Applications and impact of nanocellulose based adsorbents." *Cellulose* 27 (2020): 2967–2990. doi: 10.1007/s10570-020-03011-1.

[59] Hokkanen, Sanna, Amit Bhatnagar, and Mika Sillanpaa. "A review on modification methods to cellulose-based adsorbents to improve adsorption capacity." *Water Research* 91 (2016): 156–173. doi: 10.1016/j.waters.2016.01.008.

[60] Peng, Yucheng, Douglas J. Gardner, Yousoo Han, Zhiyong Cai, and Mandla A. Tshabalala. "Influence of drying method on the surface energy of cellulose nanofibrils determined by inverse gas chromatography." *Journal of Colloid Interface Science* 405 (2013): 85–95. doi: 10.1016/j.jcis.2013.0.

[61] Qu, Xiaolei, Pedro J. J. Alvarez, and Qilin Li. "Applications of nanotechnology in water and wastewater treatment." *Water Research* 47 (2013): 3931–3946. doi: 10.1016/j.waters.2012.09.058.

[62] Shaaban, M. F., Mohamed A. El-Khateeb, and Mohamed Saad. "Water desalination using cellulosic nanofiltration membrane composed of nano-scale polytetraflouroethylene." *Egyptian Journal of Chemistry* 62 (2019): 15–20. doi: 10.21608/EJCHEM.2018.3738.1327.

[63] Barud, Hernane S., Adalberto M. de A. Junior, Daniele B. Santos, Rosana M. N. de Assunção, Carla S. Meireles, Daniel A. Cerqueira, and Guimes R. Filho et al. "Thermal behavior of cellulose acetate produced from homogeneous acetylation of bacterial cellulose." *Thermochimica Acta* 471 (2008): 61–69. doi: 10.1016/j.tca.2008.02.009.

[64] Ferjani, Ezdine, Ramzi H. Lajimi, Andre Deratani, and Mohamed S. Roudesli. "Bulk and surface modification of cellulose diacetate based RO/NF membranes by polymethylhydrosiloxane preparation and characterization." *Desalination* 146 (2002): 325–330. doi: 10.1016/S0011-9164(02)00505-2.

[65] Basta, Altaf H., and Houssni El-Saied, "Enhanced transport properties and thermal stability of agro-based RO-membrane for desalination of brackish water." *Journal of Membrane Science* 310 (2008): 208–218. doi: 10.1016/j.memsci.2007.10.043.

[66] El-Saied, Houssni, Altaf H. Basta, Barsoum N. Barsoum, and Mohamed M. Elberry, "Cellulose membranes for reverse osmosis part I. RO cellulose acetate membranes including a composite with polypropylene." *Desalination* 159 (2003): 171–181. doi: 10.1016/S0011-9164(03)90069-5.

[67] Candido, R.G., G. G. Godoy, and Adilson R. Gonçalves. "Characterization and application of cellulose acetate synthesized from sugarcane bagasse." *Carbohydrate Polymers* 167 (2017): 280–289. doi: 10.1016/j.carbpol.2017.03.057.

[68] Bhardwaj, Nandana, and Subhas C. Kundu. "Electrospinning: a fascinating fiber fabrication technique." *Biotechnology Advances* 28 (2010): 325–347. doi: 10.1016/j.biotechadv.2010.01.004.

[69] Ahmad, Adnan, Sidra Waheed, Shahzad M. Khan, Sabad e-Gul, Muhammad Shafiq, Muhammad Farooq, Khairuddin Sanaullah, and Tahir Jamil. "Effect of silica on the properties of cellulose acetate/polyethylene glycol membranes for reverse osmosis." *Desalination* 355 (2015): 1–10. doi: 10.1016/j. desal.2014.10.004.

[70] P. Suja, C. Reshmi, P. Sagitha, and A. Sujith, "Electrospun nanofibrous membranes for water purification." *Polymer Reviews* 57 (2017): 467–504. doi: 10.1080/15583724.2017.1309664.

[71] Halim, N., M. Wirzal, M. Bilad, A. Yusoff, N. Nordin, Z. Putra, and J. Jaafar, "Effect of solvent vapor treatment on electrospun Nylon 6, 6 nanofiber membrane." *IOP Conference Series: Materials Science and Engineering, IOP Publishing* 429 (2018): 012019. doi: 10.1088/1757-899X/429/1/012019.

[72] Puspasari, Tiara, Neelakanda Pradeep, and Klaus-Viktor Peinemann. "Crosslinked cellulose thin film composite nanofiltration membranes with zero salt rejection." *Journal of Membrane Science* 491 (2015): 132–137. doi: 10.1016/j.memsci.2015.05.002.

[73] Wu, Qing-Yun, Xiao-Yan Xing, Yuan Yu, Lin Gu, and Zhi-Kang Xu. "Novel thin film composite membranes supported by cellulose triacetate porous substrates for high-performance forward osmosis." *Polymer* 153 (2018): 150–160. doi: 10.1016/j.polymer.2018.08.017.

[74] Asempour, Farhad, Daryoush Emadzadeh, Takeshi Matsuura, and Boguslaw Kruczek. "Synthesis and characterization of novel cellulose nanocrystals-based thin film nanocomposite membranes for reverse osmosis applications." *Desalination* 439 (2018): 179–187. doi: 10.1016/j.desal.2018.04.009.

[75] Hyung Park, Chan, Sung K. Jeon, Sang-Hee Park, Min G. Shin, Min S. Park, Sun-Young Lee, and Jung-Hyun Lee. "Cellulose nanocrystal-assembled reverse osmosis membranes with high rejection performance and excellent antifouling." *Journal of Material Chemistry A* 7 (2019): 3992–4001. doi: 10.1039/C8TA10932J.

[76] Smith, Ethan D., Keith D. Hendren, James V. Haag IV, E. Johan Foster, and Stephen M. Martin. "Functionalized cellulose nanocrystal nanocomposite membranes with controlled interfacial transport for improved reverse osmosis performance." *Nanomaterials* 9 (2019): 125. doi: 10.3390/nano9010125.

[77] Liu, Shasha, Ze-Xian Low, Hanaa M. Hegab, Zongli Xie, Ranwen Ou, Guang Yang, and George P. Simon et al. "Enhancement of desalination performance of thin-film nanocomposite membrane by cellulose nanofibers." *Journal of Membrane Science* 529 (2019): 117363. doi: 10.1016/j.memsci.2019.117363.

12

Nanocellulose-Based Materials for the Removal of Inorganic Toxicants

Rekha Sharma
Banasthali Vidyapith, Banasthali, India

Sapna Nehra
Dr. K.N. Modi University, Newai, India

Dinesh Kumar
Central University of Gujarat, Gandhinagar, India

CONTENTS

12.1 Introduction

The rapid growth in nanotechnology is attributed to the development of materials' diversity, showing excellent ecological and sustainability features. For a long 150 years, cellulose polymer has been discovered as the primary source of renewable energy and biodegradability. It is used in construction materials and clothing. Over the years, researchers have made tremendous efforts to disrupt fossil fuels' use by replacing it with cellulosic polymer [1, 2]. Therefore, cellulose and its corresponding derivatives have attracted attention due to their excessive abundance, eco-friendly nature, and biocompatible and renewable characteristics [3]. This material displayed remarkable features in terms of large surface area, low density, maximum aspect ratio, excellent mechanical convenience, minimal cost, and tunable adaptable surface features. There are about 1 trillion tons on global market as of 2020 because of the growing demand for cellulose in various medical and pharmaceutical industries [4]. The primary source of cellulose formation is agricultural waste like cotton stalks, crop waste, corn husk, wheat straw, rice straw, fruit waste, and so forth [5–8]. Therefore, the need for nanocellulosic materials is growing substantially in the industrial and medical fields [9]. The nanocellulose materials have been used as fillers to enhance the nanomaterials' mechanical strength, for instance, physical, thermophysical, and barricade features of the carboxymethyl cellulose (CMC) agar, chitosan, starch derivatives, alginate films, and poly(lactic acid).

Sensors are used in technology [10], enzyme stabilization [11], and automotive [12] and electronic devices [13–15]. Polymer films are also used for the functionalization of materials and to improve the mechanical strength of nanocellulose. A diverse range of metal and metal oxides were functionalized with nanocellulose materials such as zinc oxide, titanium oxide, magnesium oxide, copper oxide, silver, and iron oxide and were further used to remove toxins present in wastewater.

12.2 Types of Nanocellulose

Different types of nanocellulose components are extracted from the biomass of cellulose. The biomass of nanocellulose generally had some crystalline areas that attributed the strength to cellulose fibers and elasticity to the biomass cell wall [16]. Many hydroxyl groups and strong hydrogen bonding are found in cellulosic biomass, supporting excellent physical and mechanical features. Nanocellulose refers to cellulosic fibers' size in the range of 1–100 nm in diameter derived from natural cellulose [17]. Primarily, nanocellulose is divided into three sections known as cellulose nanocrystals (CNC), cellulose nanofibrillations (CNF), and bacterial nanocellulose (BNC). CNFs are formed through mechanical treatment, for example, ultrasonication, wet grinding, and ultra-homogenization grinding [18–21].

12.2.1 Cellulose Nanofibrils (CNFs) in Removal of Heavy Metals

CNF represents the natural cellulosic material, which contains a large surface-to-high-volume ratio and diameter in the range of 20–60 nm [22, 23]. The methods used in the abstraction of CNF show the disruption in the cellulosic fibers' arrangement laterally on their axis through the mechanical strengths' implication, for instance, high-pressure homogenization, wet grinding, cryogenic cooling, ultrasonic treatments, ultrahigh friction grinding, and combined chemical and enzymatic pretreatments. Various processes have been performed and mechanical methods used for the pretreatment of the cellulosic fiber, such as enzymatic hydrolysis, carboxymethylation phosphorylation, and 2,2,6,6-tetramethylpiperidine-1-oxyl (TEMPO) oxidation [24–26]. TEMPO oxidation pretreatment approach was used to extract CNF. The obtained carbon nanocellulose fiber product showed the characteristic properties of gel-type in the aqueous suspension medium. Fundamental features of cellulose material rely on raw materials. Similarly, the approach was rooted in high fibrillation crystal intensity levels and showed morphological and structural characteristics [27–29]. The carbon nanofiber derived from the primary cell wall was thinner and longer than the secondary cell wall. These fibers can be obtained from various cellulose sources like cotton stalks, corn husk, banana, sugar beet pulp, softwood and hardwood, rice straw, and many more [30, 31].

Qin et al. [32] treated wastewater using carboxymethylated CNF to remove heavy metal ions. They used approximately 2.7 mmol/g amount of carboxymethylated CNFs (CMCNFs). Various factors affect the adsorbent's efficacy in absorbing toxic ions of copper, including the carboxylate amount, the initial copper concentration, and contact time. The maximum adsorption capacity was reported to be 115.3 mg/g at pH 5.0. Besides, the depiction of copper adsorption mechanisms was explained through kinetics with fitting several models such as Langmuir isotherm and pseudo-second-order (PSO) models. In conclusion, CMCNF-2.7 is considered efficient at high carboxylate amounts in converting copper-contaminated water into drinking water. CMCNFs provide a new selection for the design of novel nanocellulose-based materials for water treatments [32].

Najib and Christodoulatos [33] showed arsenate As(V) adsorption in an aqueous medium using trimethylammonium chloride incorporated CNF. The adsorption process was speedy, which attained equilibrium within 2 hours. The modified CNF was found very efficient for both laboratories and field-level samples. Arsenic adsorbed on the modified surface of CNF through electrostatic interaction between As(V) and available active sites supported by the PSO kinetic model. The total absorption capacity was estimated at around 25.5 mg/g, even in the presence of other interfering ions such as nitrate, nitrite, and sulfate. The interfering ions showed a negligible effect on the adsorption capacity except for phosphate ion, which reduced the capacity. Thermodynamic parameters revealed that arsenic adsorption was temperature-dependent, exothermic, and spontaneous. According to all results, it is supposedly a promising alternative to arsenic adsorption from aqueous solutions such as modified-CNF [33].

Wang et al. [34] used three-dimensional metal-organic frameworks (MOFs) by changing metal ions and organic ligands. Translucent EuMOF@TEMPO-oxidized cellulose nanofibrils (Eu-MOF@TOCNF) materials have photoluminescence characteristics to recognize copper ions fabricated through the *in situ* hydroalcoholic medium. However, in other intervention ions, the functionalized film showed better selectivity for copper ions. Later on, it altered the copper ion's concentration to check the functionalized film's fluorescence intensity. A slow decrement in the film's fluorescence intensity by increasing copper ion amount and observed I0/I-1 excellent connection with copper are attributed to the promising material for detecting copper in an aqueous system [34].

12.2.2 Nanocellulosic Hydrogel in Removal of Heavy Metal

Guo et al. [35] used CNFs to remove heavy metal ions and various other fields because of their high aspect ratio and exhibited excellent luminescence performance. This nanocellulosic hydrogel showed fluorescent and removal behavior against the heavy metals. Several other parameters, such as the effect of contact time and initial concentration, were demonstrated to check the absorbing capacity of nanocellulosic hydrogel. A range of adsorption capacities was observed: 769, 212, 2056, and 1246 mg/g for iron, barium, lead, and copper, correspondingly. The synthesized hydrogel attributed the intense blue color to fluorescence containing 23.6% of fluorescence quantum yield and showed the high sensitivity in recognizing the iron. Carbon nanocellulose fibril modified with carbon dot provides several active hydroxyl sites that easily adsorb heavy metal. It reflects the rapidly visible changes that make it a robust sensor for heavy metals and increase heavy metals' removal. Fluorescence increased sensitivity to determine the signals' stability and heavy metals concentration [35].

Geng et al. [36] developed an eco-friendly bioadsorbent, which can be recycled and reused in removing toxic mercury heavy metal in the water bodies. They synthesized the aerogel by modifying nanocellulose with the thiol group for efficient uptake of Hg(II) ion. The freeze-drying method was used to incorporate the bamboo-derived TEMPO-oxidized NFC denoted as TO-NFC suspension in the presence of 3-mercaptopropyltrimethoxysilane (MPTs) sols. Functionalized aerogel showed the selectivity and high removal toward the Hg(II) up to 92% in the presence of other interfering ions. The concentration of Hg(II) varied from 0.01 mg/L to 85 mg/L [36].

Zarei et al. [17] extracted cellulose from the algal species *Cystoseira myrica*. They prepared iron oxide, nanocellulose, and Fe_3O_4-nanocellulose compounds using co-precipitation, acid hydrolysis, and sol-gel approaches. The adsorption data were fitted with 14 types of well-absorption isotherm models and 12 kinetics models. The density functional theory (DFT) calculation was explained in different circumstances: the interaction between polymer and mercury ions. Achieved simulation results showed a relationship with experimental results obtained [17].

12.2.3 Nanocellulose Hybrid Membranes in Removal of Heavy Metal

Zhang et al. [37] used a straightforward and practical approach based on the aggregation of nanocellulose and nanochitin. They prepared the biohybrid hydrogel (BHH) and biohybrid aerogel (BHA) adsorbents for water remediation. Self-aggregation was regulated through the electrostatic interactions among the 1D negatively and positively charged TOCNF and partly deacetylated chitin nanofiber (PDChNF). The BHA showed maximum adsorption capacity compared to other conventional, biobased adsorbents for heavy metals and dyes such as methylene blue. The maximum adsorption of dye in neutral and alkaline pH conditions was 217 and 531 mg/g. The BHA maintained the adsorption capacity of up to 505 mg/g even after processing five consecutive adsorption-desorption cycles [37].

Luo et al. [38] developed a hybrid nanoflower by creating a series of substrates to overcome the shortcomings of the flower-based structure. Three-dimensional hierarchically porous nanofibrous PVA-*co*-PE membranes (HPNM) were synthesized via a simple template strategy to fabricate the laccase-$Cu_2(PO_4)_3 \cdot 3H_2O$ hybrid nanoflowers efficiently. The mixed nanoflower composite as HNF-HPNM showed outstanding catalytic behavior in the degradation of various kinds of textile dyes, including the acid blue 25, acid yellow 76, indigo carmine, and reactive blue 2 with the highest degradation efficacy of 99.5% for indigo carmine. Further, a reusability study was done with HNFHPNM up to 14 times to degrade the indigo carmine, and a very slight degradation in the efficiency was obtained from 99.48% to 98.52% [38].

12.2.4 Nanocellulose Composite in Removal of Heavy Metal

Anirudhan et al. [39] fabricated a magnetic nanocellulose composite using sulfhydryl and carboxyl [(MB-IA)-*g*-MNCC] through co-polymerization of itaconic acid over the magnetite nanocellulose (MNCC), involving EGDMA and $K_2S_2O_8$ as crosslinker and free radical initiator, respectively. The higher adsorption capacity occurred at pH 6.5, and its kinetics was optimized with the pseudo-second-order model. Therefore, the mechanism of adsorption of cobalt on the surface of MB-IA-g-MNCC was ion exchange. Langmuir isotherm was best fitted with the adsorption result. In addition to other coexistence ions, the adsorbing rate was examined in water samples collected from nuclear power plant coolant. The adsorption-desorption cycle in the presence of an acidic medium (0.1 M HCl) was performed six times and observed the absorption capacity of 97.5% and 84.7%, respectively, in the first and sixth cycles [39].

Liu et al. [40] reported the adsorption efficacy of cellulose fabricated by enzymes to remove silver, copper, and iron from the industrial effluent. The inclusion of phosphate on nanocellulose significantly increased the rate of metal-adsorbing efficiency. The higher surface area determined the adsorption capacity and their characteristic application illustrated by the DFT calculation. Usually, the tested sample contains the single-type metal ions and the order of selectivity followed was $Ag(I) > Cu(II) > Fe(III)$, whereas in multicomponent system the selectivity order was $Ag(I) > Fe(III) > Cu(II)$ altered unrelatedly to the surface functionality of the nanocellulose. The industrial effluent collected from the mirror manufacturing industry, 99% exclusion of the copper and iron through the phosphorylated nanocellulose was achieved. The adsorbent so obtained is known as highly efficient nanocellulosic adsorbent for scavenging the multiple metal ions and concurrently from industrial sewages [40].

Mautner et al. [41] demonstrated the membrane filtration method in removing copper from the contaminated water. They chose the TCNF and reported the more significant attraction for the heavy metal ions, while nanopaper film exhibited the minimum permeability. The nanopaper-based membrane is still not modified to achieve the maximum adsorption of metal ions. The fabrication of the bare fiber with TCNF attributed the more adsorption tendency toward the copper ion with higher water permeability. In this study, raw nanocellulose composite filters were extracted from flax and agave fibers, specifically from tequila residues, and later on, fabricated with TCNF. The efficiency of synthesized nanofilters was calculated according to their adsorption capacity for copper. This unique filter obtained from the industrial crop residues could adsorb the maximum amount of copper ion at the continuous filtration method with high permeances allowed by the extremely porous nature [41].

Septevani et al. [42] reported water remediation using the EFB-based NC as super-adsorbent through acid hydrolysis of sulfuric and phosphoric acid as NCS and NCP, respectively, modified with AC. The generation of the sulfonic functionality over the surface of the sulfonic acid-modified nanocellulose is assigned larger than the empty fruit bunch phosphoric acid-functionalized nanocellulose. The prepared superabsorbent of 2 wt% displayed 24.94 mg/g adsorption capacity and selectivity toward the lead and 86% of removal tendency, which was just double compared to the rice-straw nanocellulose. This adsorbent also showed the removal tendency even after the second reusability test. The inconsequential upgrading of the removal ability of sulfonic acid-incorporated nanocellulose activated carbon that provided distributed surface functionality of micrometer diameter activated carbon. It displayed the outstanding organic deprivation ability for reducing the COD to 93%. The other factors are also maintained by many super adsorbents like sulfate, phosphate TDS, pH, and permissible limits according to the government standards [42].

Bhattacharya et al. [43] presented the BNC application derived from the *Komagataeibacter xylinus* strain SGP8 and well characterized through many techniques. Around 1.82 g/L of cellulose was collected as pellicle in the usual HS Hiveg medium. Further, the crystallinity, morphological structure, and thermal stability of BNC were described through the X-ray diffraction (XRD) and scanning electron microscopy (SEM) techniques. As noticed, approximately 86% of crystallinity index, nanofibrillar structure, and stability up to 280 °C by rising the temperature till 350 °C were attributed to depolymerization. To check the application in removal, first functionalize with $CaCO_3$. During the adsorption experiment with the optimum condition, like 10 mg/L of the cadmium's initial concentration with pH 5, 7, and 9 after the contact time of 12 gained more than 99% removal. The composite showed the removal efficacy of 69–70% at the initial concentration of 25 and 50 mg/L. Finally, cellulose derived from the *K. xylinus* strain SGP8 exhibited outstanding material properties and showed exceptional adsorption performance toward the cadmium ion [43].

Hokkanen et al. [44] showed the exclusion of zinc, nickel, copper, cobalt, and cadmium from aqueous medium by the employment of succinic anhydride for functionalization and mercerization. Further, the physicochemical properties were determined by using a few techniques like Fourier transform infrared (FTIR) and SEM. The carboxyl group's presence was determined by the FTIR study and the crystallinity enhanced, which was clearly observed in the SEM images. Several factors were studied to investigate the batch mode's adsorption performance, such as contact time, regeneration, effects of pH, and metal ion concentration. In the range of 0.72–1.95 mmol/g, the maximum adsorption rate was obtained and the selectivity order followed was Cd > Cu > Zn > Co > Ni. The Langmuir and Sips adsorption isotherms models have illustrated the monolayer type of adsorption over the surface. The regeneration study explained the reusability tendency of the modified nanocellulose, which was done by using the nitric acid in the presence of ultrasonic treatment [44].

Li et al. [45] reported the NFC- and PEI-based aerogel preparation through electrostatic interaction without involving chemical crosslinking. Attenuated total reflection Fourier transform infrared (ATR-FTIR) and thermogravimetric analysis (TGA) studies confirmed the physical crosslinking between the nanofibrillated cellulose and the polyethyleneimine. The promising aerogel showed excellent structural permanency and maintained the shape in the water. Many amine groups' availability and porous structure in the aerogel structure easily adsorb the heavy metal at their surface. The Langmuir adsorption isotherm with higher regression coefficient of more than 0.999 showed a better fitting with the adsorption data. The pseudo-second-order kinetics determined the kinetics data. The overall adsorption capacity of copper and lead calculated from the Langmuir adsorption isotherm were 175.44 and 357.44 mg/g, respectively. The aerogel-based adsorbent was regenerated by performing the successive adsorption-desorption cycles using ethylenediaminetetraacetic acid (EDTA). Therefore, the composite aerogel adsorbent showed outstanding adsorption results and behaved as the most promising adsorbent for treating industrial effluents containing heavy metal [45].

12.2.5 Nanocellulose Nanofiber in Removal of Heavy Metal

Dwivedi et al. [46] introduced the CNF as a green candidate because of its outstanding aspects: enormous availability, robustness, and high porosity to remove arsenic and chromium. In this preparation, using waste coffee-filters, four different CNFs were formed to adsorb arsenic and chromium in their corresponding +5 and +6 oxidation states found in different water systems. In between all modified hybrids, the iron crosslinked $CNF-Fe_2O_3$ (CNF-FF) and mussel-inspired dopamine-conjugated CNF (DP) showed a better removal than the commercial activated adsorbents for the arsenic and chromium in the aqueous system. X-ray absorption near edge structure (XANES) and extended X-ray absorption fine structure (EXAFS) studies determined the adsorption mechanism of arsenic and chromium to the carbon nanocellulose and their bindings through bidentate-binuclear complex and the chromium binds to catecholic −OH in the form of trinuclear complex. For the significant practical application, chromium was interconverted from hexavalent to trivalent states using dopamine. A systematic representation of the formation of CNF is shown in Fig. 12.1 [46].

Sharma et al. [47] investigated cadmium removal from the aqueous system using carboxycellulose nanofibers. The prepared nanofiber exhibited low crystallinity, high surface charge, and hydrophilicity as 26.50%, −68 mV, static contact angle 38°. These outstanding characteristics were obtained through

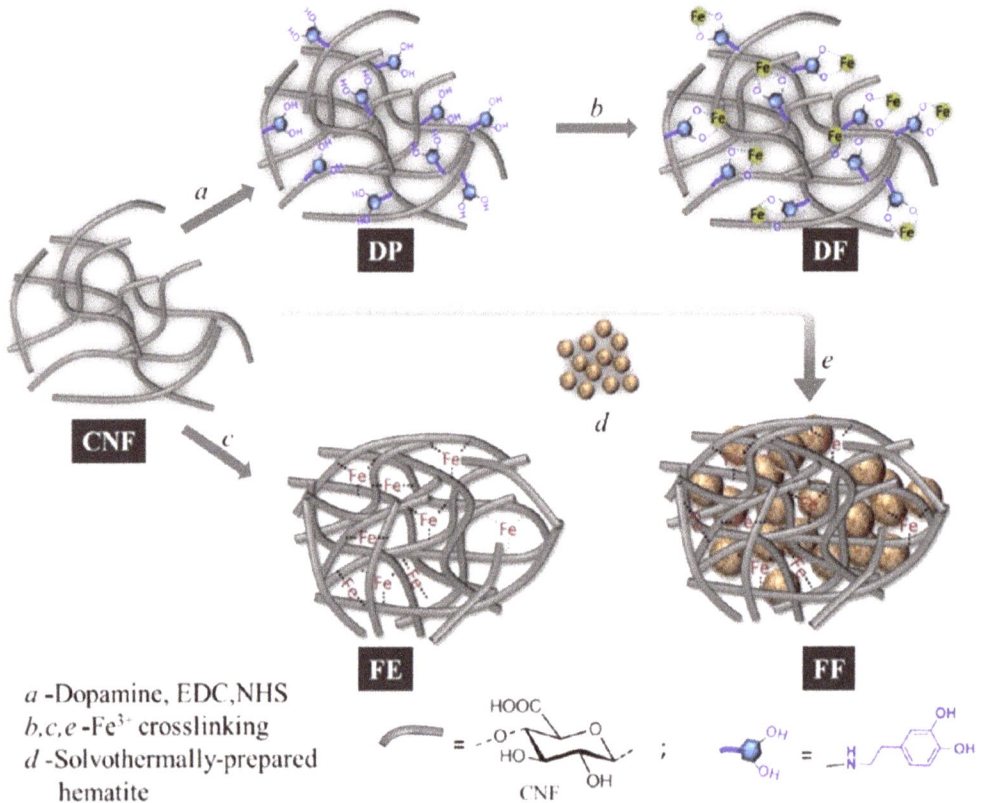

FIGURE 12.1   Synthesis scheme of CNF derivatives (Reproduced with permission from Ref. [46], © American Chemical Society 2017).

nitro-oxidation method from the raw Australian spinifex grass in nitric acid and sodium nitrite. For instance, a minimum concentration of 0.20 wt% of NOCNF suspension will remove cadmium ions in the large concentration range of 50–5000 ppm in a brief interval of time, like less than 5 minutes. The adsorption experiment results suggested that in a lower concentration below 500 ppm, the removal tendency was maximum because of the strong interactions amid the carboxylate groups over the surface of nanofiber and the cadmium ion. In higher concentrations above 1000 ppm, the adsorption process was governed by the mineralization process by forming the cadmium hydroxide nanocrystals depicted through the transmission electron microscope (TEM) and wide-angle X-ray diffraction (WAXD). The present nanofiber's adsorption capacity was reported much higher than previously reported adsorbents, approximately 2550 mg/g estimated by Langmuir adsorption isotherms. In the presence of 250 ppm concentration, it showed 84% removal efficacy. The current study recognized it as a workable strategy for converting the biomass into valuable nanomaterials that potentially adsorb the cadmium from the aqueous system [47].

Bandi et al. [48] prepared the sliver-functionalized holocellulose nanofibrils known as AgNPs/HCNF through the microwave-assisted approach in smaller time intervals without the involvement of a reducing agent. As a result, adsorbents were well elucidated by sophisticated analytical tools. Here silver nanoparticles work as the probe, which selectively detects the mercury in the range of detection limit of 10–200 μg/L. The developed probe showed mercury detection in the various forms of the water, such as spiked tap, bore, and lake water. Further paper strips were designed to detect the mercury at the site process. Additionally, under freeze-drying conditions, the sliver-functionalized nanocomposite produced the aerogel. Further, it showed the outstanding catalytic property for reducing the Congo red and methylene blue dyes [48].

TABLE 12.1

Various Nanocellulosic Adsorbents for the Removal of Inorganic Toxicants.

Adsorbent	AC (mg/g)	Adsorbate	Reference
TO-CNF/TMPTAP/PEI composite	485.44	Cu(II)	[51]
CS/PEO/PNC nanofibers	232.55	Cd(II)	[52]
TZFNC	40	Pb(II)	[53]
NCNB	2749.68, 916.65, 1937.49	Cr(VI), Co(III), Cu(II)	[54]
SNC	404.95	Hg(II)	[55]
MoS$_2$ NPs	160.04	Hg(II)	[56]
NOCNF	1470	U(IV)	[57]
TOCNF+GO	63.5	Cu(II)	[58]
MDAC-cys NDAC-cys	344.82, 357.14	As(III)	[59]
Cationic dialdehyde cellulose (c-DAC)	80.5	Cr(VI)	[60]
NC-PEI/GA	255.19	As(V)	[61]
Cellulose nanofibrils (CNFs) and lysozyme nanofibrils (LNFs)	55	Hg(II)	[62]
SP-Zr-La composites	61.5	PO$_4^{3-}$	[63]
Sugarcane bagasse	9.1	PO$_4^{3-}$	[64]
Cellulose/metal hybrid macrogels	22.25	PO$_4^{3-}$	[65]

12.3 Removal of Fluoride and Other Inorganic Toxicants

Mukherjee et al. [49] used modified nanofiber with polyaniline (PANI)-supported ferrihydrite nanocomposite to study fluoride removal. In the green synthesis, iron did not act as an oxidant but served as a precursor. The AgNPs were incorporated in the blend of the cellulose-PANI. The granular nanocomposite also exhibited the dual sites by giving active sites for adsorption and robustness that also inhibit the leaching of NPs. The synergistic active adsorption sites in doped PANI and FeOOH favored the adsorption of fluoride, which enhanced the adsorption efficacy. Further, the mechanism followed in the adsorption process was investigated by several analytical tools [49].

Sehaqui et al. [50] demonstrated that the cationic CNF exhibited three different positively charged quaternary ammonium groups synthesized using the waste pulp residue through a water-based modification method. The pulp was esterified with glycidyl trimethylammonium chloride tailed by mechanical disintegration. The resulted nanofiber was examined through the SEM, TGA, and zeta potential analysis. Further, the feasibility of the reaction was investigated by performing the conductometric titration. The amount of 1.2 mmol/g showed a higher charge and remained positively charged at different pH values. The positively charged surface easily adsorb the negatively charged ions like fluoride, nitrate, sulfate, and phosphate ions. The higher adsorption capacity of the CNF was attained at approximately 0.6 mmol/g. The cationic CNF was more selective toward multivalent ions like phosphate and sulfate than the single-valent ions like fluoride and nitrate. The synthesized cationic CNF can be transformed into the selective membranes accomplished by dynamic nitrate adsorption by utilizing a simple paper-making process [50]. Various nanocellulosic adsorbents for the removal of inorganic toxicants are listed in Table 12.1.

12.4 Conclusions and Future Perspectives

For thousands of years, natural cellulose-based materials delivered the vast worldwide market of industries in textiles, paper, forest products, and further what has been used in our society as engineering materials. The nanocellulose shaped from various naturally occurring cellulose sources has readily exploded since 2005, which partakes one or two dimensions in the nanometer range. Cellulose nanomaterials show eminent properties—low cost, low density, sustainability, renewability, non-toxicity/biocompatibility, abundance, biodegradability, and obtainability in various forms and provides opportunity to produce energy after burning at the end of their life cycle. When switching to the nanoscale, some other goods are exacerbated, for example, anisotropic shape, mechanical properties, tailorability, and specific surface area of the surface chemistry. Nanocellulose materials possess the possible share of being truly green nanomaterials with many unexpected and useful properties. For a variety of applications relevant to biomedical engineering and material science, cellulose nanomaterials' properties make them exciting materials, and they possibly partake in a high potential for developing manufacturing. Anionic or cationic surface groups can precisely match with necessities for treatment of wastewater, i.e., high mechanical properties, enhanced specific surface area, tailorability, and hydrophilicity of the surface chemistry. To treat wastewater, nanocellulose can be used as flexible membranes and effective adsorbents or be involved in the grounding of hybrid materials. The separation of cellulose nanomaterials includes energy-consuming procedures and is expensive. Encounters are primarily associated with cost-effective surface modification and inexpensive upscale manufacturing.

Acknowledgments

Rekha Sharma gratefully acknowledges the support from the Ministry of Human Resource Development Department of Higher Education, Government of India, under the scheme of Establishment of Centre of Excellence for Training and Research in Frontier Areas of Science and Technology (FAST), and for providing the financial support to perform this study vide letter No., F. No. 5–5/201 4–TS. Vll.

REFERENCES

[1] Thakur, Vijay Kumar, and Stefan Ioan Voicu. "Recent advances in cellulose and chitosan-based membranes for water purification: a concise review." *Carbohydrate Polymers* 146 (2016): 148–165. doi.org/10.1016/j.carbpol.2016.03.030.

[2] Voicu, Stefan Ioan, Roxana Mihaela Condruz, Valentina Mitran, Anisoara Cimpean, Florin Miculescu, Corina Andronescu, Marian Miculescu, and Vijay Kumar Thakur. "Sericin covalent immobilization onto cellulose acetate membrane for biomedical applications." *ACS Sustainable Chemistry & Engineering* 4 (2016): 1765–1774. doi.org/10.1021/acssuschemeng.5b01756.

[3] Park, Nae-Man, Sukyung Choi, Jee Eun Oh, and Dae Youn Hwang. "Facile extraction of cellulose nanocrystals." *Carbohydrate Polymers* 223 (2019): 115114. doi.org/10.1016/j.carbpol.2019.115114.

[4] Mugaanire, Innocent Tendo, Hua Wang, and Junfen Sun. "Fibrous microcrystalline cellulose from *Ficus natalensis* barkcloth." *European Journal of Wood and Wood Products* 77 (2019): 483–486. doi.org/10.1007/s00107-019-01382-2.

[5] Zhang, Xi-Feng, Zhi-Guo Liu, Wei Shen, and Sangiliyandi Gurunathan. "Silver nanoparticles: synthesis, characterization, properties, applications, and therapeutic approaches." *International Journal of Molecular Sciences* 17 (2016): 1534. doi.org/10.3390/ijms17091534.

[6] Hiasa, Shou, Shinichiro Iwamoto, Takashi Endo, and Yusuke Edashige. "Isolation of cellulose nanofibrils from mandarin (*Citrus unshiu*) peel waste." *Industrial Crops and Products* 62 (2014): 280–285. doi.org/10.1016/j.indcrop.2014.08.007.

[7] Bettaieb, Fedia, Ramzi Khiari, Mohammad L. Hassan, Mohamed Naceur Belgacem, Julien Bras, Alain Dufresne, and Mohamed Farouk Mhenni "Preparation and characterization of new cellulose nanocrystals from marine biomass *Posidonia oceanica*." *Industrial Crops and Products* 72 (2015): 175–182. doi.org/10.1016/j.indcrop.2014.12.038.

[8] Reddy, Narendra, and Yiqi Yang. "Properties of high-quality long natural cellulose fibers from rice straw." *Journal of Agricultural and Food Chemistry* 54 (2006): 8077–8081. doi.org/10.1021/jf0617723.

[9] Deepa, B., Eldho Abraham, Laly A. Pothan, Nereida Cordeiro, Marisa Faria, and Sabu Thomas. "Biodegradable nanocomposite films based on sodium alginate and cellulose nanofibrils." *Materials* 9 (2016): 50. doi.org/10.3390/ma9010050.

[10] Almasi, Hadi, Babak Ghanbarzadeh, Jalal Dehghannya, Ali Akbar Entezami, and Asghar Khosrowshahi Asl. "Novel nanocomposites based on fatty acid modified cellulose nanofibers/poly

(lactic acid): morphological and physical properties." *Food Packaging and Shelf Life* 5 (2015): 21–31. doi.org/10.1016/j.fpsl.2015.04.003.

[11] Oun, Ahmed A., and Jong-Whan Rhim. "Preparation and characterization of sodium carboxymethyl cellulose/cotton linter cellulose nanofibril composite films." *Carbohydrate Polymers* 127 (2015): 101–109. doi.org/10.1016/j.carbpol.2015.03.073.

[12] Annamalai, Pratheep K. and Depan, Dilip "Nano-cellulose reinforced chitosan nanocomposites for packaging and biomedical applications." In *Green biorenewable biocomposites: from knowledge to industrial applications*, eds. V. K. Thakur and M. R. Kessler, 489–506. Boca Raton, FL: Apple Academic Press, 2015.

[13] Reddy, Jeevan Prasad, and Jong-Whan Rhim. "Characterization of bionanocomposite films prepared with agar and paper-mulberry pulp nanocellulose." *Carbohydrate Polymers* 110 (2014): 480–488. doi.org/10.1016/j.carbpol.2014.04.056.

[14] Mahmoud, Khaled A., Keith B. Male, Sabahudin Hrapovic, and John H. T. Luong. "Cellulose nano-crystal/gold nanoparticle composite as a matrix for enzyme immobilization." *ACS Applied Materials & Interfaces* 1 (2009): 1383–1386. doi.org/10.1021/am900331d.

[15] Ashori, Alireza. "Wood–plastic composites as promising green-composites for automotive industries." *Bioresource Technology* 99 (2008): 4661–4667. doi.org/10.1016/j.biortech.2007.09.043.

[16] Shefa, Anha Afrin, Mirana Taz, Monir Hossain, Yong Sik Kim, Sun Young Lee, and Byong-Taek Lee. "Investigation of efficiency of a novel, zinc oxide loaded TEMPO-oxidized cellulose nanofiber-based hemostat for topical bleeding." *International Journal of Biological Macromolecules* 126 (2019): 786–795. doi.org/10.1016/j.ijbiomac.2018.12.079.

[17] Zarei, Saeid, Mahmood Niad, and Hossein Raanaei. "The removal of mercury ion pollution by using Fe_3O_4-nanocellulose: synthesis, characterizations and DFT studies." *Journal of Hazardous Materials* 344 (2018): 258–273. doi.org/10.1016/j.jhazmat.2017.10.009.

[18] Farshchi, Elnaz, Sajad Pirsa, Leila Roufegarinejad, Mohammad Alizadeh, and Mahmoud Rezazad. "Photocatalytic/biodegradable film based on carboxymethyl cellulose, modified by gelatin and TiO_2-Ag nanoparticles." *Carbohydrate Polymers* 216 (2019): 189–196. doi.org/10.1016/j.carbpol.2019.03.094.

[19] Zheng, Ming, Peng-Li Wang, Si-Wei Zhao, Yuan-Ru Guo, Li, Fu-Long Yuan, and Qing-Jiang Pan. "Cellulose nanofiber induced self-assembly of zinc oxide nanoparticles: theoretical and experimental study on interfacial interaction." *Carbohydrate Polymers* 195 (2018): 525–533. doi.org/10.1016/j.carbpol.2018.05.016.

[20] Pandey, Jitendra K., M. S. A. Bistamam, and Hitoshi Takagi. "Cellulose nanofibers from waste newspaper." *Journal of Biobased Materials and Bioenergy* 6 (2012): 115–118. doi.org/10.1166/jbmb.2012.1195.

[21] Dufresne, Alain. "Nanocellulose: potential reinforcement in composites." *Natural Polymers* 2 (2012): 1–32. doi:10.1039/9781849735315-00001.

[22] Lavoine, Nathalie, Isabelle Desloges, Alain Dufresne, and Julien Bras. "Microfibrillated cellulose—its barrier properties and applications in cellulosic materials: a review." *Carbohydrate Polymers* 90 (2012): 735–764. doi.org/10.1016/j.carbpol.2012.05.026.

[23] Anwar, Zahid, Muhammad Gulfraz, and Muhammad Irshad. "Agro-industrial lignocellulosic biomass a key to unlock the future bio-energy: a brief review." *Journal of Radiation Research and Applied Sciences* 7 (2014): 163–173. doi.org/10.1016/j.jrras.2014.02.003.

[24] Herrick, Franklin W., Ronald L. Casebier, J. Kelvin Hamilton, and Karen R. Sandberg. "Microfibrillated cellulose: morphology and accessibility." *Journal of Applied Polymer Science: Applied Polymer Symposia* 37 (1983). Available at: https://www.osti.gov/biblio/5039044.

[25] Szabo, Orsolya Erzsebet, and Emilia Csiszar. "Some factors affecting efficiency of the ultrasound-aided enzymatic hydrolysis of cotton cellulose." *Carbohydrate Polymers* 156 (2017): 357–363. doi.org/10.1016/j.carbpol.2016.09.039.

[26] Desmaisons, Johanna, Elisa Boutonnet, Martine Rueff, Alain Dufresne, and Julien Bras. "A new quality index for benchmarking of different cellulose nanofibrils." *Carbohydrate Polymers* 174 (2017): 318–329. doi.org/10.1016/j.carbpol.2017.06.032.

[27] Dilamian, Mandana, and Babak Noroozi. "A combined homogenization-high intensity ultrasonication process for individualization of cellulose micro-nano fibers from rice straw." *Cellulose* 26 (2019): 5831–5849. doi.org/10.1007/s10570-019-02469-y.

[28] Rajinipriya, Malladi, Malladi Nagalakshmaiah, Mathieu Robert, and Saïd Elkoun. "Importance of agri-cultural and industrial waste in the field of nanocellulose and recent industrial developments of wood

based nanocellulose: a review." *ACS Sustainable Chemistry & Engineering* 6 (2018): 2807–2828. doi. org/10.1021/acssuschemeng.7b03437.

[29] Li, Meng, Li-jun Wang, Dong Li, Yan-Ling Cheng, and Benu Adhikari. "Preparation and characterization of cellulose nanofibers from de-pectinated sugar beet pulp." *Carbohydrate Polymers* 102 (2014): 136–143. doi.org/10.1016/j.carbpol.2013.11.021.

[30] Meng, Fanrong, Guoqing Wang, Xueyu Du, Zhifen Wang, Shuying Xu, and Yucang Zhang. "Extraction and characterization of cellulose nanofibers and nanocrystals from liquefied banana pseudo-stem residue." *Composites Part B: Engineering* 160 (2019): 341–347. doi.org/10.1016/j.compositesb.2018.08.048.

[31] Kumar, Ritesh, Sanju Kumari, Shivani Singh Surah, Bhuvneshwar Rai, Rakesh Kumar, Sidhharth Sirohi, and Gulshan Kumar. "A simple approach for the isolation of cellulose nanofibers from banana fibers." *Materials Research Express* 6 (2019): 105601. doi.org/10.1088/2053-1591/ab3511.

[32] Qin, Famei, Zhiqiang Fang, Jie Zhou, Chuan Sun, Kaihuang Chen, Zixian Ding, Guanhui Li, and Xueqing Qiu. "Efficient removal of Cu^{2+} in water by carboxymethylated cellulose nanofibrils: performance and mechanism." *Biomacromolecules* 20 (2019): 4466–4475. doi.org/10.1021/acs.biomac.9b01198.

[33] Najib, Nadira, and Christos Christodoulatos. "Removal of arsenic using functionalized cellulose nanofibrils from aqueous solutions." *Journal of Hazardous Materials* 367 (2019): 256–266. doi.org/10.1016/j. jhazmat.2018.12.067.

[34] Wang, Haiping, Yujia Pei, Xueren Qian, and Xianhui An. "Eu-metal organic framework@ TEMPO-oxidized cellulose nanofibrils photoluminescence film for detecting copper ions." *Carbohydrate Polymers* 236 (2020): 116030. doi.org/10.1016/j.carbpol.2020.116030.

[35] Guo, Xin, Dong Xu, Hanmeng Yuan, Qiuyan Luo, Shiyao Tang, Liu, and Yiqiang Wu. "A novel fluorescent nanocellulosic hydrogel based on carbon dots for efficient adsorption and sensitive sensing in heavy metals." *Journal of Materials Chemistry A* 7 (2019): 27081–27088. doi.org/10.1039/C9TA11502A.

[36] Geng, Biyao, Haiying Wang, Shuai Wu, Jing Ru, Congcong Tong, Yufei Chen, Hongzhi Liu, Shengchun Wu, and Xuying Liu. "Surface-tailored nanocellulose aerogels with thiol-functional moieties for highly efficient and selective removal of Hg(II) ions from water." *ACS Sustainable Chemistry & Engineering* 5 (2017): 11715–11726. doi.org/10.1021/acssuschemeng.7b03188.

[37] Zhang, Xuefeng, Islam Elsayed, Chanaka Navarathna, Gregory T. Schueneman, and E. I. Barbary Hassan. "Biohybrid hydrogel and aerogel from self-assembled nanocellulose and nanochitin as a high-efficiency adsorbent for water purification." *ACS Applied Materials & Interfaces* 11 (2019): 46714–46725. doi.org/10.1021/acsami.9b15139.

[38] Luo, Mengying, Mufang Li, Shan Jiang, Hao Shao, Joselito Razal, Dong Wang, and Jian Fang. "Supported growth of inorganic-organic nanoflowers on 3D hierarchically porous nanofibrous membrane for enhanced enzymatic water treatment." *Journal of Hazardous Materials* 381 (2020): 120947. doi.org/10.1016/j.jhazmat.2019.120947.

[39] Anirudhan, T. S., F. Shainy, and J. R. Deepa. "Effective removal of Cobalt (II) ions from aqueous solutions and nuclear industry wastewater using sulfhydryl and carboxyl functionalised magnetite nanocellulose composite: batch adsorption studies." *Chemistry and Ecology* 35 (2019): 235–255. doi.org/10.1080 /02757540.2018.1532999.

[40] Liu, Peng, Pere Ferrer Borrell, Mojca Božič, Vanja Kokol, Kristiina Oksman, and Aji P. Mathew. "Nanocelluloses and their phosphorylated derivatives for selective adsorption of Ag^+, Cu^{2+} and Fe^{3+} from industrial effluents." *Journal of Hazardous Materials* 294 (2015): 177–185. doi.org/10.1016/j. jhazmat.2015.04.001.

[41] Mautner, Andreas, Yosi Kwaw, Kathrin Weiland, Mlando Mvubu, Anton Botha, Maya Jacob John, Asanda Mtibe, Gilberto Siqueira, and Alexander Bismarck. "Natural fibre-nanocellulose composite filters for the removal of heavy metal ions from water." *Industrial Crops and Products* 133 (2019): 325–332. doi.org/10.1016/j.indcrop.2019.03.032.

[42] Septevani, Athanasia Amanda, Annisa Rifathin, Ajeng Arum Sari, Yulianti Sampora, Gita Novi Ariani, and Dewi Sondari. "Oil palm empty fruit bunch-based nanocellulose as a super-adsorbent for water remediation." *Carbohydrate Polymers* 229 (2020): 115433. doi.org/10.1016/j.carbpol.2019.115433.

[43] Bhattacharya, Amrik, Ayesha Sadaf, Swati Dubey, Rajesh P. Singh, and Sunil Kumar Khare. "Production and characterization of *Komagataeibacter xylinus* SGP8 nanocellulose and its calcite-based composite for removal of Cd ions." *Environmental Science and Pollution Research International* (2020). doi. org/10.1007/s11356-020-08845-7.

[44] Hokkanen, Sanna, Eveliina Repo, and Mika Sillanpaa. "Removal of heavy metals from aqueous solutions by succinic anhydride modified mercerized nanocellulose." *Chemical Engineering Journal* 223 (2013): 40–47. doi.org/10.1016/j.cej.2013.02.054.

[45] Li, Jian, Keman Zuo, Weibing Wu, Zhaoyang Xu, Yonggang Yi, Yi Jing, Hongqi Dai, and Guigan Fang. "Shape memory aerogels from nanocellulose and polyethyleneimine as a novel adsorbent for removal of Cu(II) and Pb(II)." *Carbohydrate Polymers* 196 (2018): 376–384. doi.org/10.1016/j.carbpol.2018.05.015.

[46] Dwivedi, Amarendra Dhar, Naresh D. Sanandiya, Jitendra Pal Singh, Syed M. Husnain, Keun Hwa Chae, Dong Soo Hwang, and Yoon-Seok Chang. "Tuning and characterizing nanocellulose interface for enhanced removal of dual-sorbate (As(V) and Cr(VI)) from water matrices." *ACS Sustainable Chemistry & Engineering* 5 (2017): 518–528. doi.org/10.1021/acssuschemeng.6b01874.

[47] Sharma, Priyanka R., Aurnov Chattopadhyay, Sunil K. Sharma, Lihong Geng, Nasim Amiralian, Darren Martin, and Benjamin S. Hsiao. "Nanocellulose from spinifex as an effective adsorbent to remove cadmium (II) from water." *ACS Sustainable Chemistry & Engineering* 6 (2018): 3279–3290. doi.org/10.1021/acssuschemeng.7b03473.

[48] Bandi, Rajkumar, Madhusudhan Alle, Chan-Woo Park, Song-Yi Han, Gu-Joong Kwon, Jin-Chul Kim, and Seung-Hwan Lee. "Rapid synchronous synthesis of Ag nanoparticles and Ag nanoparticles/holocellulose nanofibrils: Hg (II) detection and dye discoloration." *Carbohydrate Polymers* (2020): 116356. doi.org/10.1016/j.carbpol.2020.116356.

[49] Mukherjee, Sritama, Haritha Ramireddy, Avijit Baidya, A. K. Amala, Chennu Sudhakar, Biswajit Mondal, Ligy Philip, and Thalappil Pradeep. "Nanocellulose-reinforced organo-inorganic nanocomposite for synergistic and affordable defluoridation of water and an evaluation of its sustainability metrics." *ACS Sustainable Chemistry & Engineering* 8 (2019): 139–147. doi.org/10.1021/acssuschemeng.9b04822.

[50] Sehaqui, Houssine, Andreas Mautner, Uxua Perez de Larraya, Numa Pfenninger, Philippe Tingaut, and Tanja Zimmermann. "Cationic cellulose nanofibers from waste pulp residues and their nitrate, fluoride, sulphate and phosphate adsorption properties." *Carbohydrate Polymers* 135 (2016): 334–340. doi.org/10.1016/j.carbpol.2015.08.091.

[51] Mo, Liuting, Huiwen Pang, Yi Tan, Shifeng Zhang, and Jianzhang Li. "3D multi-wall perforated nanocellulose-based polyethylenimine aerogels for ultrahigh efficient and reversible removal of Cu (II) ions from water." *Chemical Engineering Journal* 378 (2019): 122157. doi.org/10.1016/j.cej.2019.122157.

[52] Brandes, Ricardo, Dan Belosinschi, François Brouillette, and Bruno Chabot. "A new electrospun chitosan/phosphorylated nanocellulose biosorbent for the removal of cadmium ions from aqueous solutions." *Journal of Environmental Chemical Engineering* 7 (2019): 103477. doi.org/10.1016/j.jece.2019.103477.

[53] Alipour, A., S. Zarinabadi, A. Azimi, and M. Mirzaei. "Adsorptive removal of Pb(II) ions from aqueous solutions by thiourea-functionalized magnetic ZnO/nanocellulose composite: optimization by response surface methodology (RSM)." *International Journal of Biological Macromolecules* 151 (2020): 124–135. doi.org/10.1016/j.ijbiomac.2020.02.109.

[54] Shahnaz, Tasrin, Vivek Sharma, Senthilmurugan Subbiah, and Selvaraju Narayanasamy. "Multivariate optimisation of Cr(VI), Co(III) and Cu(II) adsorption onto nanobentonite incorporated nanocellulose/chitosan aerogel using response surface methodology." *Journal of Water Process Engineering* 36 (2020): 101283. doi.org/10.1016/j.jwpe.2020.101283.

[55] Ram, Bhagat, and Ghanshyam S. Chauhan. "New spherical nanocellulose and thiol-based adsorbent for rapid and selective removal of mercuric ions." *Chemical Engineering Journal* 331 (2018): 587–596. doi.org/10.1016/j.cej.2017.08.128.

[56] Haseen, U., and Ahmad, H. "Preconcentration and determination of trace Hg(II) using a cellulose nanofiber mat functionalized with MoS_2 nanosheets." *Industrial & Engineering Chemistry Research* 59 (2020): 3198–3204. doi.org/10.1021/acs.iecr.9b06067.

[57] Sharma, Priyanka R., Aurnov Chattopadhyay, Sunil K. Sharma, and Benjamin S. Hsiao. "Efficient removal of UO_2^{2+} from water using carboxycellulose nanofibers prepared by the nitro-oxidation method." *Industrial & Engineering Chemistry Research* 56 (2017): 13885–13893. doi.org/10.1021/acs.iecr.7b03659.

[58] Zhu, Chuantao, Peng Liu, and Aji P. Mathew. "Self-assembled TEMPO cellulose nanofibers: graphene oxide-based biohybrids for water purification." *ACS Applied Materials & Interfaces* 9 (2017): 21048–21058. doi.org/10.1021/acsami.7b06358.

[59] Chen, Hui, Sunil K. Sharma, Priyanka R. Sharma, Heidi Yeh, Ken Johnson, and Benjamin S. Hsiao. "Arsenic (III) removal by nanostructured dialdehyde cellulose-cysteine microscale and nanoscale fibers." *ACS Omega* 4 (2019): 22008–22020. doi.org/10.1021/acsomega.9b03078.

[60] Huang, Xiangyu, Guilherme Dognani, Pejman Hadi, Mengying Yang, Aldo E. Job, and Benjamin S. Hsiao. "Cationic dialdehyde nanocellulose from sugarcane bagasse for efficient chromium (VI) removal." *ACS Sustainable Chemistry & Engineering* 8 (2020): 4734–4744. doi.org/10.1021/acssuschemeng.9b06683.

[61] Chai, Fei, Runkai Wang, Lili Yan, Guanghui Li, Yiyun Cai, and Chunyan Xi. "Facile fabrication of pH-sensitive nanoparticles based on nanocellulose for fast and efficient As(V) removal." *Carbohydrate Polymers* (2020): 116511. doi.org/10.1016/j.carbpol.2020.116511.

[62] Silva, Nuno H. C. S., Paula Figueira, Elaine Fabre, Ricardo J. B. Pinto, Maria Eduarda Pereira, Armando J. D. Silvestre, Isabel M. Marrucho, Carla Vilela, and Carmen S. R. Freire. "Dual nanofibrillar-based bio-sorbent films composed of nanocellulose and lysozyme nanofibrils for mercury removal from spring waters." *Carbohydrate Polymers* 238 (2020): 116210. doi.org/10.1016/j.carbpol.2020.116210.

[63] Du, Weiyan, Yaru Li, Xing Xu, Yanan Shang, Baoyu Gao, and Qinyan Yue. "Selective removal of phosphate by dual Zr and La hydroxide/cellulose-based bio-composites." *Journal of Colloid and Interface Science* 533 (2019): 692–699. doi.org/10.1016/j.jcis.2018.09.002.

[64] Shang, Yanan, Kangying Guo, Peng Jiang, Xing Xu, and Baoyu Gao. "Adsorption of phosphate by the cellulose-based biomaterial and its sustained release of laden phosphate in aqueous solution and soil." *International Journal of Biological Macromolecules* 109 (2018): 524–534. doi.org/10.1016/j.ijbiomac.2017.12.118.

[65] Lei, Xiaojuan, Xuehai Dai, Sihui Long, Ning Cai, Zhaocheng Ma, and Xiaogang Luo. "Facile design of green engineered cellulose/metal hybrid macrogels for efficient trace phosphate removal." *Industrial & Engineering Chemistry Research* 56 (2017): 7525–7533. doi.org/10.1021/acs.iecr.7b00587.

13

Nanocellulose in Sensing Application for Water Purification

Sapna Nehra
Dr. K. N. Modi University, Newai, India

Rekha Sharma
Banasthali Vidyapith, Banasthali, India

Dinesh Kumar
Central University of Gujarat, Gandhinagar, India

CONTENTS

13.1 Introduction

Availability of freshwater is essential for maintaining ecological nature [1]. However, many industrial and textile hazardous substances adversely affect the water quality, which has become a primary concern, particularly in developing countries [2]. Many pollutants such as petroleum by-product metal, pharmaceutical drugs, organic dyes, and heavy metals are found in water sources [3]. The toxic levels and their bioaccumulation have demonstrated adverse effects on human health and other life on earth [4]. Leaching of heavy metals is one factor responsible for the contamination of drinking water. People use contaminated water without knowing their concentration. Therefore, the health of users is at risk from various toxic components that are present in high concentration. Therefore, to inhibit the long-time exposure is essential to detect metal contamination in the water resources. However, to determine the concentration of metal contaminants, several techniques have been used to monitor the contaminated water. Only the standard methods were employed, i.e., those permitted by the environmental protection agency (EPA), like inductively coupled plasma mass spectrometry (ICPMS) [5].

The leaching of the metal occurred unexpectedly high and there is need for efficient materials and techniques. Several researchers have developed efficient techniques, such as electrochemical, potentiometric, colorimetric, to determine the metal concentration at the domestic level [6]. The potential sensors exhibited a smaller life duration because of its reference electrodes—another limitation in the previously used colorimetric sensors in the single-use direction. Hence, the urgent need to develop a sensor with a long-time working range without the involvement of manpower. So, by using the nanocellulose and its derivatives, many researchers have developed the ideal sensor that exhibited long-time durability, required no further examination after installation, and is ecofriendly in nature [7]. The cellulosic material in nanoform exhibited one dimension in high mechanical strength, stability, and surface area [8]. A different form obtained from the cellulose like bacterial cellulose (BC), cellulose nanocrystals (CNC), nanofibrillated cellulose (NFC) is determined by its isolation methods [9]. Because of unique potential characteristics, nanocellulose is treated as outstanding support in different nanocomposites such as ceramics, non-metals, reduced graphene oxides, and CNT for the implication of sensing [10]. This chapter elaborates on the various nanocellulose-related composites employed for the on-site sensing of hazardous substances present in water. It also deals with ongoing tasks and modifications on nanocellulose and future outcomes are also analyzed.

13.2 Properties and Structure of Nanocellulose

The term cellulose is derived from the French word "cellule," which is related to living cell and glucose. The term was given by Anselme Pen, a French chemist, in 1838 [11]. Cellulosic polymers are found in many plant materials such as hemp, wood, grain straws, and cotton [12]. About 1012 tons of cellulose are formed every year, and about 6×10^9 tons are manufactured through several industries like textiles, paper, etc. [13]. The materiality of cellulose is semicrystalline, so both crystalline and inconsistent cellulose remain, determined by the extracted source of cellulose [14].

It is known as a linear biopolymer, made up of β-D-glucopyranose units connected to β-1,4-glycosidic connections [15]. Additionally, the hydroxyl functional group's hydrogen bonding formed the crystal structure because of strong interactions [16]. The cellulose structure comprises three types of hydroxyl groups: primary, secondary, and hydrophilic at C6, C2, and C3 positions, respectively [17]. However, it reflects the insolubility in water and some other solvents. Because of the availability of many hydroxyl functional groups', it is easy to associate with inorganic toxins such as dyes, heavy metals, and other toxic anions [18].

13.3 Types of Nanocellulose

Nanocellulose is nanostructured cellulose and is available in roughly three categories:

  i. BC, which is produced by bacteria and referred to as microbial cellulose (MC).
 ii. Cellulose nanofibers (CNFs), also known as nanofibrillated cellulose or microfibrillated cellulose, is composed of nanosized cellulose fibrils 5–20 nm in width and several micrometers in length. The fibrils are derived from wooden fiber pulp at high temperature, high pressure, high velocity, and by microfluidization.
iii. CNC, also known as nanocrystalline cellulose or cellulose nanowhiskers (CNW), can be made from 100–1000 nm sized native fibers by homogenization and microfluidization.

The shape, properties, structure, and morphology of NC are always influenced by the methods used to extract them [19].

13.4 Sensing Application of Nanocellulose

13.4.1 Cellulose Nanofibrils (CM-CNF) in Sensing of Heavy Metal

Yuan et al. [20] showed the detrimental effects of the hexavalent chromium—Cr(VI)—on eyes, skin, and respiratory system and considered occupational carcinogens may cause cancer. To remove and recognize Cr(VI) from solutions, a cobweb three-dimensional net-like framework, fluorescent aerogel, was used and prepared with photoluminescent carbon and renewable CM-CNF. The fluorescence properties, microstructure, chemical elements, and surface morphology of this gel were demonstrated and experimented with to know the adsorption properties of Cr(VI) onto this gel. The PSO and Freundlich model was used to illustrate the adsorption mechanism and its corresponding efficacy was 433.5 mg/g. When applied, the prepared aerogel to sense Cr(VI) attained 11.8% quantum yield and emitted a bright blue fluorescence. Besides all these displayed, a linear relationship was found between Cr(VI) and the quantum yield. This fluorescence aerogel was found favorable adsorbent for the exclusion of the Cr(VI) [20].

Guo et al. [21] simultaneously detected and removed the heavy metals. A fluorescent nanocellulosic hydrogel was synthesized and found to be an optical sensor for detecting heavy metals. The synthesized nanocellulosic fluorescent hydrogel exhibited the intense blue color fluorescence with 23.6% quantum yield and detected the presence of Fe(III). At the time of hydrogel synthesis, the CNFs fabricated with carbon dots (CDs), assisted as adsorption-aggregator, and which permitted the fast visible changes toward the heavy metals as the optical sensor result from the enhanced efficacy, stability of signal, and sensitivity [21].

Teodoro et al. [22] developed an electrochemical sensor to detect Hg(II) from an aqueous system even in minimal concentrations. They utilized the green nanoarchitecture, which comprised CNW, reduced graphene oxide (rGO), and polyamide 6 (PA6) electrospun nanofibers. Various characterization methods were used to characterize the synthesized nanosensor, such as thermogravimetric analysis (TGA), scanning transmission electron microscopy (STEM), UV-vis, and FTIR. These methods were also used to describe the stability, morphology, sensitivity, and functional moieties of the hybrid composite. Further, the cyclic voltammetry (CV) analysis confirmed the charge transference characteristic of PA6/CNW:rGO. Because of the graphene's outstanding electrical behavior, the hybrid composite could detect the very minimal concentration of Hg(II) and enhanced the sensing ability [22].

Fu et al. [23] worked on minimum stability because of the formation of metallophilic Hg(II)-Au(I) interaction. Further, the solid-state transparent CNF matrix-reinforced luminescent gold nanoparticles (AuNPs) with intense red fluorescence were incorporated. The nanostructured CNF matrix provides higher scale immobilizing spots to the AuNPs' sensor blocks. The highly clear CNF support makes AuNPs easy to escape the fluorescence light without scattering loss. The fluorescence signal's medium was altered successively from the signal's origin to transmission to quenching, which results in maximum sensitivity, stability, and selectivity. The obtained recognition value of 1.0 nm was less than the 10.0 nm standard mark recommended by the US EPA [23].

Zhang et al. [24] employed cellulose and dyes film composite as a colorimetric sensor. When the cellulose was regenerated in the ethanol, the films were further recovered by imbibing it into the dye ethanol solution trailed by hot-pressing. They used the 1-(2-pyridylazo)-2-naphthol (PAN), which resulted in a very tough mechanical strength CP film. Further, the film showed light transmittance, thermodynamic stability, and 52.9 MPa of tensile strength. PAN nanoparticles were dispersed on cellulose because of the interaction between the nitrogen of the azo group of PAN and cellulose. The robust nanosensor was utilized to detect the Zn(II) by changing the color after testing with tap water and demonstrated the minimal amount of 100 parts per billion (ppb). The two more dyes were incorporated over the cellulose: 1-(2-thiazolylazo)-2-naphthol (TAN) and dithizone (Dith), and the obtained composite film works as a colorimetric sensor for the exclusion of Zn(II) and Cu(II) [24].

13.4.2 Cellulose Composite in Sensing

Padmalaya et al. [25] used a co-precipitation approach to synthesize the zinc oxide nanoparticle. Further, ZnO was functionalized with the significant ester as cellulose acetate (CA). Several techniques were utilized to investigate the functionalization of the bare ZnO and CA. By the help of the XRD technique, the reduction in the diameter after the functionalization was determined. The SEM images demonstrated the hexagonal rod shape obtained due to agglomeration when the CA fabricated over the ZnO. The FTIR analytical tools also confirmed the formation of NC. Additionally, the CV technique was attributed to its electrochemical behavior in the direction of Cd(II) and also gave its low detection limit of 0.41 μM with the sensitivity of 3.11 μA/(μM). In the range of 0.1–0.5 μM, it exhibited the 0.9937 regression coefficient [25].

Lei et al. [26] fabricated MOF UiO-66 and UiO-66-NH$_2$ over the elastic cellulose aerogel through *in situ* approach at room temperature. The metal ion adsorbed over the surface is equal to the sum of MOF cellulose aerogels, which shows that the cellulose aerogel's fabrication did not restrict the MOF's active sites. The decomposition temperature of MOF composite was enhanced to 62.1 °C [26].

Khalilzadeh et al. [27] detected venlafaxine using an electrochemical sensor based on Fe$_3$O$_4$@ cellulose nanocrystals/copper nanocomposite. They used green method to synthesize by involving the leaf extract of *Petasites hybridus*, which acted as a stabilizing and reducing agent. The electrochemical behavior of venlafaxine on the modified electrode was showed via the cyclic and differential pulse voltammetry and chronoamperometry. Through the cyclic voltammeter, the sensor's performance was checked and venlafaxine was recognized. In the presence of optimum conditions, the linear alteration range of venlafaxine concentrations and LOD was found to be 0.05–600 μM and 0.01 μM, respectively. Therefore, the synthesized nanocomposite was an efficient electrode sensor for checking the venlafaxine in urine, water, and pharmaceutical formulation samples even in traces [27].

Zhu et al. [28] detected Hg(II) with high sensitivity and selectivity. They developed functionalized silver nanoparticles (AgNPs) for colorimetric sensing. AgNPs were linearly stabilized by thermoresponsive poly(nisopropyl acrylamide) (PNIPAm). The synthesized composite showed the color differentiation from yellow to taupe brown because Ag-Hg generated core-shell of nanoparticles. Later on, the composite displayed the better selectivity and maximum sensitivity with 75 nm of LOD and the regression coefficient of $R^2 = 0.9903$ and 0.9905 for the ultrapure and tap water, respectively. The optimization of the specific parameters of the sensor depends on the thermosensitive PNIPAm. The sensor might be used for the parting of the Ag-Hg composites with the highest efficacy of 97.60 and 95.40% corresponds to sliver and mercury without producing the secondary hazardous pollutant. Therefore, this dual nanocomposite showed the potential in the recombination and separation of Hg(II) [28].

Song et al. [29] developed a new carbon quantum dot (CDQ) NFC composite aerogel through green chemical synthesis and utilized for the adsorption and recognition of the Cr(III) from the aqueous system. The synthesized CQDs adsorbed the Cr(III) ion from the aqueous system with different pH value alterations. The fluorescence effects diminished by the enhancing heavy metal concentration might be adsorbed and it showed real-time assessment in the water system [29].

13.4.3 Bacterial Nanocellulose (BC) in Sensing

Mohammed et al. [30] detected the hazardous metal ions present even in traces. A very accurate gold-coated bovine serum albumin nanocrystal as Au@BSA NCs was used to sense and scavenge the heavy metal ions, especially Hg(II) from the aqueous system. The leaching effect and adsorption experiment showed the sensing and scavenging nature of toxic ions as 1:2 ratio from the water system. With other ions, the nanocomposite ions were most selective toward the Hg(II) [30].

13.4.4 Membrane-Based Nanocellulose in Sensing

Jiang et al. [31] detected aqueous Cd(II) using a highly sensitive cellulose membrane-based colorimetric sensor strip (CCSS). It was successfully developed by creating intermolecular H- bonding between Victoria blue B (VBB) and cellulose membrane (CM) through the tap casting method. Comparing with the CM, the CCSS showed the uniform dense skin film confirmed by the SEM images. The intermolecular type of interaction among the CB and VBB was confirmed through the FTIR analysis. The pore volume and pore size of CCSS were reduced after incorporating the VBB and adsorption of the Cd(II). During the adsorption process, the strip's color was altered from yellow to blue-green, seen with naked eyes. The LOD value for the CCSS was found to be very low 0.01 mg/L, and the response was attained within 1 minute. Further, the CCSS was utilized to recognize cadmium in the field-collected polluted water

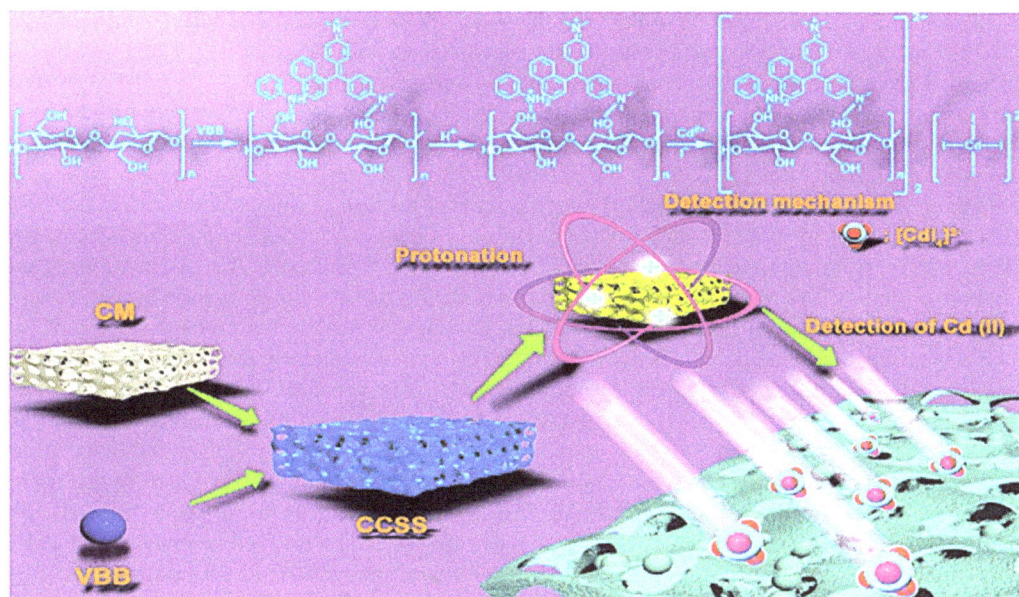

FIGURE 13.1 The mechanism for the design of the sensor strips and the detection of cadmium ions (Reproduced with permission from Ref. [31], © American Chemical Society 2020).

samples with strong anti-interference nature compared to the old filter paper [31]. The highly pure nature of CM was found favorable to increase the recognition tendency of the sensor. The adsorption mechanism of the Cd(II) and the formation of CCSS are schematically shown in Fig. 13.1 [31].

Li et al. [32] prepared the fluorogenic bioadsorbent and showed high sensitivity with the recognition boundary of 84 ppb that lies beneath the higher disposal of standard enterprise in China. Further, fluorescent quenching features were noticed because of the photoinduced electron transfer mechanism. The higher efficiency of 143.88 mg/g was obtained and followed by the Langmuir adsorption isotherm model. The kinetic study was determined through the PSO kinetics model [32].

Yu et al. [33] performed the semiquantitative visual analysis of the Cu(II) and the Hg(II) from the aqueous system. They employed the *in situ* microwave-assisted method to prepare the sponge cellulose fluorescence spherical, renamed as CS-CDs. During the preparation as a base material, nitrogen-doped reagents and external carbon sources were cellulose, PEI, and citric acid. Its fluorescent nature and porous structure contributed to its selectivity and the adsorption behavior toward the Hg(II) and showed the detection limit of 26 nM. Later on, during the fluorescence experiment, the CS's color varied from its basic pale yellow color to the deep green when the concentration of the Cu(II) was enhanced from 3 to 60 µM. The 0.11 and 3 µM were the fluorescence and visual detection limits, which were considered lower as compared to the recommended limit of 20 µM of Cu(II) in drinking water. The discoloration mechanism was attributed to the formation of the Cu(II) amine complexes because of aggregation over the cellulose sphere substrate's surface. In conclusion, we can say that the CS-CDs were employed for checking the tap and river water, depending on the greatly reproducible results. The present CS-CD-related device was a very appropriate probe from environmental intervention and remarkably stable up to the tiniest 22 cycles of revelation to EDTA [33].

13.5 Sensing of Dyes

Albukhari et al. [34] synthesized the AgNPs using a green method in which the reducing agent is extracted from the *Duranta erecta* leaves. With the biological method's help, the AgNPs were fabricated over the cellulose acetate filter paper (Ag@CAF) and titanium dioxide (Ag@TiO$_2$) to investigate the application in the decontamination of the harmful organic substances and dyes such as 2-nitrophenol (2-NP), 2-nitroaniline (2-NA), 4-nitrophenol (4-NP), trinitrophenols (TN rhodamine BP), and rhodamine B (RhB), respectively. Some outstanding features were investigated in the CAF and TiO_2-related Ag nanocrystals such as catalytic activity, facile separation, and high stability. The rough and porous structure behaves as eco-friendly, stable, and renewable, which can universally implement for the decontamination of drinking water [34].

Gu et al. [35] developed the bio-enabled hybrid adsorbent through the interaction between the CNF and faceted silver nanoparticles (AgFNPs) to recognize the organic dyes from the aqueous system. The faceted AgNPs were prepared in the presence of eco-friendly environmental circumstances using the TEMPO-oxidized CNFs to regulate the shape and the reducing and stabilizing agent. First, the spherical shape of the small zero-valent AgNPs 2–16 nm in diameter was formed over the surface of the CNF, which later converted into the larger form 100–200 nm in diameter and faceted AgNPs when followed the Ostwald ripening procedure by reacting with H_2O_2. The prepared CNF/AgFNP was fabricated over the surface-enhanced Raman scattering (SERS) substrate showing the outstanding increment in the methylene blue by showing the signal up to 6.8×10^3 at 448 cm^{-1}. In the Raman intensities, some 3D variations showed approximately 32%, which were trailed by the Gaussian profiles assigned to the analyte's recognition as dyes at sub-PPM level. The development of discrete hot spots in the AgNPs governs the SERS increment, which can be illustrated through spatially resolved laser ablation ICPMS. Therefore, this report becomes important for designing a robust and sensitive SERS substrate [35].

13.6 Sensing of Silver

Wang et al. [36] prepared a unique cellulose-related calorimetric sensor (DAC-Tu) containing the nitrogen- and sulfur-active sites to recognize the silver ion from the water samples. This cellulose-based adsorbent showed outstanding selectivity toward the silver ion in the presence of other ions and it is very sensitive, showing the visible color from white to black. The recognition limits for the silver was 10^{-6} mol/L, which was detected by the naked eyes within 10 minutes. The functional group nitrogen and sulfur atoms, especially coordination, were chelated with silver to form N-Ag, S-Ag, and Ag_2S and augmented on the cellulose medium. So, the present sensor displayed the various color change concerning the varying amount of silver. Therefore, the synthesized colorimetric sensor displayed immense application to recognize Ag(I) by naked eyes [36].

13.7 Sensing of Other Anions

Nawaz et al. [37] demonstrated the iron's rapid detection in an aqueous system and showed the good detection limit. They developed the cellulose-related fluorescent sensor by involving the crosslinker using 4,4′-methylene diphenyl diisocyanate (MDI) for the chemical interaction among the 1,10-phenanthroline-5-amine (Phen) onto CA and formed Phen-MDI-CA. The fabrication and adulterating impact of cellulose framework resulted in the outstanding fluorescence features solution and solid phase. Specifically, its polymeric chain remarkably enhanced the sensitivity of phenanthroline. By the interaction with the Fe$^{(II)}$, insoluble and non-fluorescent complex was formed as Fe-(Phen-MDI-CA). So, this complex behaves as the divergent mode of chromogenic sensor for the fast detection of iron. The recognition range was varied in the instrumental free visual, and fluorescence mode was 50 and 2.6 ppb. It was first time reported that sensors were utilized to detect iron in an aqueous system via the naked eye. Phen-MDI-CA showed better solubility and processability in common organic solvents, enabling the application of various material forms, such as coating, film, and printing ink. The detection mechanism is shown in Fig. 13.2 [37].

Nawaz et al. [38] demonstrated the microsensitive and multiresponsive sensors for detecting heavy metal in an aqueous system. They synthesized diverse, responsive fluorescent sensors by functionalizing phenanthroline over the CA via the crosslinker as MDI. The functionalized composite showed the visuals and microlevel detection of the traces of anions because of the functional amine moiety. The continuous exposure to different amines in the solution medium as well as to the vapor state the composite

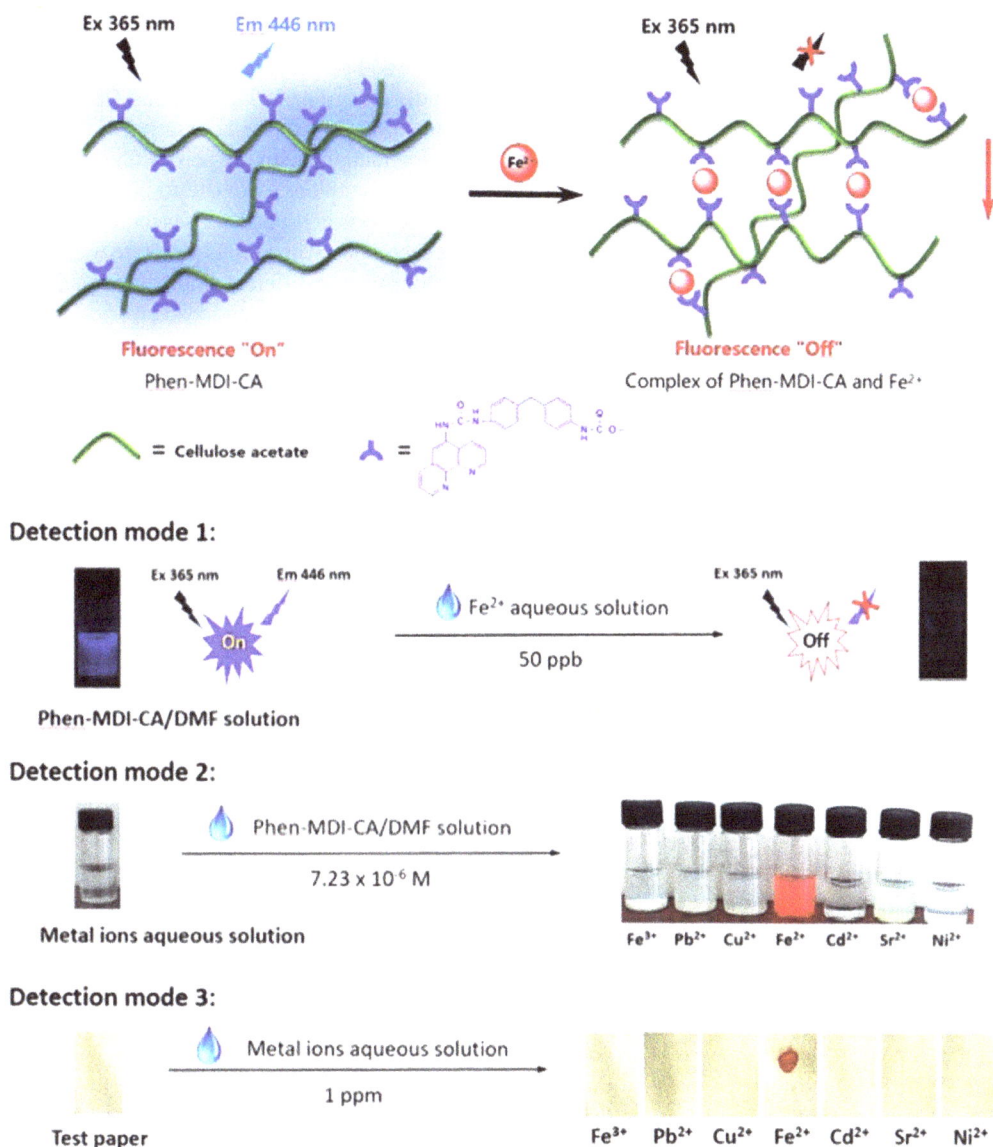

FIGURE 13.2 Detection of Fe²⁺ ions in multimode (reproduced with permission from Ref. [37], © American Chemical Society 2017).

attributed to the various fluorescence colors might distinguish the triethylamine, pyrrolidine methylamine, hydrazine aniline, ethylenediamine, and from other amines. The combination of the composite and the radiometric system of malachite green precisely detects the $B_4O_7^{2-}$, PO_4^{3-}, and CO_3^{2-} ions by displaying the different visible and fluorescence colors. Each ion showed a different fluorescent behavior at different minimal molar concentration like 0.18, 0.69, 0.86 nmol of $B_4O_7^{2-}$, PO_4^{3-}, and CO_3^{2-}, correspondingly. Remarkably, the present sensor showed the qualitative and quantitative recognition of specific anions in the combination of other anions because of the cellulose polymer chain and the distinguished association among the sensor and analytes [38].

Park et al. [39] showed the sensing of other ions than the heavy metal present in the water like iodine by using the cesium lead bromide perovskite PQDs. At present, the PQDs attract more attention of the researchers owing to their outstanding optoelectronic features like quantum efficiency, maximum photoluminescence, tunable wavelength, color purity, and narrow emission. Although besides its remarkable features, its unstable nature in the presence of long-time exposure concerning the pH, temperature, and moisture inhibits its application. They introduced the facile and portable cellulose-containing colorimetric sensor incorporated with $CsPbBr_3$ PQDs for the nakedly recognized Cl^- and I^- ions. During composite, the preparation used the hot injection strategy, which attributed to the homogeneous formation of $CsPbBr_3$PQDs having great sensitivity, providing strong interactions with the porous fibers resulting in the excellent stability and durability under different moisture conditions, with end-to-end outstanding photoluminescence characteristics. Initially, the synthesized composite used as the sensor in the sample solution showed specific color alteration faster, even within 5 seconds. The porous structure was mainly responsible for the fast diffusion, which results in the quantification of different colors through the red, green, and blue imaging analysis. The unique composite behaves as an efficient colorimetric sensor to detect the trace elements from the aqueous system [39]. Table 13.1 gives brief idea of miscellaneous of nanocellulosic sensors.

TABLE 13.1

Miscellaneous of Nanocellulosic Sensors.

Sensor	Target Ions	Detection Limit	Equilibrium Time	Reference
Fluorescence-modified cellulose (FMC)	F^-	10 mg/L	10 minutes	[40]
Graphene/multiwalled carbon nanotubes (G-MWNTs)	Pb(II), Cd(II)	0.2 μg/L	5 minutes	[41]
Spherical nanocellulose chemosensor	Cr (VI)	30 ppb	5 minutes	[42]
TEMPO-oxidized cellulose	F^-	0.005 M	–	[43]
Cellulose-graphene oxide-based nanocomposite	Ni(II)	0.01–0.1 ppm	–	[44]
Cellulose-grafted 2,5-dithiourea	Ag(I),Cu (II)	10^{-3} and 10^{-4} mol/L	5 seconds	[45]
Anthracene-modified cellulose (AMC)	Hg (II)	84 ppb	–	[32]
Dialdehyde cellulose (DAC), 2,6-pyridine dihydrazide (PDH)	Cu(II)	10–7 mol/L	30 seconds	[46]
Rhodamine-based fluorescence cellulose nanocrystals (RhB-CNCs)	Hg(II)	746 nM	–	[47]
Rhodamine B derivative (RBD)/cellulose acetate (CA)	Cu(II)	18 ppm	–	[48]
BC/PANI-DBSA/PAMPS	NH_3	200 ppb	10.2 seconds	[49]
CDs-CMC-HA c	Fe(III)	55 nm	–	[50]
Cellulose filter paper	Hg(II), Cr(VI)	5 ppb, 5 ppm	–	[51]
Novel cellulose-Schiff base	Cu(II)	1.054	30 seconds	[52]
Fluorescent cellulose nanocrystals	Pb(II)	1.5×10^{-7} mol/L	–	[53]
Cellulose acetate nanofibers/Fe/carbon dot	Hg(II), Pb(II)	–	–	[54]
Microcrystalline cellulose (MCC)	Fe(III)	0.01 ppm	–	[55]
Porous graphene/carboxymethyl cellulose/ fondaparinux nanocomposite	Cd(II), Pb(II)	0.28 and 0.17 nM	–	[56]
Cellulose-graphene nanocomposite	Ni(II)	10 ppm	–	[57]
Nanocellulosic fiber-modified carbon paste	Cd(II), Pb(II)	88 and 33 μg/L	10 minutes	[58]
PEI-PAN cellulose membrane	Cu(II)	0.5–2.0 mg/ L	–	[59]
Rhodamine B derivative (RBD)/cellulose acetate (CA)	Cu(II)	18 ppm	33 minutes	[48]
D-Penicillamine-anchored nanocellulose (DPA-NC)	Cu(II)	0.048 pM	–	[60]
Cellulose-based solid amine	Cr (IV)	0.5 mg/L	20 minutes	[61]
Ternary cellulosic nanocomposites	Pb(II)	10 nM/L	–	[62]
Carboxy methyl cellulose/poly-vinyl-alcohol and chitosan	Cu(II)	1.3 ppm	–	[63]

13.8 Conclusion and Future Outcomes

This chapter studied properties of nanocellulose, like high surface-area-to-volume ratio, functionalization ability, high strength, and sustainability. The vital role of nanocellulose in sensing and eliminating hazardous ions in the aqueous system are discussed. In water treatment technologies, the capability of the nanocellulose as a durable exchange of nanofabricated form of carbon nanotube, composites, nanocomposites, and filtration membranes with a functionalized form of these is displayed. The chapter deals with various cellulose-based nanomaterials that sustain reasonable future endeavors in water treatment using these nanomaterials.

Acknowledgments

Dinesh Kumar is thankful to DST, New Delhi, for financial support to this work (sanctioned vide project Sanction Order F. No. DST/TM/WTI/WIC/2K17/124(C).

REFERENCES

[1] Ummartyotin, Sarute, and Hathaikarn Manuspiya. "A critical review on cellulose: from fundamental to an approach on sensor technology." *Renewable and Sustainable Energy Reviews* 41 (2015): 402–412. doi.org/10.1016/j.rser.2014.08.050.

[2] Kumari, Sapana, and Ghanshyam S. Chauhan. "New cellulose-lysine Schiff-base-based sensor-adsorbent for mercury ions." *ACS Applied Materials & Interfaces* 6 (2014): 5908–5917. doi.org/10.1021/am500820n.

[3] Ogundare, Segun A., and Werner van Zyl. "Amplification of SERS 'hot spots' by silica clustering in a silver-nanoparticle/nanocrystalline-cellulose sensor applied in malachite green detection." *Colloids and Surfaces A: Physicochemical and Engineering Aspects* 570 (2019): 156–164. doi.org/10.1016/j.colsurfa.2019.03.019.

[4] Khattab, Tawfik A., Sawsan Dacrory, Hussein Abou-Yousef, and Samir Kamel. "Development of microporous cellulose-based smart xerogel reversible sensor via freeze drying for naked-eye detection of ammonia gas." *Carbohydrate Polymers* 210 (2019): 196–203. doi.org/10.1016/j.carbpol.2019.01.067.

[5] Nunez, Cristina, Mario Diniz, Alcindo A. Dos Santos, Jose Luis Capelo, and Carlos Lodeiro. "New rhodamine dimer probes for mercury detection via color changes and enhancement of the fluorescence emission: fast recognition in cellulose supported devices." *Dyes and Pigments* 101 (2014): 156–163. doi.org/10.1016/j.dyepig.2013.09.019.

[6] US EPA. Table of Regulated Drinking Water Contaminants. Pennsylvania Avenue N.W. Washington, DC. Available at: https://www.epa.gov/ground-water-and-drinking-water/tableregulated-drinking-water-contaminants (accessed January 18, 2017).

[7] Rull-Barrull, Jordi, Martin d'Halluin, Erwan Le Grogneca, and François-Xavier Felpin. "Chemically-modified cellulose paper as smart sensor device for colorimetric and optical detection of hydrogen sulfate in water." *Chemical Communications* 52 (2016): 2525–2528. doi.org/10.1039/C5CC09889K.

[8] Lucio, Melone, Bonafede Simone, Tushi Dorearta, Punta Carlo, and Cametti Massimo. "Dip in colorimetric fluoride sensing by a chemically engineered polymeric cellulose/bPEI conjugate in the solid state." *RSC Advances* 5 (2015): 83197–83205. doi: 10.1039/C5RA16764G.

[9] Bi, Lei, Yi-Ping Chen, Chen Wang, Jing Su, and Gang Pan. "Microalgae-derived cellulose/inorganic nanocomposite rattle-type microspheres as an advanced sensor for pollutant detection." *Chemical Engineering Journal* (2020): 125073. doi.org/10.1016/j.cej.2020.125073.

[10] Cho, Soo-Yeon, Hayoung Yu, Junghoon Choi, Hohyung Kang, Seoungwoong Park, Ji-Soo Jang, Hye-Jin Hong, Il-Doo Kim, Seoung-Ki Lee, Hyeon Su Jeong and Hee-Tae Jung. "Continuous meter-scale synthesis of weavable tunicate cellulose/carbon nanotube fibers for high-performance wearable sensors." *ACS Nano* 13 (2019): 9332–9341. doi.org/10.1021/acsnano.9b03971.

[11] Ng, Hon-Meng, Lee Tin Sin, Tiam-Ting Tee, Soo-Tueen Bee, David Hui, Chong-Yu Low, and A. R. Rahmat. "Extraction of cellulose nanocrystals from plant sources for application as reinforcing agent in polymers." *Composites Part B: Engineering* 75 (2015): 176–200. doi.org/10.1016/j.compositesb.2015.01.008.

[12] Tang, Juntao, Jared Sisler, Nathan Grishkewich, and Kam Chiu Tam. "Functionalization of cellulose nanocrystals for advanced applications." *Journal of Colloid and Interface Science* 494 (2017): 397–409. doi.org/10.1016/j.jcis.2017.01.077.

[13] Ye, Jun, Zhang, Mingming and Xiong, Jian. "Fluorescence probe based carboxymethyl cellulose/Tb(III) nanocomposites for detection of Mn^{2+} with simpleness, rapidness and high sensitivity." *Carbohydrate Polymers* 190 (2018): 156–161. doi.org/10.1016/j.carbpol.2018.02.042.

[14] Si, Junhui, Zhixiang Cui, Qianting Wang, Qiong Liu, and Chuntai Liu. "Biomimetic composite scaffolds based on mineralization of hydroxyapatite on electrospun poly(ε-caprolactone)/nanocellulose fibers." *Carbohydrate Polymers* 143 (2016): 270–278. doi.org/10.1016/j.carbpol.2016.02.015.

[15] Lavoine, Nathalie, Isabelle Desloges, Alain Dufresne, and Julien Bras. "Microfibrillated cellulose—its barrier properties and applications in cellulosic materials: a review." *Carbohydrate Polymers* 90 (2012): 735–764. doi.org/10.1016/j.carbpol.2012.05.026.

[16] Khalil, HPS Abdul, Y. Davoudpour, Chaturbhuj K. Saurabh, Md. S. Hossain, A. S. Adnan, R. Dungani, M. T. Paridah, et al. "A review on nanocellulosic fibres as new material for sustainable packaging: process and applications." *Renewable and Sustainable Energy Reviews* 64 (2016): 823–836. doi.org/10.1016/j.rser.2016.06.072.

[17] Yahya, Mazlita, You Wei Chen, Hwei Voon Lee, and Wan Hasamudin Wan Hassan. "Reuse of selected lignocellulosic and processed biomasses as sustainable sources for the fabrication of nanocellulose via Ni(II)-catalyzed hydrolysis approach: a comparative study." *Journal of Polymers and the Environment* 26 (2018): 2825–2844. doi.org/10.1007/s10924-017-1167-2.

[18] Marett, Josh, Alex Aning, and E. Johan Foster. "The isolation of cellulose nanocrystals from pistachio shells via acid hydrolysis." *Industrial Crops and Products* 109 (2017): 869–874. doi.org/10.1016/j.indcrop.2017.09.039.

[19] Habibi, Youssef. "Key advances in the chemical modification of nanocelluloses." *Chemical Society Reviews* 43 (2014): 1519–1542. doi.org/10.1039/C3CS60204D.

[20] Yuan, Hanmeng, Guanhua Yang, Qiuyan Luo, Teng Xiao, Yingfeng Zuo, Xin Guo, Dong Xu, and Yiqiang Wu. "A 3D net-like structured fluorescent aerogel based on carboxy-methylated cellulose nanofibrils and carbon dots as a highly effective adsorbent and sensitive optical sensor of Cr (VI)." *Environmental Science: Nano* 7 (2020): 773–781. doi.org/10.1039/C9EN01394F.

[21] Guo, Xin, Dong Xu, Hanmeng Yuan, Qiuyan Luo, Shiyao Tang, Liu, and Yiqiang Wu. "A novel fluorescent nanocellulosic hydrogel based on carbon dots for efficient adsorption and sensitive sensing in heavy metals." *Journal of Materials Chemistry A* 7 (2019): 27081–27088. doi.org/10.1039/C9TA11502A.

[22] Teodoro, Kelcilene B. R., Fernanda L. Migliorini, Murilo H. M. Facure, and Daniel S. Correa. "Conductive electrospun nanofibers containing cellulose nanowhiskers and reduced graphene oxide for the electrochemical detection of mercury (II)." *Carbohydrate Polymers* 207 (2019): 747–754. doi.org/10.1016/j.carbpol.2018.12.022.

[23] Fu, Junjun, Jiayi Zhu, Yan Tian, Kui He, Huang Yu, Linlin Chen, Dongjun Fang, et al. "Green and transparent cellulose nanofiber substrate-supported luminescent gold nanoparticles: a stable and sensitive solid-state sensing membrane for Hg (II) detection." *Sensors and Actuators B: Chemical* (2020): 128295. doi.org/10.1016/j.snb.2020.128295.

[24] Zhang, Meng, Lina Zhang, Huafeng Tian, and Ang Lu. "Universal preparation of cellulose-based colorimetric sensor for heavy metal ion detection." *Carbohydrate Polymers* 236 (2020): 116037. https://doi.org/10.1016/j.carbpol.2020.116037.

[25] Padmalaya, G., B. S. Sreeja, P. Dinesh Kumar, S. Radha, V. Poornima, M. Arivanandan, Sujan Shrestha, and T. S. Uma. "A facile synthesis of cellulose acetate functionalized zinc oxide nanocomposite for electrochemical sensing of cadmium ions." *Journal of Inorganic and Organometallic Polymers and Materials* 29 (2019): 989–999. doi.org/10.1007/s10904-018-0989-2.

[26] Lei, Chao, Junkuo Gao, Wenjing Ren, Yuanbo Xie, Somia Yassin Hussain Abdalkarim, Shunli Wang, Qingqing Ni, and Juming Yao. "Fabrication of metal-organic frameworks@ cellulose aerogels composite materials for removal of heavy metal ions in water." *Carbohydrate Polymers* 205 (2019): 35–41. doi.org/10.1016/j.carbpol.2018.10.029.

[27] Khalilzadeh, Mohammad A., Somayeh Tajik, Hadi Beitollahi, and Richard A. Venditti "Green synthesis of magnetic nanocomposite with iron oxide deposited on cellulose nanocrystals with copper (Fe_3O_4@ CNC/Cu): investigation of catalytic activity for the development of a venlafaxine electrochemical sensor." *Industrial & Engineering Chemistry Research* 59 (2020): 4219–4228. doi.org/10.1021/acs.iecr.9b06214.

[28] Zhu, Chenxue, Jie Bian, Yuxi Li, Junshen Liu, Xunyong Liu, Xuezhen Gao, Guiying Li, and Yi Liu. "A novel and ultrasensitive yellow to taupe brown colorimetric sensing and removal method for Hg(II) based on the thermosensitive poly (*N*-isopropyl acrylamide) stabilized silver nanoparticles." *New Journal of Chemistry* 43 (2019): 15879–15885. https://doi.org/10.1039/C9NJ03955D.

[29] Song, Zihui, Xueqi Chen, Xinchao Gong, Xing Gao, Qian Dai, Tat Thang Nguyen, and Minghui Guo. "Luminescent carbon quantum dots/nanofibrillated cellulose composite aerogel for monitoring adsorption of heavy metal ions in water." *Optical Materials* 100 (2020): 109642. https://doi.org/10.1016/j.optmat.2019.109642.

[30] Mohammed, Nishil, Avijit Baidya, Vasanthanarayan Murugesan, Avula Anil Kumar, Mohd Azhardin Ganayee, Jyoti Sarita Mohanty, Kam Chiu Tam, and Thalappil Pradeep. "Diffusion-controlled simultaneous sensing and scavenging of heavy metal ions in water using atomically precise cluster-cellulose nanocrystal composites." *ACS Sustainable Chemistry & Engineering* 4 (2016): 6167–6176. doi.org/10.1021/acssuschemeng.6b01674.

[31] Jiang, Xiangyang, Jian Xia, and Xiaogang Luo. "Simple, rapid, and highly sensitive colorimetric sensor strips from a porous cellulose membrane stained with Victoria blue b for efficient detection of trace Cd(II) in water." *ACS Sustainable Chemistry & Engineering* 8 (2020): 5184–5191. doi.org/10.1021/acssuschemeng.9b07614.

[32] Li, Ming, Bo Li, Li Zhou, Yuling Zhang, Qianyong Cao, Rujie Wang, and Huining Xiao. "Fluorescence-sensitive adsorbent based on cellulose using for mercury detection and removal from aqueous solution with selective "on-off" response." *International Journal of Biological Macromolecules* 132 (2019): 1185–1192. doi.org/10.1016/j.ijbiomac.2019.04.048.

[33] Yu, Shujuan, Wei Li, Yuki Fujii, Taro Omura, and Hideto Minami. "Fluorescent spherical sponge cellulose sensors for highly selective and semiquantitative visual analysis: detection of Hg^{2+} and Cu^{2+} ions." *ACS Sustainable Chemistry & Engineering* 7 (2019): 19157–19166. doi.org/10.1021/acssuschemeng.9b05142.

[34] Albukhari, Soha M., Muhammad Ismail, Kalsoom Akhtar, and Ekram Y. Danish. "Catalytic reduction of nitrophenols and dyes using silver nanoparticles@ cellulose polymer paper for the resolution of wastewater treatment challenges." *Colloids and Surfaces A: Physicochemical and Engineering Aspects* 577 (2019): 548–561. doi.org/10.1016/j.colsurfa.2019.05.058.

[35] Gu, Jin, and Anthony Dichiara. "Hybridization between cellulose nanofibrils and faceted silver nanoparticles used with surface enhanced Raman scattering for trace dye detection." *International Journal of Biological Macromolecules* 143 (2020): 85–92. https://doi.org/10.1016/j.ijbiomac.2019.12.018.

[36] Wang, Lei, Cunzhi Zhang, Hui He, Hongxiang Zhu, Wei Guo, Shile Zhou, Shuangfei Wang, Joe R. Zhao, and Jian Zhang. "Cellulose-based colorimetric sensor with N, S sites for Ag^+ detection." *International Journal of Biological Macromolecules* (2020). https://doi.org/10.1016/j.ijbiomac.2020.07.018.

[37] Nawaz, Haq, Weiguo Tian, Jinming Zhang, Ruonan Jia, Zhangyan Chen, and Jun Zhang. "Cellulose-based sensor containing phenanthroline for the highly selective and rapid detection of Fe^{2+} ions with naked eye and fluorescent dual modes." *ACS Applied Materials & Interfaces* 10 (2018): 2114–2121. doi.org/10.1021/acsami.7b17342.

[38] Nawaz, Haq, Jinming Zhang, Weiguo Tian, Kunfeng Jin, Ruonan Jia, Tiantian Yang, and Jun Zhang. "Cellulose-based fluorescent sensor for visual and versatile detection of amines and anions." *Journal of Hazardous Materials* 387 (2020): 121719. doi.org/10.1016/j.jhazmat.2019.121719.

[39] Park, Bumjun, Sung-Min Kang, Go-Woon Lee, Cheol Hwan Kwak, Muruganantham Rethinasabapathy, and Yun Suk Huh. "Fabrication of $CsPbBr_3$ perovskite quantum dots/cellulose-based colorimetric sensor: dual-responsive onsite detection of chloride and iodide ions." *Industrial & Engineering Chemistry Research* 59 (2019): 793–801. doi.org/10.1021/acs.iecr.9b05946.

[40] Li, Meng, Zhijiang Liu, Hui-Chen Wang, Adam C. Sedgwick, Jordan E. Gardiner, Steven D. Bull, Hui-Ning Xiao, and Tony D. James. "Dual-function cellulose composites for fluorescence detection and removal of fluoride." *Dyes and Pigments* 149 (2018): 669–675. https://doi.org/10.1016/j.dyepig.2017.11.033.

[41] Wang, Hui, Junfeng Wang, Gang Liu, Zhankuan Zhang, and Xiaopeng Hou. "Electrochemical sensing of Pb(II) and Cd(II) in decorative material of wood panel using nano-cellulose paper-based electrode modified using graphene/multi-walled carbon nanotubes/bismuth film." *International Journal of Electrochemical Science* 14 (2019): 11253–11266. doi: 10.20964/2019.12.13.

[42] Ram, Bhagat, Ghanshyam S. Chauhan, Akshita Mehta, Reena Gupta, and Kalpana Chauhan. "Spherical nanocellulose-based highly efficient and rapid multifunctional naked-eye Cr(VI) ion chemosensor and

adsorbent with mild antimicrobial properties." *Chemical Engineering Journal* 349 (2018): 146–155. doi. org/10.1016/j.cej.2018.05.085.

[43] Riva, Laura, Andrea Fiorati, Aurora Sganappa, Lucio Melone, Carlo Punta, and Massimo Cametti. "Naked-eye heterogeneous sensing of fluoride ions by co-polymeric nanosponge systems comprising aromatic-imide-functionalized nanocellulose and branched polyethyleneimine." *ChemPlusChem* 84 (2019): 1512–1518. https://doi.org/10.1002/cplu.201900348.

[44] Daniyal, Wan Mohd Ebtisyam Mustaqim Mohd, Yap Wing Fen, Jaafar Abdullah, Silvan Saleviter, and Nur Alia Sheh Omar. "Preparation and characterization of hexadecyltrimethylammonium bromide modified nanocrystalline cellulose/graphene oxide composite thin film and its potential in sensing copper ion using surface plasmon resonance technique." *Optik* 173 (2018): 71–77. doi.org/10.1016/j. ijleo.2018.08.014.

[45] Guo, Wei, Hui He, Hongxiang Zhu, Xudong Hou, Xingjuan Chen, Shile Zhou, Shuangfei Wang, Lingtao Huang, and Jiehan Lin. "Preparation and properties of a biomass cellulose-based colorimetric sensor for Ag+ and Cu²⁺." *Industrial Crops and Products* 137 (2019): 410–418. https://doi.org/10.1016/j. indcrop.2019.05.044.

[46] Zhou, Shile, Hui He, Wei Guo, Hongxiang Zhu, Fei Xue, Meixiao Cheng, Jiehan Lin, Lei Wang, and Shuangfei Wang. "Structural design of a high sensitivity biomass cellulose-based colorimetric sensor and it's *in situ* visual recognition mechanism for Cu²⁺." *Carbohydrate Polymers* 219 (2019): 95–104. doi. org/10.1016/j.carbpol.2019.04.094.

[47] Ye, Xiu, Yanhui Kang, and Jinping Zhou. "Rhodamine labeled cellulose nanocrystals as selective "naked-eye" colorimetric and fluorescence sensor for Hg²⁺ in aqueous solutions." *Cellulose* 27 (2020): 5197–5210. https://doi.org/10.1007/s10570-020-03126-5.

[48] Tungsombatvisit, Nilobol, Thitirat Inprasit, Dini Rohmawati, and Penwisa Pisitsak. "Rhodamine derivative-based cellulose acetate electrospun colorimetric sensor for Cu²⁺ sensing in water: effects of alkaline treatment." *Fibers and Polymers* 20 (2019): 481–489. https://doi.org/10.1007/s12221-019-8907-y.

[49] Yang, Luyu, Lei Yang, Shuaining Wu, Feng Wei, Ying Hu, Xuran Xu, Lei Zhang, and Dongping Sun. "Three-dimensional conductive organic sulfonic acid co-doped bacterial cellulose/polyaniline nanocomposite films for detection of ammonia at room temperature." *Sensors and Actuators B: Chemical* (2020): 128689. https://doi.org/10.1016/j.snb.2020.128689.

[50] Sarkar, Chandrani, Angshuman Ray Chowdhuri, Amit Kumar, Dipranjan Laha, Subhadra Garai, Jui Chakraborty, and Sumanta Kumar Sahu. "One pot synthesis of carbon dots decorated carboxymethyl cellulose-hydroxyapatite nanocomposite for drug delivery, tissue engineering and Fe³⁺ ion sensing." *Carbohydrate Polymers* 181 (2018): 710–718. https://doi.org/10.1016/j.carbpol.2017.11.091.

[51] Ismail, Muhammad, M. I. Khan, Kalsoom Akhtar, Jongchul Seo, Murad Ali Khan, Abdullah M. Asiri, and Sher Bahadar Khan. "Phytosynthesis of silver nanoparticles: naked eye cellulose filter paper dual mechanism sensor for mercury ions and ammonia in aqueous solution." *Journal of Materials Science: Materials in Electronics* 30 (2019): 7367–7383. doi.org/10.1007/s10854-019-01049-x.

[52] Zhang, Lianwei, Ruijia Wang, Rui Liu, Xiaolin Du, Ranju Meng, Lin Liu, and Juming Yao. "Rapid capture and visual detection of copper ions in aqueous solutions and biofluids using a novel cellulose-Schiff base." *Cellulose* 25 (2018): 6947–6961. https://doi.org/10.1007/s10570-018-2083-x.

[53] Song, Ruyuan, Qing Zhang, Youlu Chu, Lei Zhang, Hongqi Dai, and Weibing Wu. "Fluorescent cellulose nanocrystals for the detection of lead ions in complete aqueous solution." *Cellulose* 26 (2019): 9553–9565. https://doi.org/10.1007/s10570-019-02760-y.

[54] Ahmadian-Fard-Fini, Shahla, Davood Ghanbari, Omid Amiri, and Masoud Salavati-Niasari. "Electro-spinning of cellulose acetate nanofibers/Fe/carbon dot as photoluminescence sensor for mercury (II) and lead (II) ions." *Carbohydrate Polymers* 229 (2020): 115428. https://doi.org/10.1016/j. carbpol.2019.115428.

[55] Fan, Jiang, Sufeng Zhang, Yongshe Xu, Ning Wei, Ben Wan, Liwei Qian, and Ye Liu. "A polyethylenimine/salicylaldehyde modified cellulose Schiff base for selective and sensitive Fe³⁺ detection." *Carbohydrate Polymers* 228 (2020): 115379. doi.org/10.1016/j.carbpol.2019.115379.

[56] Priya, T., N. Dhanalakshmi, V. Karthikeyan, and N. Thinakaran. "Highly selective simultaneous trace determination of Cd²⁺ and Pb²⁺ using porous graphene/carboxymethyl cellulose/fondaparinux nanocomposite modified electrode." *Journal of Electroanalytical Chemistry* 833 (2019): 543–551. doi. org/10.1016/j.jelechem.2018.12.039.

[57] Daniyal, Wan Mohd Ebtisyam Mustaqim Mohd, Yap Wing Fen, Jaafar Abdullah, Amir Reza Sadrolhosseini, Silvan Saleviter, and Nur Alia Sheh Omar. "Label-free optical spectroscopy for characterizing binding properties of highly sensitive nanocrystalline cellulose-graphene oxide-based nanocomposite towards nickel ion." *Spectrochimica Acta Part A: Molecular and Biomolecular Spectroscopy* 212 (2019): 25–31. doi.org/10.1016/j.saa.2018.12.031.

[58] Rajawat, Deepak Singh, Abhishek Kardam, Shalini Srivastava, and Soami Piara Satsangee. "Nanocellulosic fiber-modified carbon paste electrode for ultra-trace determination of Cd(II) and Pb(II) in aqueous solution." *Environmental Science and Pollution Research* 20 (2013): 3068–3076. doi. org/10.1007/s11356-012-1194-4.

[59] Danwittayakul, Supamas, and Phitchaya Muensri. "Polyethyleneimine coated polyacrylonitrile cellulose membrane for colorimetric copper (II) determination." *Journal of Water Chemistry and Technology* 42 (2020): 22–29. doi: 10.3103/S1063455X20010075.

[60] Taheri, M., F. Ahour, and S. Keshipour. "Sensitive and selective determination of Cu^{2+} at D-penicillamine functionalized nano-cellulose modified pencil graphite electrode." *Journal of Physics and Chemistry of Solids* 117 (2018): 180–187. doi.org/10.1016/j.jpcs.2018.02.035.

[61] Xue, Fei, Hui He, Hongxiang Zhu, Huanhuan Huang, Qi Wu, and Shuangfei Wang. "Structural design of a cellulose-based solid amine adsorbent for the complete removal and colorimetric detection of Cr(VI)." *Langmuir* 35 (2019): 12636–12646. doi.org/10.1021/acs.langmuir.9b01788.

[62] Teodoro, Kelcilene BR, Flávio M. Shimizu, Vanessa P. Scagion, and Daniel S. Correa. "Ternary nanocomposites based on cellulose nanowhiskers, silver nanoparticles and electrospun nanofibers: use in an electronic tongue for heavy metal detection." *Sensors and Actuators B: Chemical* 290 (2019): 387–395. doi.org/10.1016/j.snb.2019.03.125.

[63] Wu, Qian, Xiaojie Wang, Sefiu Abolaji Rasaki, Tiju Thomas, Chuanxi Wang, Chi Zhang, and Minghui Yang. "Yellow-emitting carbon-dots-impregnated carboxy methyl cellulose/poly-vinyl-alcohol and chitosan: stable, freestanding, enhanced-quenching Cu^{2+}-ions sensor." *Journal of Materials Chemistry C* 6 (2018): 4508–4515. doi.org/10.1039/C8TC00660A.

14

Nanocellulose-Based Materials for Photocatalysis in Water Treatment Applications

Meena Nemiwal
Malaviya National Institute of Technology, Jaipur, India

Ankita Dhillon
Banasthali Vidyapith, Banasthali, India

CONTENTS

14.1 Introduction

Cellulose is the most abundantly naturally existing biopolymer exploited as a reinforcing agent for composites of fiber and thermoplastic materials [1, 2]. Nanocelluloses (NCs) are extracted from the celluloses in nanoscale dimensions in various forms. NC is a material that can be renewed and shows a high surface area, high strength, unique surface chemistry, and chemical inertness [3]. It exists as cellulose nanocrystals (CNCs), and another form is cellulose nanofibrils (CNFs) with a rod-like shape. Its lengths vary ranging from 100 nm to 2000 nm, and diameter varies between 2 and 20 nm [4]. The various forms and sizes of NC depend on the source of the cellulose and different preparation routes [5]. Hydrochloric or sulfuric acid is used in acid hydrolysis during preparation processes. Electrostatically stable CNCs are developed by the interaction of sulfuric acids with hydroxyl groups that lie on the surface of cellulose to form esters of sulfate, which makes the surface negatively charged. Whereas hydrochloric acid is used, the CNCs produced will be with low surface charge density. Therefore, depending on the preparation routes, various chemical modifications can be carried out [6]. CNCs are bio-based nanoscale materials, which have attracted researchers' interest in academia and industry because of its excellent physicochemical and structural properties like high mechanical strength, optical transparency, tunable surface chemistry, and low-density renewability, biodegradability, and biocompatibility [7]. Thus, CNCs are emerging as a potential tool for various applications in electronics [8], pharmaceuticals, biomedical, supercapacitors, membranes [9], and nanocomposites. Another class is bacterial nanocelluloses (BNCs), synthesized from different bacteria and possess a very high crystallinity [10].

Owing to the specific properties such as hydrophilic surface chemistry, chemical inertness, high strength, and high surface area, NC has been emerging as a promising material for water remediation application [11]. Although several approaches have been explored for water remediation, photocatalysis has been emerging as a promising field due to its greener approach. Semiconductor materials have been explored as photocatalysts, but their agglomeration and reusability limit their application [12]. Therefore, their composites with cellulose material are gaining attention because of improved electron-hole pair separation. Cellulose materials immobilize the photocatalysts by providing structural support and reduce the contamination of photocatalysts, and hence the performance for photocatalysis is enhanced and becomes recyclable [13]. Researchers have explored several semiconductor materials to develop semiconductor/cellulose composites such as ZnO, TiO_2, WO_3, ZnS, CdS, BiOI, BiOBr, BiOCl, Ag_3PO_4, $AgVO_4$, AgI, AgBr, and $AgCrO_4$ [14]. It has been observed that composites of semiconductors and cellulose improved photocatalytic activity by 1.3–3.5 times than that of the individual semiconductor or cellulose. Nowadays, non-metallic semiconductors like carbon nitride and graphite have also been employed to modify photocatalysis [15] further. In this view, this chapter covers the basic mechanism of photocatalysis and cellulose-based photocatalysts used for the degradation of organic pollutants in water and wastewater.

14.2 Fundamentals of Photocatalytic Oxidation

Utilization of sunlight as a source of renewable energy for photocatalytic degradation of organic pollutant in water by exploiting heterogeneous photocatalysts has been emerging as a promising method [16]. Photoreactions are accelerated due to photocatalysts' availability, and it should be non-toxic, cheap, photoactive, and able to use solar light efficiently. In the advanced oxidation process (AOP), water molecules are generally oxidized into •OH radical species, and this process is carried out on the catalyst surface. Another active species produced from the oxygen is the radical of superoxide ($O_2^{•-}$) ion [17]. For a photocatalyst to be active for photochemical reaction, its valence band (VB) redox potential should be more negative than that is required for the generation of hydroxyl radical. The conduction band (CB) should be more positive than that needed to generate superoxide radical from oxygen. Although many metal oxides and composites of metal oxides have been explored for photocatalysts, TiO_2 and ZnO are the highest used semiconductor materials [18].

The basic mechanism includes the excitation of electrons from VB to CB upon irradiation of photons bearing energy higher than the semiconductor bandgap. As a result, holes are generated in the VB, and the CB has the availability of electrons [19]. Thus, the e^-/h^+ pair generated either recombine or interact with water and O_2 in VB and CB, respectively [20]. Value of chemical potential for e^- is + 0.5 to −1.5 V to Normal Hydrogen Electrode (NHE), making it feasible to interact electron-rich species due to strong reduction potential, and h^+ has chemical potential value +1.0 to +3 V to NHE that shows its strong oxidative power for interaction with electron-deficient species. They produced highly reactive species that can degrade organic pollutants into other harmless products [21]. Excited electrons are exploited for producing superoxide anion ($O_2^{•-}$) from molecular oxygen, and water molecules take up holes for the production of hydroxyl radical (•OH) [22]. These active species superoxide anion ($O_2^{•-}$) and hydroxyl radical (•OH) exhibit the potential to degrade the structures of organic compounds, lipids, proteins, carbohydrates, biomolecules, inorganic compounds, nucleic acids, etc. and act as disinfectant in water (Fig. 14.1) [23]. However, e^-/h^+ pair recombination causes instability in semiconductors, and by forming their composites with cellulose-based material as substrate, electron availability could be increased.

14.3 Nanocellulose-Based Materials for Photocatalysis
in Water Treatment Applications

Nowadays, composites of organic- and inorganic-based nanostructures are being exploited as photocatalysts with specific thermal, mechanical, electrical, and optical properties. Since using environment-friendly biomaterials for replacing petroleum-based feed stokes is the only way for a clean and

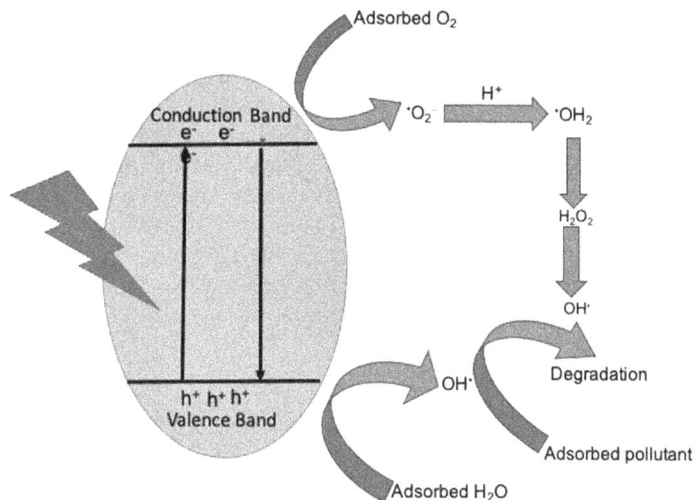

FIGURE 14.1 Mechanistic view showing photocatalysis and degradation of dye (from Ref. [23]).

sustainable environment, cellulose-based materials have shown potential application in this area because of its abundance and biocompatibility. Material based on cellulose exhibits exceptional properties such as hydrophilicity, chirality, large surface area, and wide chemical variability [24].

NCs have been employed as photocatalysts under irradiation of UV or visible light for organic pollutants degradation in water. Their photocatalytic performances have been found to enhance by incorporating nanoparticles (NPs) with semiconductor properties. Cellulose provides excellent host properties to other NPs by giving stability and controlling the NPs' growth without disturbing their morphology [25]. Tremendous research has been carried out using NC-based materials like fibers, membrane, hybrid materials, and thin film to act as photocatalyst for water remediation [26].

In hybrid materials of cellulose and NP (TiO_2, ZnO, graphene oxide (GO), Fe_2O_3), the cellulose is adsorbed on metal oxide. It incorporates its hydroxyl (—OH) groups, resulting in increased surface area and widening of the wavelength response extended to the visible range. Similarly, BNCs are also employed for composite formation with metal oxide NPs because of their high surface-to-volume ratio, high mechanical strength, and excellent hydrophilic properties [27]. Based on NPs used for hybrid formation, cellulose-based materials may be categorized as follows.

14.3.1 TiO_2/Nanocellulose Composite as Photocatalyst

TiO_2 metal oxide is a semiconductor that has been extensively employed for various applications like the splitting of water for hydrogen evolution reactions, photocatalysis, etc., owing to high thermal stability, nontoxicity, low cost, and easy to handle [28]. In this view, Sun et al. developed cellulose/TiO_2 composite monolith for methylene blue (MB) degradation through photocatalysis and observed 99% degradation within 60 minutes under UV light irradiation with 90% retention of efficiency after ten cycles (Fig. 14.2) [29].

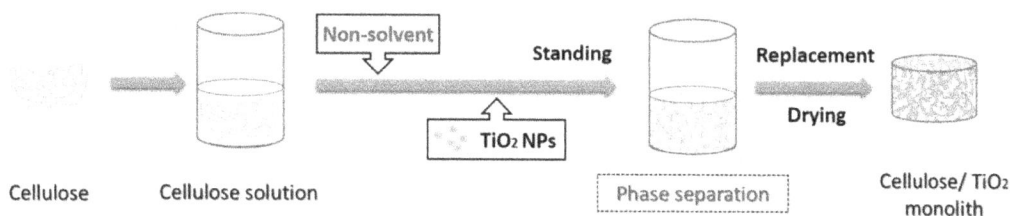

FIGURE 14.2 Schematic diagram showing synthesis of cellulose/TiO_2 monoliths. NPs: nanoparticles (from Ref. [29]).

In further similar research, microfibrillated cellulose (MFC)-polyamide-amine-epichlorohydrin (PAE)-TiO_2 NPs composite was developed by a two-step mixing process. The resultant composite was reusable, reproducible, uniform, and flexible. This composite exhibited a 95% photocatalytic degradation of methyl orange within 150 minutes under UV light irradiation. The role of polyamide-amine was to keep NPs and restrict their release into the environment [25].

TiO_2 has also been explored with BNC for methyl orange degradation in water. TiO_2 was supported on the BNC, and the composite with rare earth elements was doped into it by the sol-gel method. The results showed 100% dye degradation with (IV)-doped composite, whereas 79.7% with (III)-doped composite and the optimal dosages were 2 and 5 mmol/L for (IV) and (III), respectively [30]. Besides this, Brandes et al. synthesized spherical BC/TiO_2 nanocomposites for efficient MB degradation from the water. They were found to degrade MB 70.83 and 89.58% after 35 minutes *in situ* and *ex situ*, respectively [31].

14.3.2 ZnO/Nanocellulose Composite as Photocatalyst

CNCs prepared from natural cellulose by acid hydrolysis are rod-shaped of various nanosize ranging from 1 to 100 nm, mainly depending on the origin [32]. CNCs have shown applications in photocatalysis and the biomedical field, but researchers have modified their properties by chemical modifications and doping with other NPs such as ZnO.

ZnO NPs exhibit high catalytic activity, optoelectronic properties, and disinfectant properties for pathogens of a wide variety [33]. The bandgap value for ZnO is 3.3 eV, which resembles that of TiO_2 (3.2 eV for anatase), making it a promising photocatalyst and may be exploited for many applications such as solar cells, gas sensors, UV photodetectors, optical devices, and batteries [34, 35]. Recently, ZnO NPs have shown application to treat wastewater that is the degradation of organic pollutants and other effluents [36, 37]. However, ZnO's high surface energy and high surface area lead to aggregation of NPs that limits its use. Thus, nanoscale dispersion of ZnO is improved by incorporating nanofibrillar materials (cellulose) during their synthesis and showed excellent applications in many areas [38].

Guan et al. synthesized hybrids of ZnO/CNC with varying morphologies of flower-like shape, thin-sheet, and nearly spherical using bamboo CNC as a precursor (Fig. 14.3) [39]. The resultant hybrid showed excellent adsorption for cationic dyes in the treatment of wastewater. ZnO/CNC 8.5 showed high

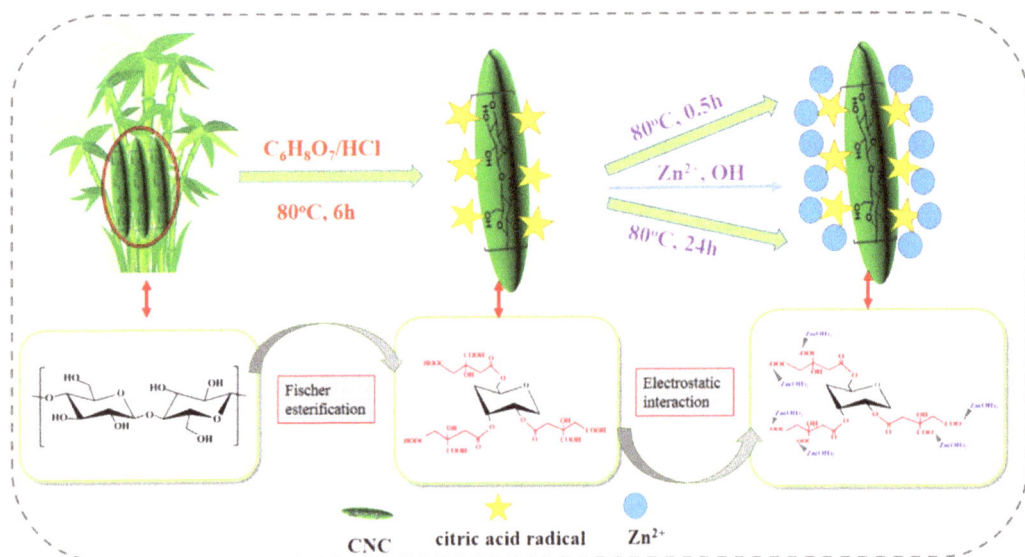

FIGURE 14.3 Schematic presentation of synthesis of ZnO/CNC hybrids with formation mechanism [Reproduced with permission from Ref. [39]. © Elsevier 2019].

and facile dye removal of malachite green, 99.02% and MB, 93.55%, and adsorption capacity reached up to 49.51 mg/g for MG and 46.77 mg/g for MB [39].

Similarly, the composite of NC with ZnO has been developed by using oil palm empty fruit bunches as a source of NC by alkaline and acid treatment. These NCs have acted as host polymers for the fabrication of ZnO NPs. The percentage crystalline index was improved from 35.7% to 43.3% and 53.3% by alkaline and acid treatment. The observed photocatalytic properties of ZnO/CN composite were more than that of the pure ZnO [40]. The composites CNC/ZnO have been investigated to show antibacterial activity with improved photocatalytic performances [41]. Besides CNC, BNC has also attracted researchers for its application as a photocatalyst for dye degradation [42].

Wei et al. developed hexagonal and crystallized wurtzite ZnO with NC using the hydrothermal method and employed it to degrade MB. The degradation efficiency was 94.4% within 30 minutes under irradiation of UV light of 6 W, and the reusability of the photocatalyst was also very good up to eight cycles. Cotton was used as a source of NC, which is renewable and environment-friendly [43]. Cellulose material provides the base as a substrate to immobilize the semiconductor NPs. Xiao et al. exploited TEMPO-oxidized (2,2,6,6-tetramethyl-1-piperidinyloxy as TEMPO) cellulose as a reactive template carbon/ZnO composite investigated for MB degradation up to 96.11% within 120 minutes [44].

14.3.3 Graphene Oxide/Nanocellulose Composite as Photocatalyst

GO has been widely used for the composite formation for improving the transmission capacity of the e^-/h^+ pairs. The benzene ring's conjugate effect in the GO increases the rate of photodegradation performance. Tu et al. developed a composite film of regenerated cellulose (RC) and GO incorporated with

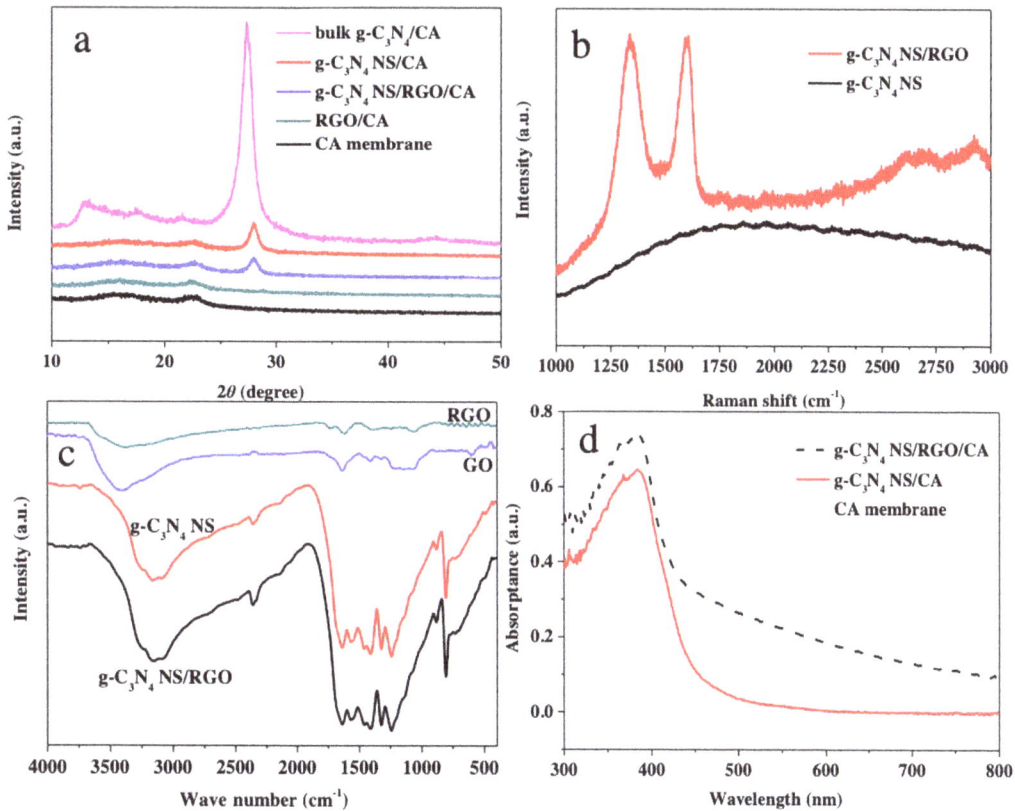

FIGURE 14.4 g-C_3N_4 NS/RGO/CA composite: (a) XRD patterns, (b) Raman spectra, (c) FTIR spectra, and (d) UV-vis absorption [Reproduced with permission from Ref. [15]. © Elsevier 2016].

Cu_2O NPs in this film's micropores. The resultant nanocomposite was employed for the methyl orange dye degradation under visible-light irradiation. The rate of degradation was found to be improved from 2.0 mg/h/g/cat to 6.5 mg/h/g/cat [45]. Different solvents have been exploited as dispersing agents to enhance these films' electrical, thermal, and mechanical properties. *N*-methylmorpholine-*N*-oxide (NMMO) monohydrate solvent was employed as monohydrate solvent for the fabrication of GO/cellulose composite films, and rheological properties were enhanced [46].

14.3.4 g-C_3N_4/Nanocellulose Composite as Photocatalyst

Graphitic carbon nitride is a non-metallic, polymeric semiconductor with a bandgap of 2.7 eV, which is suitable for fulfilling the thermodynamic need for hydrogen and oxygen production through water splitting [47, 48]. Many research groups are working for cleaner production of hydrogen by exploiting g-C_3N_4 [49]. Although the photocatalytic activity of the g-C_3N_4 for degradation of organic pollutants is better than that of the traditional photocatalysts like TiO_2, their composites with cellulose have shown remarkable performances. Zhao et al. synthesized composite membrane of graphitic carbon nitride nanosheet, reduced GO, and cellulose acetate (g-C_3N_4 NS/RGO/CA). They exploited for rhodamine B dye degradation under visible light irradiation (Fig. 14.4) [15]. The observed photocatalytic performances were better than observed for g-C_3N_4 alone [15].

14.4 Conclusion

This chapter reported the photocatalytic activity of cellulose-based materials and their composites with semiconductors such as TiO_2, ZnO, etc. The investigations showed that doping of cellulose exhibited enhanced photocatalytic activity with thermal stability and appreciable hydrophilicity. Although these studies showed the fundamental understanding of the photocatalysis by composites of cellulose and semiconductors, there is still room to explore this area. Though it is assumed that the basic mechanism to explain cellulose and semiconductor composite stability is chemical bond or electrostatic force between cellulose and semiconductor, different binding energies show that there should be some other properties for fixing of cellulose and semiconductor. Thus, a deep understanding is required to explain the mechanism that can be explored by combining the Density Functional Theory (DFT) with experimental results. Photostability of cellulose in the presence of semiconductors is still an overlooked area that needs in-depth study. In summary, for efficient photocatalysis by cellulose-based material, smart, robust, and sustainable composites should be developed and commercialized.

Acknowledgment

The author Meena Nemiwal is thankful to the Malaviya National Institute of Technology (MNIT), Jaipur, for its support.

REFERENCES

[1] Dufresne, Alain. "Cellulose nanomaterial reinforced polymer nanocomposites." *Current Opinion in Colloid & Interface Science* 29 (2017): 1–8. doi: https://doi.org/10.1016/j.cocis.2017.01.004.

[2] Fardioui, Meriem, Mohamed El Mehdi Mekhzoum, and Rachid Bouhfid. "Bionanocomposite materials based on chitosan reinforced with nanocrystalline cellulose and organo-modified montmorillonite." In *Nanoclay reinforced polymer composites*, eds. M. Jawaid, A. el K. Qaiss, and R. Bouhfidpp, 167–194. Singapore: Springer. 2016.

[3] Mondal, Subrata. "Preparation, properties and applications of nanocellulosic materials." *Carbohydrate Polymers* 163 (2017): 301–316. doi: https://doi.org/10.1016/j.carbpol.2016.12.050.

[4] Xie, Hongxiang, Haishun Du, Xianghao Yang, and Chuanling Si. "Recent strategies in preparation of cellulose nanocrystals and cellulose nanofibrils derived from raw cellulose materials." *International Journal of Polymer Science* 2018 (2018). doi: https://doi.org/10.1155/2018/7923068.

[5] Isogai, Akira, and Yaxin Zhou. "Diverse nanocelluloses prepared from TEMPO-oxidized wood cellulose fibers: nanonetworks, nanofibers, and nanocrystals." *Current Opinion in Solid State and Materials Science* 23 (2019): 101–106. doi: https://doi.org/10.1016/j.cossms.2019.01.001.

[6] Niu, Fuge, Mengya Li, Qi Huang, Xiuzhen Zhang, Weichun Pan, Jiansheng Yang, and Jianrong Li. "The characteristic and dispersion stability of nanocellulose produced by mixed acid hydrolysis and ultrasonic assistance." *Carbohydrate Polymers* 165 (2017): 197–204. doi: https://doi.org/10.1016/j.carbpol.2017.02.048.

[7] Li, Ya-Yu, Bin Wang, Ming-Guo Ma, and Bo Wang. "Review of recent development on preparation, properties, and applications of cellulose-based functional materials." *International Journal of Polymer Science* 2018 (2018). doi: https://doi.org/10.1155/2018/8973643.

[8] Chen, Chaoji, and Liangbing Hu. "Nanocellulose toward advanced energy storage devices: structure and electrochemistry." *Accounts of Chemical Research* 51 (2018): 3154–3165. doi: https://doi.org/10.1021/acs.accounts.8b00391.

[9] Voisin, Hugo, Lennart Bergstrom, Peng Liu, and Aji P. Mathew. "Nanocellulose-based materials for water purification." *Nanomaterials* 7 (2017): 57. doi: https://doi.org/10.3390/nano7030057.

[10] Faria-Tischer, Paula C. S., Renato M. Ribeiro-Viana, and Cesar Augusto Tischer. "Bio-based nanocomposites: strategies for cellulose functionalization and tissue affinity studies." In *Materials for biomedical engineering*, eds. V. Grumezescu and A. M. Grumezescu, 205–244. Oxfordshire, UK: Elsevier. 2019.

[11] Jimenez Saelices, Clara, and Isabelle Capron. "Design of pickering micro-and nanoemulsions based on the structural characteristics of nanocelluloses." *Biomacromolecules* 19 (2018): 460–469. doi: https://doi.org/10.1021/acs.biomac.7b01564.

[12] Lee, Seul-Yi, and Soo-Jin Park. "TiO_2 photocatalyst for water treatment applications." *Journal of Industrial and Engineering Chemistry* 19 (2013): 1761–1769. doi: https://doi.org/10.1016/j.jiec.2013.07.012.

[13] Kim, KeeHong, Pravin G. Ingole, SangHee Yun, WonKil Choi, JongHak Kim, and HyungKeun Lee. "Water vapor removal using CA/PEG blending materials coated hollow fiber membrane." *Journal of Chemical Technology & Biotechnology* 90 (2015): 1117–1123. doi: https://doi.org/10.1002/jctb.4421.

[14] Zhao, Yilin, Yaoqiang Wang, Gang Xiao, and Haijia Su. "Fabrication of biomaterial/TiO_2 composite photocatalysts for the selective removal of trace environmental pollutants." *Chinese Journal of Chemical Engineering* 27 (2019): 1416–1428. doi: https://doi.org/10.1016/j.cjche.2019.02.003.

[15] Zhao, Huanxin, Shuo Chen, Xie Quan, Hongtao Yu, and Huimin Zhao. "Integration of microfiltration and visible-light-driven photocatalysis on g-C_3N_4 nanosheet/reduced graphene oxide membrane for enhanced water treatment." *Applied Catalysis B: Environmental* 194 (2016): 134–140. doi: https://doi.org/10.1016/j.apcatb.2016.04.042.

[16] Ahmed, Syed Nabeel, and Waseem Haider. "Heterogeneous photocatalysis and its potential applications in water and wastewater treatment: a review." *Nanotechnology* 29 (2018): 342001. doi: https://doi.org/10.1088/1361-6528/aac6ea.

[17] Kanakaraju, Devagi, Beverley D. Glass, and Michael Oelgemöller. "Advanced oxidation process-mediated removal of pharmaceuticals from water: a review." *Journal of Environmental Management* 219 (2018): 189–207. doi: https://doi.org/10.1016/j.jenvman.2018.04.103.

[18] Byrne, Ciara, Gokulakrishnan Subramanian, and Suresh C. Pillai. "Recent advances in photocatalysis for environmental applications." *Journal of Environmental Chemical Engineering* 6 (2018): 3531–3555. doi: https://doi.org/10.1016/j.jece.2017.07.080.

[19] Brunet, Léna, Delina Y. Lyon, Ernest M. Hotze, Pedro J. J. Alvarez, and Mark R. Wiesner. "Comparative photoactivity and antibacterial properties of C60 fullerenes and titanium dioxide nanoparticles." *Environmental Science & Technology* 43 (2009): 4355–4360. doi: https://doi.org/10.1021/es803093t.

[20] Han, F., V. S. R. Kambala, R. Dharmarajan, Y. Liu, and R. Naidu. "Photocatalytic degradation of azo dye acid orange 7 using different light sources over Fe^{3+}-doped TiO_2 nanocatalysts." *Environmental Technology & Innovation* 12 (2018): 27–42. doi: https://doi.org/10.1016/j.eti.2018.07.004.

[21] Magalhaes, Pedro, Luisa Andrade, Olga C. Nunes, and Adélio Mendes. "Titanium dioxide photocatalysis: fundamentals and application on photoinactivation." *Reviews on Advanced Materials Science* 51 (2017): 91–129.

[22] Lin, Hsiu-Fen, Shih-Chieh Liao, and Sung-Wei Hung. "The DC thermal plasma synthesis of ZnO nanoparticles for visible-light photocatalyst." *Journal of Photochemistry and Photobiology A: Chemistry* 174 (2005): 82–87. doi: https://doi.org/10.1016/j.jphotochem.2005.02.015.

[23] Singh, Pardeep, Pooja Shandilya, Pankaj Raizada, Anita Sudhaik, Abolfazl Rahmani-Sani, and Ahmad Hosseini-Bandegharaei. "Review on various strategies for enhancing photocatalytic activity of graphene-based nanocomposites for water purification." *Arabian Journal of Chemistry* 13 (2020): 3498–3520. doi: https://doi.org/10.1016/j.arabjc.2018.12.001.

[24] Osong, Sinke H., Sven Norgren, and Per Engstrand. "Processing of wood-based microfibrillated cellulose and nanofibrillated cellulose, and applications relating to papermaking: a review." *Cellulose* 23 (2016): 93–123. doi: https://doi.org/10.1007/s10570-015-0798-5.

[25] Garusinghe, Uthpala M., Vikram S. Raghuwanshi, Warren Batchelor, and Gil Garnier. "Water resistant cellulose-titanium dioxide composites for photocatalysis." *Scientific Reports* 8 (2018): 1–13. doi: https://doi.org/10.1038/s41598-018-20569-w.

[26] Nevo Y., Peer N., Yochelis S., Igbaria M., Meirovitch S., and Shoseyov O. et al. "Nano bio optically tunable composite nanocrystalline cellulose films." *RSC Advances* 5 (2015): 7713–9. doi: https://doi.org/10.1039/c4ra11840e.

[27] Xu, Mengya, He Wang, Gang Wang, Lin Zhang, Gang Liu, Zhixiang Zeng, Tianhui Ren, Wenjie Zhao, Xuedong Wu, and Qunji Xue. "Study of synergistic effect of cellulose on the enhancement of photocatalytic activity of ZnO." *Journal of Materials Science* 52 (2017): 8472–8484. doi: https://doi.org/10.1007/s10853-017-1106-6.

[28] Wittmar, Alexandra, Hanna Thierfeld, Steffen Köcher, and Mathias Ulbricht. "Routes towards catalytically active TiO_2 doped porous cellulose." *RSC Advances* 5 (2015): 35866–35873. doi: https://doi.org/10.1039/c5ra03707g.

[29] Sun, Xiaoxia, Kunpeng Wang, Yu Shu, Fangdong Zou, Boxing Zhang, Guangwu Sun, Hiroshi Uyama, and Xinhou Wang. "One-pot route towards active TiO_2 doped hierarchically porous cellulose: highly efficient photocatalysts for methylene blue degradation." *Materials* 10 (2017): 373. doi: https://doi.org/10.3390/ma10040373.

[30] Zhang, Xiuju, Wenbin Chen, Zhidan Lin, Jia Yao, and Shaozao Tan. "Preparation and photocatalysis properties of bacterial cellulose/TiO_2 composite membrane doped with rare earth elements." *Synthesis and Reactivity in Inorganic, Metal-Organic, and Nano-Metal Chemistry* 41 (2011): 997–1004. doi: https://doi.org/10.1080/15533174.2011.591334.

[31] Brandes, Ricardo, Elizabeth C. A. Trindade, Daniel F. Vanin, Vanessa M. M. Vargas, Claudimir A. Carminatti, Hazim A. Al-Qureshi, and Derce O. S. Recouvreux. "Spherical bacterial cellulose/TiO_2 nanocomposite with potential application in contaminants removal from wastewater by photocatalysis." *Fibers and Polymers* (2018): 1861–1868. doi: https://doi.org/10.1007/s12221-018-7798-7.

[32] Trache, Djalal, M. Hazwan Hussin, M. K. Mohamad Haafiz, and Vijay Kumar Thakur. "Recent progress in cellulose nanocrystals: sources and production." *Nanoscale* 9 (2017): 1763–1786. doi: https://doi.org/10.1039/c6nr09494e.

[33] Lee, Kian Mun, Chin Wei Lai, Koh Sing Ngai, and Joon Ching Juan. "Recent developments of zinc oxide based photocatalyst in water treatment technology: a review." *Water Research* 88 (2016): 428–448. doi: https://doi.org/10.1016/j.watres.2015.09.045.

[34] Singh, Sonal, Ravi Pendurthi, Manika Khanuja, S. S. Islam, Suchitra Rajput, and S. M. Shivaprasad. "Copper-doped modified ZnO nanorods to tailor its light assisted charge transfer reactions exploited for photo-electrochemical and photocatalytic application in environmental remediation." *Applied Physics A* 123 (2017): 184. doi: https://doi.org/10.1007/s00339-017-0806-8.

[35] Awan, Fatima, Muhammad Shahidul Islam, Yeyu Ma, Cindy Yang, Zengqian Shi, Richard M. Berry, and Kam C. Tam. "Cellulose nanocrystal-ZnO nanohybrids for controlling photocatalytic activity and UV protection in cosmetic formulation." *ACS Omega* 3 (2018): 12403–12411. doi: https://doi.org/10.1021/acsomega.8b01881.

[36] Samadi, Morasae, Mohammad Zirak, Amene Naseri, Elham Khorashadizade, and Alireza Z. Moshfegh. "Recent progress on doped ZnO nanostructures for visible-light photocatalysis." *Thin Solid Films* 605 (2016): 2–19. doi: https://doi.org/10.1016/j.tsf.2015.12.064.

[37] Sharma, Priyanka R., Sunil K. Sharma, Richard Antoine, and Benjamin S. Hsiao. "Efficient removal of arsenic using zinc oxide nanocrystal-decorated regenerated microfibrillated cellulose scaffolds." *ACS Sustainable Chemistry & Engineering* 7 (2019): 6140–6151. doi: https://doi.org/10.1021/acssuschemeng.8b06356.

[38] Ul-Islam, Mazhar, Waleed Ahmad Khattak, Muhammad Wajid Ullah, Shaukat Khan, and Joong Kon Park. "Synthesis of regenerated bacterial cellulose-zinc oxide nanocomposite films for biomedical applications." *Cellulose* 21 (2014): 433–447. doi: https://doi.org/10.1007/s10570-013-0109-y.

[39] Guan, Ying, Hou-Yong Yu, Somia Yassin Hussain Abdalkarim, Chuang Wang, Feng Tang, Jaromir Marek, Wei-Lai Chen, Jiri Militky, and Ju-Ming Yao. "Green one-step synthesis of ZnO/cellulose nanocrystal hybrids with modulated morphologies and superfast absorption of cationic dyes." *International Journal of Biological Macromolecules* 132 (2019): 51–62. doi: https://doi.org/10.1016/j.ijbiomac.2019.03.104.

[40] Lefatshe, Kebadiretse, Cosmas M. Muiva, and Lemme P. Kebaabetswe. "Extraction of nanocellulose and *in-situ* casting of ZnO/cellulose nanocomposite with enhanced photocatalytic and antibacterial activity." *Carbohydrate Polymers* 164 (2017): 301–308. doi: https://doi.org/10.1016/j.carbpol.2017.02.020.

[41] Yu, Hou-Yong, Guo-Yin Chen, Yi-Bo Wang, and Ju-Ming Yao. "A facile one-pot route for preparing cellulose nanocrystal/zinc oxide nanohybrids with high antibacterial and photocatalytic activity." *Cellulose* 22 (2015): 261–273. doi: https://doi.org/10.1007/s10570-014-0491-0.

[42] Zheng, Wei-li, Wei-li Hu, Shi-yan Chen, Yi Zheng, Bi-hui Zhou, and Hua-ping Wang. "High photocatalytic properties of zinc oxide nanoparticles with amidoximated bacterial cellulose nanofibers as templates." *Chinese Journal of Polymer Science* 32 (2014): 169–176. doi: https://doi.org/10.1007/s10118-014-1386-0.

[43] Wei, Guangyu, Hong-Fen Zuo, Yuan-Ru Guo, and Qing-Jiang Pan. "Synthesis of ZnO with enhanced photocatalytic activity: a novel approach using nanocellulose." *BioResources* 11 (2016): 6244–6253. doi: https://doi.org/10.15376/biores.11.3.6244-6253.

[44] Xiao, He, Weibo Zhang, Yicui Wei, and Lihui Chen. "Carbon/ZnO nanorods composites templated by TEMPO-oxidized cellulose and photocatalytic activity for dye degradation." *Cellulose* 25 (2018): 1809–1819. doi: https://doi.org/10.1007/s10570-018-1651-4.

[45] Tu, Kai, Qiyang Wang, Ang Lu, and Lina Zhang. "Portable visible-light photocatalysts constructed from Cu_2O nanoparticles and graphene oxide in cellulose matrix." *The Journal of Physical Chemistry C* 118 (2014): 7202–7210. doi: https://doi.org/10.1021/jp412802h.

[46] Kim, Chan-Jun, Waliullah Khan, Dong-Hun Kim, Kwang-Soo Cho, and Soo-Young Park. "Graphene oxide/cellulose composite using NMMO monohydrate." *Carbohydrate Polymers* 86 (2011): 903–909. doi: https://doi.org/10.1016/j.carbpol.2011.05.041.

[47] Su, Fangzheng, Smitha C. Mathew, Grzegorz Lipner, Xianzhi Fu, Markus Antonietti, Siegfried Blechert, and Xinchen Wang. "mpg-C_3N_4-catalyzed selective oxidation of alcohols using O_2 and visible light." *Journal of the American Chemical Society* 132 (2010): 16299–16301. doi: https://doi.org/10.1021/ja102866p.

[48] Bellardita, Marianna, Elisa I. García-López, Giuseppe Marcì, Igor Krivtsov, José R. García, and Leonardo Palmisano. "Selective photocatalytic oxidation of aromatic alcohols in water by using P-doped g-C_3N_4." *Applied Catalysis B: Environmental* 220 (2018): 222–233. doi: https://doi.org/10.1016/j.apcatb.2017.08.033.

[49] Wang X., Maeda K., Thomas A., Takanabe K., Xin G., and Carlsson J. M., et al. "A metal-free polymeric photocatalyst for hydrogen production from water under visible light." *Nature Materials* 8 (2009): 76–80. doi: https://doi.org/10.1038/nmat2317.

15

Nanocellulose-Based Materials for the Removal of Organic Toxicants and Antibacterial Applications

Pallavi Jain
SRM Institute of Science and Technology, Delhi-NCR Campus, Modinagar, India

A. Geetha Bhavani
Noida International University, Greater Noida, India

CONTENTS

15.1 Introduction

Water is the most important requirement for the survival of life on Earth. Because of rapid growth and industrialization in the population, water scarcity is increasing day by day. These actions pollute crust water with toxic contamination, which is vulnerable to both humans and the ecosystem. Contamination in water can be broadly classified into organic and inorganic compounds, which are carcinogenic. These organic compounds comprise dyes (coming from coloring, tanning, and respective manufacturing industries), pesticides (coming from agricultural fields and individual manufacturing units), pharmaceutical drugs, aromatic hydrocarbons, and volatile organic compounds, the toxicity of which affects all the living organisms and the environment. Removing these organic contaminants from water bodies is our crucial role [1]. Removing toxic pollutants with low concentration is challenging, and technical updates toward efficiency are necessary to determine these contaminants.

Water treatment was carried out on nanocellulose (NC) using membrane filtration through absorbing and disinfection methods. The adsorption was the well-practiced method for eliminating toxic contaminants using porous materials from water [2]. The porous material plays the adsorbent role, and the pollutants were absorbed on the adsorbent's surface; finally, contaminants are removed. The adsorption methods have many advantages, such as unit operational parameters; the process was quite covenanting for handling, abundant adsorbents, economical operations, and compatibility in bulk-scale operations [3]. Conventional activated carbons are swiped with NC, displaying high adsorption capacities (ACs) with a low-cost process. The products from the agricultural and industrial units are collected and prepared in environment-friendly ways to reduce NC's cost [4].

Bacterial diseases have played an essential role in human life and deaths throughout history. Bacteria are also accountable for several contagious infections: wound infections, sepsis, gonorrhea (sexually transmitted infections), pneumonia, and many other infections. Many pathogenic bacteria such as *Escherichia coli* (E.coli) and *Campylobacter* can be transmitted and spread to the disease through food and water. Bacteria often spread intentionally or unintentionally and spread the disease. Therefore, it is imperative to remove such bacterial pathogens properly. NC has gained tremendous popularity for antibacterial use due to its exceptional physical properties, unique surface chemistry, and excellent biological properties [5]. The NC has an excess of hydroxyl groups, making them water stable. Antibacterial NC materials developed via nanomaterials combination, modification of the surface, antibiotics, and antibacterial polymers combination [6]. Most metal/metal oxide (MO) nanoparticles (NPs), such as Au, Ag, Cu, magnesium oxide (MgO), copper oxide (CuO), titanium dioxide (TiO_2), and zinc oxide (ZnO) NPs, and Cu/TiO_2 nanocomposites are excellent contenders for antibacterial applications [7]. Besides, other antimicrobial polymers are used for antibacterial applications. For example, chitosan and its derivatives are strong antibacterial agents [8]. Antimicrobial products based on NC may be utilized in various applications. This chapter aims to compress the NC as adsorbents to effectively eliminate organic contaminants such as dyes, pesticides, drug residues, and other chemical target compounds. The specific applications of the antimicrobial materials based on NC are further discussed.

15.2 Nanocellulose

Cellulose is found in abundant renewable sources such as algae, some bacteria, plants, and tunicates. The interesting polymer has a comprehensive precursor that holds potential functional groups and alters by specific modifications per industrial and commercial requirements [9]. The advantage of cellulose was stretched by bundling bundles with mixing cellulose chains with highly ordered NPs called cellulose nanomaterial or NC. These cellulose nanomaterials' outstanding physicochemical properties are useful as future materials [10]. Besides the above advantages, NCs are also supplying the combined chemical inertness, excellent strength, high filtration ability, low coefficient of thermal expansion, low density, high porosity, dimensional stability, and a possibility to modify the surface properties (Fig. 15.1) [11, 12]. The NC was divided into nanofibers and nanostructured materials. The first category of nanofibers includes cellulose nanofiber (CNF), cellulose nanocrystals (CNCs), and bacterial cellulose (BC), while the second nanostructured materials are microcrystalline cellulose and cellulose microfibrils.

The adsorbents' upgradation creates new adsorbents with high efficiency and toward the greenery process for removing contaminants [13]. It is significant to tailor any adsorbent's desired properties for precise applications like removing specific contaminants wastewater. The adsorbent adsorption depends upon the polarity and pH of the effluent. The polar interactions' effectiveness is more prominent because the surface is holding negative charges, which is not effective at low pH levels as it affects the protonation by removing specific interactions by non-covalent bonds [14]. Parameters like the adsorbate and surface interaction duration, ion-exchange capacity, concentration of the targeted molecules, adsorbent quantity, and temperature affect the total adsorbent performance [15]. The new adsorbents process the larger porosity with a high surface area of NCs, leading to logical commercial purposes. The other researcher also reported that NC materials serve as a biorenewable, ecological, and low cost to fulfill the industrial demand in bulk scale with a profile of less carbon footprint [16].

The many reports featured the cellulose nanomaterials characteristics like high aspect ratio [17], high active surface areas [18], high rigorousness [18], renewability [19], high mechanical strength [20], and low toxicity [21] made it highly acceptable in worldwide for commercial unit operations. The unique feature of adsorptive processes is the simplicity and diversity of accessing the functional groups (hydroxyl, carboxyl, and amino groups) introduced into the NC framework with fabrication methods that will alter

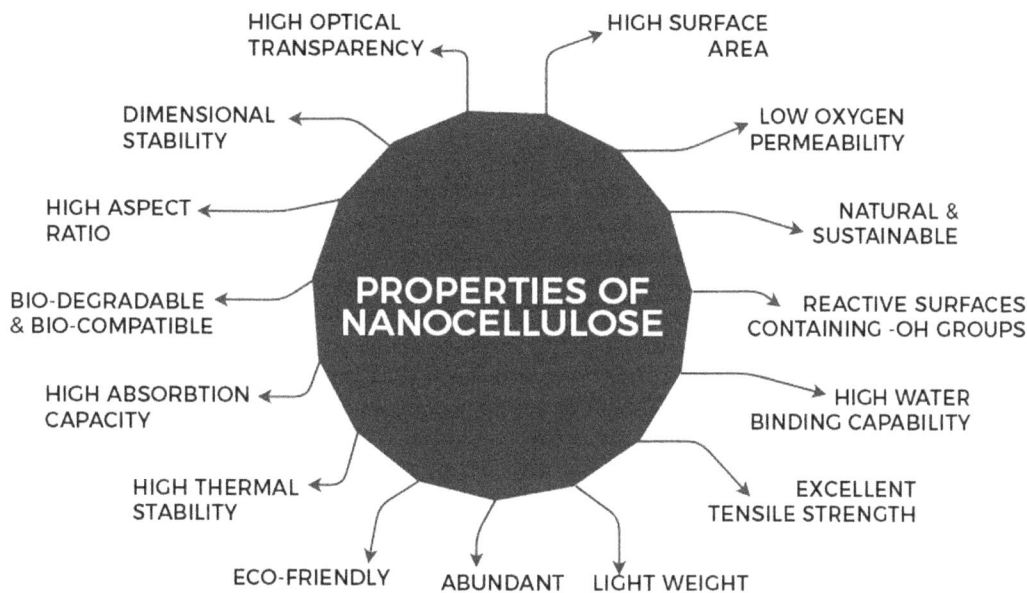

FIGURE 15.1   Advantages of NC.

the surface chemistry for any specific applications [22]. The fabrication of surface functional groups insertion in polymer matrices of NC leads to be used for compatibilization and adsorption [23]. Finally, the initiation of new methods with a combination of low cost and quick processing has reduced NC materials' capital expenditure over other materials like graphene etc. [24]. NC showed many outstanding factors like adsorption method for water treatment, disinfection, and membrane filtration are the main contributors. Many areas are needed to explore and advance low-cost adsorbents without by-products formation and high ACs [25].

The absorbents based on NC can be developed via three different techniques. The foremost technique was by functional group structure alteration into NC. These functional groups are composed of the combinations of cationic groups, sulfate-half ester groups, carbonyl groups, carboxyl groups, aldehyde groups, etc. [26]. The second technique was the amalgamation of other polymeric structures (PLA,PEG, GMA) straightly over NC adsorbent, also known as "grafting" [27]. The third technique was through polymerization using various monomer species in the presence of NC adsorbent structures. These techniques are intended to improve the surface area by inserting micro-polymeric, nanosized structures, and increase the porosity of the application like cryogels and aerogels are created [28]. Binary systems of NC adsorbents are needed to study to increase the efficiency of NC's binding synergy with other materials. NC-based adsorbents are well commercialized due to their excellent properties, like extremely polar negative charges on the surface (interdependent on zeta potential and pH) allow the electrostatic relations between surface functional groups and positively charged molecules through the polarity [29]. The main aim of modifying any adsorbents with polymer or NC or its derivatives is to provide high density functional groups, leading to the increased surface area. These enhanced NC features meet the desired absorbing ability of adsorption for specific target molecules [30].

15.3   Separation of Organic Contaminants Using NC Materials

Water pollutants are being released naturally or domestically from the human requirement. For example, polycyclic aromatic hydrocarbons, heavy metals, pesticides, pharmaceutical compounds, and biomolecular contaminants. Contaminant removal is being done with a variety of methods. Many local units' adsorption method was still well-practiced due to low cost and high efficiency for municipal wastewater [31]. The setback was kinetics of the targeted pollutants can change the adsorption's perfor-mance, and a single pollutant could not measure. To overcome these setbacks, the adsorbent system must design to give exceptional results.

15.3.1 Separation of Petroleum Products Using NC-Based Materials

Among many applications of NC-based materials, the separation of petroleum compounds from water using modified with ultra-porous aerogels is also well considered with the property of hydrophobicity. The hydrophobicity enhances the wetting phenomena because of the misbalance between surface energy. For example, the spongelike structured adsorbent was prepared through vapor deposition technique using hydrophobic natured silanes on CNF aerogels [32]. These pores are 40 mm in size and able to adsorb 45 times higher than their capacity, known as super-hydrophobicity. The material having super-hydrophobic property leads to the ultimate absorption performance of non-polar liquids with the reusability of adsor-bent [33]. CNC nanopapers were prepared with deterioration in water flux even after 12 runs and with great salt resistance to separate the crude oil spills from seawater. CNC structures with cellulose chains will restrict the mobility of hydrophilic groups, which leads to the stability of the hydrophilicity, which is important for any adsorbent applications [34]. Highly porous functionalized aerogels with the TiO_2-loaded CNF materials for an oleophilic coating were developed. The enhanced property leads to the removal of pollutants, specifically oils and organic compounds from water. It is also reported that using the freeze-drying method can increase the AC with cost-effective and large-scale production [32]. The magnetite aerogels loaded with magnetic nanocellulose was screened and modified with oleic acid to improve the AC by altering the hydrophobic property using oleic acid [35]. The screening results found to be for adsorption rate of cyclohexane was 68.06 g/g, ethyl acetate was 56.32 g/g, and vacuum pump oil was 33.24 g/g. Finally, oleic acid was able to change effectively by raising the specific surface area.

15.3.2 Separation of Pharmaceutics Using NC-Based Materials

A powerful adsorbent must be antibiotic resistant and removed even in low concentrations, affecting the wastewater and further accumulating. CNCs also prepared from green seaweed and further treated with acid hydrolysis provided the effective removal of tetracycline hydrochloride from wastewater at pH 5.5 [36]. NC materials were recycled by releasing the trapped antibiotics using NaOH or 5% HNO_3 from NC binding sites. Hu et al. [37] prepared magnetic carboxylated CNC to remove fluoroquinolones from water. This material showed the great creditable ability of separation, high surface area, and attained relative standard deviation of 0.6–7.5% with recovery ratio of 81.2–93.7% are promising results for effec-tive removal of antibiotics. This material is also reusable by desorbing ofloxacin with 30% acetic acid/methanol (v/v) from the adsorbent. By using this desorption method, 70% of the starting material was recovered by the seventh cycle. Abu-Danso et al. [38] found that modification with phosphate ions over NC effectively removes the diclofenac compound from water. In this report, diclofenac adsorption was initially 29.6 mg/g at pH 5.6 range and found a maximum of 107.9 mg/g AC with increasing diclofenac concentration from the experiment. The absorbent reused was done by desorption of diclofenac by elution with a mixture of 0.1 M HNO_3, 0.1 M NaOH, and H_2O. The regeneration process among HNO_3 was 80% removal efficiency compared to NaOH after final cycles. Cigarette filters are used as a NC precursor to observe the AC and aspects of waste material utilization. Hu et al. [39] showed adsorption of diclofenac sodium on amino-functionalized (bilayer) CNCs/Chitosan-chitosan composite. Ruiz-Palomero et al. [40] demonstrated β-Cyclodextrin decorated NC for selective fluorimetric determination of danofloxacin. Table 15.1 illustrates some CN-based materials used to remove pharmaceuticals [36–42].

15.3.3 Separation of Biomolecules Using NC Materials

The NC-based materials were also applied to remove the large biomolecule via adsorption process through a great affinity of the cellulose acetate (CA) nanofibers membranes prepared through electros-pun method [43]. The membrane was regenerated using the buffer solutions like hydrochloric acid/potas-sium hydrogen phthalate, with pH 3.0. Hence, the complete removal of DNA molecules was challenging as the eluent's concentration from the membrane matrix drops the performance by 11.5% for every cycle.

TABLE 15.1

Some NC-Based Absorbent Employed for the Removal of Pharmaceuticals.

NC Absorbent	Targeted Pharmaceuticals	Absorption Capacity (mg/g)	Reference
Green seaweed synthesized CNCs	Tetracycline hydrochloride	–	[36]
M-CCNs@MIP	Fluoroquinolones	45.64	[37]
HPO-CNFs	Diclofenac	107.90	[38]
Amino-functionalized (bilayer) CNCs/ chitosan composite	Diclofenac sodium	444.44	[39]
Polyvinyl alcohol/sodium alginate /CNC)@ polyethyleneimine	Diclofenac sodium	418.41	[41]
CuS hollow nanospheres@N-doped CNCs hybrid composites	Ciprofloxacin	–	[42]

At the end of the fifth cycle, the total performance loss was noticed around 54%. The hydrogel as a superabsorbent was used to recover the β-casein from an aqueous solution through the interfaces among positively charged sorbent and negatively charged protein [30].

The NC and other magnetic nanomaterials are combined to prepare the hybrid nanocomposites to enhance an adsorbent's surface properties [44, 45]. The immunoglobulin was adsorbed effectively through NC modified with poly(methacrylic-acid-*co*-vinyl sulfonic acid) to prepare a magnetite composite [45]. This composite allowed the NC matrix membrane to be effective magnetic and ion-exchange properties of sorbent [46]. The maximum AC was achieved at a neutral pH of 6.8 due to developed ionic strength with the ensuing discharge of the immunoglobulin base from adsorbent in the process of regeneration. The composite membrane significantly permits selective filtration of the immunoglobulin from the proteins mixture, which was proved with amazing applications. Ishak et al. [47] found the hybrid material changed with TEMPO-magnetic CNC to adsorb the protein from the fish waste. Lombardo et al. [48] recovered the proteins (bovine serum albumin) of 90% from an aeration tank using CNC grafted with pyridinium. The report examined the effect of carboxylate and sulfate CNC on surface chemistry and found no interface between negatively charged carboxylate and sulfate groups. The adsorption feasibility of single amino acids was noticed over NC changed showed outstanding adsorption of 2.7 M concentration of amino acids [49]. Table 15.2 illustrates some NC-based materials used to remove biomolecules [44–48, 50–52].

The small-sized organic molecules released from industries are necessary for removal through adsorption-based remediation. Cyanogen bromide (CNBr) was used to modify peroxidase-conjugated CNC to eliminate chlorinated phenolic molecules from wastewater [53]. Hybrid composite materials are more effective due to their prominent synergistic effects between the surface of graphene oxide (GO) and CNC composite to remove 1-butyl-3-methylimidazolium chloride compound [54]. The addition of GO to CNC substantially increased the overall performance compared to mere CNC activity, and finally, maximum performance found over 30% of GO content. Sometimes inorganic compounds are also removed within

TABLE 15.2

Some NC-Based Absorbent Employed for the Removal of Biomolecules.

NC Absorbent	Targeted Biomolecules	Absorption Capacity	Reference
Poly[(ar-vinylbenzyl)trimethylammonium chloride)]- modified cellulose acetate nanofibers	DNA	23.51 µg/mg	[44]
Poly(methacrylic acid-*co*-vinyl sulfonic acid)-grafted- magnetite/ NC composite	Immunoglobulin	200.21 mg/g	[45]
Modified CNCs	Protein	–	[47]
Sulfated, carboxylated, and pyridinium-grafted 27 CNCs	Bovine serum albumin	–	[48]
Poly(acrylic acid)-modified poly(glycidylmethacrylate)- grafted NC hydrogel	Lysozyme	148.42 mg/g	[50]
Cellulose-based organogel	Aromatic compounds	50–400 µmol/g	[51]
γ-Cyclodextrin-modified cellulose nanofibers	Toluene	0.027 mg/g	[52]

the removal of organic industrial molecules. Sehaqui et al. [55] found a unique synthesis method using waste pulp residues to prepare cationic CNF through etherification with glycidyl trimethyl ammonium chloride. They used cationic CNF to remove multivalent ions (phosphate and sulfate), monovalent ions (fluoride and nitrate), and anions that can selectively inhibit adsorption.

15.3.4 Separation of Textile Dyes Using NC Materials

This is also important to remove dye molecules released to waters from dying sectors. These dyes are effectively removed using an adsorption process by composite materials and regenerate the materials systems through the desorption process [56]. Among many methods, the photofouling, coagulation, forward oxidation, electrochemical destruction, and membrane filtration process are mainly used [57]. Zhou et al. [58] used nanocomposite hydrogels/CNC to remove methylene blue from an aqueous solution with increasing CNC content. Similarly, methylene blue was removed using alginate hydrogels/CNC and was synthesized with extraction from TEMPO-oxidized carex meyeriana kunth (CMK) adsorbents from water [59]. Benmassaoud et al. [60] prepared hybrid composite CNC-magnetic NPs to remove Sudan dyes that emerged in food samples through the capillary liquid chromatography technique. The CN hybrid composite material effectively reduced the Sudan levels. Methanol was a widely used regeneration agent for any adsorbent, compared to other ethanol and acetone solvents. CNF was crosslinked from polyethyleneimine and polydopamine and removed methyl orange dye with AC of 265.9 mg/g [61]. Methyl orange was attained with 0.1 M NaOH, leading to a drop in AC of 89% at the end of the fourth cycle. Guan et al. [62] prepared absorbent with high AC to remove the dye over 90%, through ZnO-loaded CNC nanohybrids structures. Finally, the hybrid materials with high ZnO content display antibacterial properties besides *E. coli*. The shape of the hybrid material was pH dependent. The spherical material was obtained at a pH of 8, and the flower-like or sheet-like material was found at a pH of 10, which is beneficial for adsorbing specific target molecules. Table 15.3 tabulates NC-based adsorbents used for the removal of different dyes [25, 58, 59, 63, 64, 65–70].

TABLE 15.3

Some NC-based Adsorbents are Employed for the Removal of Dyes.

Nanocellulose Adsorbents	Targeted Dyes	Adsorption Capacity (mg/g)	Reference
CNCs after carboxylation	Methylene Blue	769	[25]
CNC-reinforced keratin	Reactive Black 5	1250	[58]
CNCs-based amino-functionalized adsorbent	Congo Red 4BS	199.5	[59]
	Acid Red GR	134.7	
	Reactive Light-Yellow K-4G	183.0	
Partially hydrolyzed and polyacrylamide-modified CNC hydrogel	Methylene Blue	326	[60]
Oxidized and hydrogen peroxide-treated CNC extract	Methylene Blue	217.4	[61]
Acid-treated CNCs	Methylene Blue	87.8	[65]
Chitosan-modified CNC composite membrane	Victoria Blue 2B	–	[66]
CNCs-based hyperbranched polyethyleneimine adsorbent	Cationic Basic Yellow 28Congo Red	1860	[67]
		2100	
CNCs-based carboxylate functionalized adsorbent	Crystal Violet, Malachite, Green, Methylene Blue, Basic Fuchsin	243.90	[68]
Fe_3O_4@cellulose core-shell magnetic biopolymer	Congo red	131	[69]
Cellulose aerogel based on cationic cellulose nanofibrils	Red Dye 180	160	[70]
	Blue Dye CR19	230	
	Orange Dye 142	560	

15.4 NC-Based Biomedical Materials

In upgrading the evolution of novel materials for many biomedical applications, polymers' thrust and preparation methods from natural precursors are very much concerned for both material and biologist's researchers. The upgradation is aimed at physical and chemical properties, which keep exceptional compaction property by cellulose materials. These cellulose materials have a history of many applications in the medical, wastewater treatment, and chemical industries. The cellulose materials coated or encapsulated with the desired drug for oral administration as a pharmaceutical application are safe and widely accepted. For advanced applications, the cellulose derivatives were modified as NC for effective and safe drug loading to target sites as carriers with potential biodegradability [71]. The researchers are creating alternative materials day by day with certain improvements in surface disinfection, antibiotic resistance, and antimicrobial (antimicrobial wound dressing). The NC-based biomaterials are intended to prepare with high porosity. These pores are the physical barriers to carrying the drugs (medicines or antibiotics) to specific woods and external infection [72]. Among many antimicrobial agents, silver (Ag) was known for anti-spoilage, and anti-infections have been well documented from ancient times. The Ag derivatives are upgraded as AgNPs and found very active probe as an antifungal, antibacterial, and antiviral property [73]. The AgNPs were prepared from $AgNO_3$ as a precursor through a chemical reduction in the hydrogen atmosphere. The prepared AgNPs were impregnated over NC and further used for therapeutic purposes.

The AgNPs/nanocellulose biomaterial showed good antimicrobial activity because of high surface area, porosity, size, and shape [74]. Poly(lactic acid)/CNC/Ag was prepared with $AgNO_3$ and was loaded over CNC polymeric matrices. It provides significant antibacterial activity was due to synergy effects and mechanical reinforcement [75]. CNC/Ag over polymer blended with poly(3-hydroxybutyrate-*co*-3-hydroxyvalerate) materials [76] and CNC/Ag over polymer blended polyurethane [77] were prepared. The CNF/Ag composite materials were also quite effective for antibacterial activity was due to macromolecular linkers of CNF and Ag and electrostatic forces [78]. The fluorescent Ag nanoclusters have also proven for the antibacterial property. These were synthesized by immersing CNF material in Ag and poly(methacrylic acid) composition [79]. Apart from silver nitrate, precursors such as silver sulfadiazine and silver chloride are also used to prepare AgNPs to develop BC/Ag bio-nanomaterials for pronounced biocompatibility, antibacterial, and antimicrobial activity [80].

15.4.1 NC as Antibacterial Agent Based on Metal Oxides

The cost of the Ag has focused the researcher on finding an alternative with the same activity. The metal oxides such as MgO, CuO, TiO_2, and ZnO show antimicrobial activity on CNC, CNF, and BC materials for antibacterial activity. The main advantage of MgO was holding potential antibacterial activity compared to other metal oxides, which was due to its basic nature [81]. It damages the cell membrane. Finally, bacterial cells will die by leaking intracellular contents.

15.4.1.1 Antibacterial Materials Based on TiO_2

CNC/TiO_2 composites were initiated through evaporation and casting methods and finally blended with wheat gluten. The biocomposite NC/TiO_2 was tested over *E. coli*, *Saccharomyces cervisiae*, and *Saccharomyces aureus*, and found promising antibacterial activity with the comparison of without TiO_2-coated material [82]. CNF/TiO_2 was synthesized in an aqueous solution to promote "green chemistry" methodology, which was relatively stable in the presence of UV irradiation and revealed antimicrobial action against *E. coli* and *S. aureus* [83].

15.4.1.2 Antibacterial Materials Based on ZnO

CNC/ZnO nanocomposite was prepared by *in situ* dispersion of ZnO in the CNC solution and found improved antibacterial activity was highest against *S. aureus* than *E. coli* with the comparison of mere ZnO [84]. CNC nanocomposites were loaded with metal oxides like CeO_2 and ZnO in the form of NPs through a binding agent (polyaniline) and found active against *E. coli* and *B. subtilis* [85]. ZnO NPs loaded over CA membrane was prepared with the electrospinning method and demonstrated antibacterial action against *S. aureus*, *E. coli*, and *Citrobacter* [86]. CA/ZnO membrane with polymer (poly(3-hydroxy-butyrateco-3-hydroxy-valerate) showed 100% antibacterial activity against *E. coli* and *S. aureus* [87].

15.4.2 Antibacterial Materials Based on Graphene Oxide (GO)

GO has grown in interest due to wide applications by easy handling and tailored nanostructural properties. CNC modified with a reduced form of GO (rGO) was synthesized through acid hydrolysis known as Hummer's method, and the final product was CNC/rGO/poly-lactic acid nanocomposite. This nanocomposite film was found antibacterial ability against *E. coli* and *S. aureus* [88]. The mechanical mixing was used to mix BC and GO to form BC/GO nanocomposites, and they exhibit remarkable antimicrobial ability on *S. cerevisiae*. This antimicrobial ability was due to nanocomposites' close contact and the *S. cerevisiae* cell surfaces [89].

15.5 Conclusion

Contaminants need to be removed from wastewater to overcome water scarcity. The wastewater can effectively be treated with NC-based materials by various modified techniques to improve the AC to remove specific targeted toxic contaminants. The AC was significantly improved through potential modifications. The NC holding unique feature of adsorptive processes is the simplicity and diversity of accessing the functional groups (hydroxyl, carboxyl, and amino groups) was altering the surface chemistry for any specific applications. The adsorbate and surface interaction duration, ion-exchange capacity, the concentration of the targeted molecules, adsorbent quantity, and temperature affect the total adsorbent performance. The study's goal is to summarize the recent developments and potential production of NC through consideration of selected examples in the removal of organic contaminants and antibacterial applications. There are primarily four processing antibacterial products based on nanocellulose, and the developed products have wide applications. These NC should be giving more importance to realizing their ability as absorbent and antibacterial materials, which will support the production of new materials.

Acknowledgment

Dr. Pallavi Jain is grateful to SRM Institute of Science and Technology, Modinagar for support and encouragement.

REFERENCES

[1] Santhosh, Chella, Venugopal Velmurugan, George Jacob, Soon Kwan Jeong, Andrews Nirmala Grace, and Amit Bhatnagar. "Role of nanomaterials in water treatment applications: a review." *Chemical Engineering Journal* 306 (2016): 1116–1137. doi: 10.1016/j.cej.2016.08.053.

[2] Saxena, Reena, Megha Saxena, and Amit Lochab. "Recent progress in nanomaterials for adsorptive removal of organic contaminants from wastewater." *Chemistry Select* 5 (2020): 335–353. doi: 10.1002/slct.201903542.

[3] Jiuhui, Q. U. "Research progress of novel adsorption processes in water purification: a review." *Journal of Environmental Sciences* 20 (2008): 1–13. doi: 10.1016/S1001-0742(08)60001-7.

[4] Mohammed, Nishil, Nathan Grishkewich, Richard M. Berry, and Kam Chiu Tam. "Cellulose nanocrystal-alginate hydrogel beads as novel adsorbents for organic dyes in aqueous solutions." *Cellulose* 22 (2015): 3725–3738.

[5] Liu, Kai, Hunan Liang, Joseph Nasrallah, Lihui Chen, Liulian Huang, and Yonghao Ni. "Preparation of the CNC/Ag/beeswax composites for enhancing antibacterial and water resistance properties of paper." *Carbohydrate Polymers* 142 (2016): 183–188. doi: 10.1016/j.carbpol.2016.01.044.

[6] Zhang, Shaobo, Xinghui Yang, Bin Tang, Lingjun Yuan, Kui Wang, Xiangyu Liu, Xingli, Zhu, et al. "New insights into synergistic antimicrobial and antifouling cotton fabrics via dually finished with quaternary ammonium salt and zwitterionic sulfobetaine." *Chemical Engineering Journal* 336 (2018): 123–132. doi: 10.1016/j.cej.2017.10.168.

[7] Heli, B., E. Morales-Narvaez, H. Golmohammadi, A. Ajji, and A. Merkoci, "Modulation of population density and size of silver nanoparticles embedded in bacterial cellulose via ammonia exposure: visual detection of volatile compounds in a piece of plasmonic nanopaper." *Nanoscale* 8 (2016): 7984–7991. doi: 10.1039/C6NR00537C.

[8] Xu, Kaimeng, Can Liu, Kunyong Kang, Zhifeng Zheng, Siqun Wang, Zhenguan Tang, and Wenxiu Yang. "Isolation of nanocrystalline cellulose from rice straw and preparation of its biocomposites with chitosan: physicochemical characterization and evaluation of interfacial compatibility." *Composites Science and Technology* 154 (2018): 8–17. doi: 10.1016/j.compscitech.2017.10.022.

[9] Mokhena, T. C, and M. J. John. "Cellulose nanomaterials: new generation materials for solving global issues." *Cellulose* 27 (2020): 1149–1194. doi: 10.1007/s10570-019-02889-w.

[10] Foster, E. Johan, Robert J. Moon, Umesh P. Agarwal, Michael J. Bortner, Julien Bras, Sandra Camarero-Espinosa, and Kathleen J. Chan et al. "Current characterization methods for cellulose nanomaterials." *Chemical Society Reviews* 47 (2018): 2609–2679. doi: 10.1039/C6CS00895J.

[11] Vineeth, S. K., Ravindra V. Gadhave, and Pradeep T. Gadekar. "Chemical modification of nano-cellulose in wood adhesive." *Open Journal of Polymer Chemistry* 9 (2019): 86–99. doi: 10.4236/ojpchem.2019.94008.

[12] Naz, Sania, Joham S. Ali, and Muhammad Zia. "Nanocellulose isolation characterization and applications: a journey from non-remedial to biomedical claims." *Bio-Design and Manufacturing* 2 (2019): 187–212. doi: 10.1007/s42242-019-00049-4.

[13] Thambiraj S, G. Sharmila, and D. Ravi Shankaran. "Green adsorbents from solid wastes for water purification application." *Material Today: Proceedings* 5 (2018): 16675–16683. doi: 10.1016/j.matpr.2018.06.029.

[14] Erol Kadir, and Kazim Kose. "Efficient polymeric material for separation of human hemoglobin." *Artificial Cells, Nanomedicine, and Biotechnology* 45 (2017): 39–45. doi: 10.1080/21691401.2016.1233112.

[15] Hokkanen, Sanna, Eveliina Repo, Terhi Suopajarvi, Henrikki Liimatainen, Jouko Niinimaa, and Mika Sillanpaa. "Adsorption of Ni(II), Cu(II) and Cd(II) from aqueous solutions by amino modified nanostructured microfibrillated cellulose." *Cellulose* 21 (2014): 1471–1487. doi: 10.1007/s10570-014-0240-4.

[16] Chenampulli, Sangeetha, G. Unnikrishnan, A. Sujith, Sabu Thomas, and Tania Francis. "Cellulose nano-particles from Pandanus: viscometric and crystallographic studies." *Cellulose* 20 (2013): 429–438. doi: 10.1007/s10570-012-9831-0.

[17] Jorfi, Mehdi, and Erika J. Foster. "Recent advances in nanocellulose for biomedical applications." *Journal of Applied Polymer Science* 132 (2015): 41719. doi: 10.1002/app.41719.

[18] Ruiz-Palomero, Celia, M. Laura Soriano, and Miguel Valcarcel. "Nanocellulose as analyte and analytical tool: opportunities and challenges." *TrAC Trends in Analytical Chemistry* 87 (2017): 1–18. doi: 10.1016/j.trac.2016.11.007.

[19] Hokkanen, Sanna, Amit Bhatnagar, and Mika Sillanpaa. "A review on modification methods to cellulose-based adsorbents to improve adsorption capacity." *Water Research* 91 (2016): 156–173. doi: 10.1016/j.watres.2016.01.008.

[20] Tang, Lirong, Biao Huang, Qilin Lu, Siqun Wang, Wen Ou, Wenyi Lin, and Xuerong Chen. "Ultrasonication-assisted manufacture of cellulose nanocrystals esterified with acetic acid." *Bioresource Technology* 127 (2013): 100–105. doi: 10.1016/j.biortech.2012.09.133.

[21] Lam, Edmond, Keith B. Male, Jonathan H. Chong, Alfred C.W. Leung, and John H.T. Luong. "Applications of functionalized and nanoparticle-modified nanocrystalline cellulose." *Trends Biotechnology* 30 (2012): 283–290. doi: 10.1016/j.tibtech.2012.02.001.

[22] Chowdhury, Shamik, and Rajashekhar Balasubramanian. "Recent advances in the use of graphene-family nano-adsorbents for removal of toxic pollutants from wastewater." *Advances in Colloid and Interface Science* 204 (2014): 35–56. doi: 10.1016/j.cis.2013.12.005.

[23] Zhao, Xiaowei, Gang Zhang, Qiong Jia, Chengji Zhao, Weihong Zhou, and Weijie Li. "Adsorption of Cu(II), Pb(II), Co(II), Ni(II), and Cd(II) from aqueous solution by poly(aryl ether ketone) containing pendant carboxyl groups (PEK-L): equilibrium, kinetics, and thermodynamics." *Chemical Engineering Journal* 171 (2011): 152–158. doi: 10.1016/j.cej.2011.03.080.

[24] Ma, Qianyun, Dongying Hu, and Lijuan Wang. "Preparation and physical properties of tara gum film reinforced with cellulose nanocrystals." *International Journal of Biological Macromolecules* 86 (2016): 606–612. doi: 10.1016/j.ijbiomac.2016.01.104.

[25] Batmaz, Rasim, Nishil Mohammed, Masuduz Zaman, Gagan Minhas, Richard M. Berry, and Kam C. Tam. "Cellulose nanocrystals as promising adsorbents for the removal of cationic dyes." *Cellulose* 21 (2014): 1655–1665. doi: 10.1007/s10570-014-0168-8.

[26] Sultan, Sahar, and Aji P. Mathew "3D printed scaffolds with gradient porosity based on a cellulose nanocrystal hydrogel." *Nanoscale* 10 (2018): 4421–4431. doi: 10.1039/C7NR08966J.

[27] Habibi, Youssef, and Alain Dufresne. "Highly filled bio-nanocomposites from functionalized polysaccharide nanocrystals." *Biomacromolecules* 9 (2008): 1974–1980. doi: 10.1021/bm8001717.

[28] Du, Haishun, Wei Liu, Miaomiao Zhang, Chuanling Si, Xinyu Zhang, and Bin Li. "Cellulose nanocrystals and cellulose nanofibrils based hydrogels for biomedical applications." *Carbohydrate Polymers* 209 (2019): 130–144. doi: 10.1016/j.carbpol.2019.01.020.

[29] Bilgin, Elif, Kadir Erol, Kazım Kose, and Dursun A. Kose. "Use of nicotinamide decorated polymeric cryogels as heavy metal sweeper." *Environmental Science and Pollution Research* 25 (2018): 27614–27627. doi: 10.1007/s11356-018-2784-6.

[30] Köse, Kazim, Miran Mavlan, and Jeffrey P. Youngblood. "Applications and impact of nanocellulose based adsorbents." *Cellulose* 27 (2020): 2967–2990. doi: 10.1007/s10570-020-03011-1.

[31] Fan, Mingyi, Jiwei Hu, Rensheng Cao, Wenqian Ruan, and Xionghui Wei. "A review on experimental design for pollutants removal in water treatment with the aid of artificial intelligence." *Chemosphere* 200 (2018): 330–343. doi: 10.1016/j.chemosphere.2018.02.111.

[32] Korhonen, Juuso T., Marjo Kettunen, Robin H. A. Ras, and Olli Ikkala. "Hydrophobic nanocellulose aerogels as floating, sustainable, reusable, and recyclable oil absorbents." *ACS Applied Materials & Interfaces* 3 (2011): 1813–1816. doi: 10.1021/am200475b.

[33] Lee, Koon-Yang, Yvonne Aitomaki, Lars A. Berglund, Kristiina Oksman, and Alexander Bismarck. "On the use of nanocellulose as reinforcement in polymer matrix composites." *Composites Science and Technology* 105 (2014): 15–27. doi: 10.1016/j.compscitech.2014.08.032.

[34] Wu, Ming-Bang, Chao Zhang, Jun-Ke Pi, Chang Liu, Jing Yang, and Zhi-Kang Xu. "Cellulose nanocrystals as anti-oil nanomaterials for separating crude oil from aqueous emulsions and mixtures." *Journal of Materials Chemistry A* 7 (2019): 7033–7041. doi: 10.1039/C9TA00420C.

[35] Gu, Hongbo, Xiaomin Zhou, Shangyun Lyu, Duo Pan, Mengyao Dong, Shide Wu, and Tao Ding, et al. "Magnetic nanocellulose-magnetite aerogel for easy oil adsorption." *Journal of Colloid and Interface Science* 560 (2020): 849–856. doi: 10.1016/j.jcis.2019.10.084.

[36] Rathod, Manali, Soumya Haldar, and Shaik Basha. "Nanocrystalline cellulose for removal of tetracycline hydrochloride from water via biosorption: equilibrium, kinetic and thermodynamic studies." *Ecological Engineering* 84 (2015): 240–249. doi: 10.1016/j.ecoleng.2015.09.031.

[37] Hu, Zhao-Hong, Yan-Fei Wang, Ahmed Mohamed Omer, and Xiao-kun Ouyang. "Fabrication of ofloxacin imprinted polymer on the surface of magnetic carboxylated cellulose nanocrystals for highly selective adsorption of fluoroquinolones from water." *International Journal of Biological Macromolecules* 107 (2018): 453–462. doi: 10.1016/j.ijbiomac.2017.09.009.

[38] Abu-Danso, Emmanuel, Afrouz Bagheri, and Amit Bhatnagar. "Facile functionalization of cellulose from discarded cigarette butts for the removal of diclofenac from water." *Carbohydrate Polymers* 219 (2019): 46–55. doi: 10.1016/j.carbpol.2019.04.090.

[39] Hu, Dalin, Haoyu Huang, Ran Jiang, Nan Wang, Hongping Xu, Yang-Guang Wang, and Xiao-kun Ouyang. "Adsorption of diclofenac sodium on bilayer amino-functionalized cellulose nanocrystals/chitosan composite." *Journal of Hazardous Materials* 369 (2019): 483–493. doi: 10.1016/j.jhazmat.2019.02.057.

[40] Ruiz-Palomero, Celia, M. Laura Soriano, and Miguel Valcarcel. "β-Cyclodextrin decorated nanocellulose: a smart approach towards the selective fluorimetric determination of danofloxacin in milk samples." *Analyst* 140 (2015): 3431–3438. doi: 10.1039/C4AN01967A.

[41] Fan, Lihong, Yuqing Lu, Li-Ye Yang, Fangfang Huang, and Xiao-Kun Ouyang. "Fabrication of polyethylenimine-functionalized sodium alginate/cellulose nanocrystal/polyvinyl alcohol core-shell microspheres ((PVA/SA/CNC)@PEI) for diclofenac sodium adsorption." *Journal of Colloid and Interface Science* 554 (2019): 48–58. doi: 10.1016/j.jcis.2019.06.099.

[42] Nekouei Farzin, Shahram Nekouei, and Hanieh Kargarzadeh. "Enhanced adsorption and catalytic oxidation of ciprofloxacin on hierarchical CuS hollow nanospheres@N-doped cellulose nanocrystals hybrid composites: kinetic and radical generation mechanism studies." *Chemical Engineering Journal* 335 (2018): 567–578. doi: 10.1016/j.cej.2017.10.179.

[43] Rojanarata, Theerasak, Samarwadee Plianwong, Kosit Su-uta, Praneet Opanasopit, and Tanasait Ngawhirunpat. "Electrospun cellulose acetate nanofibers as thin layer chromatographic media for eco-friendly screening of steroids adulterated in traditional medicine and nutraceutical products." *Talanta* 115 (2013): 208–213. doi: 10.1016/j.talanta.2013.04.078.

[44] Demirci, Serkan, Asli Celebioglu, and Tamer Uyar. "Surface modification of electrospun cellulose acetate nanofibers via RAFT polymerization for DNA adsorption." *Carbohydrate Polymers* 113 (2014): 200–207. doi: 10.1016/j.carbpol.2014.06.086.

[45] Anirudhan T. S., and S. R. Rejeena. "Poly(methacrylic acid-*co*-vinyl sulfonic acid)-grafted-magnetite/nanocellulose superabsorbent composite for the selective recovery and separation of immunoglobulin from aqueous solutions." *Separation and Purification Technology* 119 (2013): 82–93. doi: 10.1016/j.seppur.2013.08.019.

[46] Bayramoglu, Gulay, Meltem Yılmaz, Ayşegul Ulku Şene, and M. Yakup Arıca "Preparation of nano-fibrous polymer grafted magnetic poly(GMA-MMA)-*g*-MAA beads for immobilization of trypsin via adsorption." *Biochemical Engineering Journal* 40 (2008): 262–274. doi: 10.1016/j.bej.2007.12.013.

[47] Ishak N. S., K. M. Ku Ishak, Y. Bustami, and H. Rokiah. "Evaluation of cellulose nanocrystals (CNCs) as protein adsorbent in stick water." *Material Today: Proceedings* 17 (2019): 516–524. doi: 10.1016/j.matpr.2019.06.330.

[48] Lombardo, Salvatore, Samuel Eyley, Christina Schutz, Hans van Gorp, Sabine Rosenfeldt, Guy Van den, Mooter, and Wim Thielemans. "Thermodynamic study of the interaction of bovine serum albumin and amino acids with cellulose nanocrystals." *Langmuir* 33 (2017): 5473–5481. doi: 10.1021/acs.langmuir.7b00710.

[49] Zimnitsky, Dmitry S., Tatiana L. Yurkshtovich, and Pavel M. Bychkovsky. "Multilayer adsorption of amino acids on oxidized cellulose." *Journal of Colloid and Interface Science* 285 (2005): 502–508. doi: 10.1016/j.jcis.2004.12.021.

[50] Anirudhan T. S., and S. R. Rejeena. "Poly(acrylic acid)-modified poly(glycidylmethacrylate)-grafted nanocellulose as matrices for the adsorption of lysozyme from aqueous solutions." *Chemical Engineering Journal* 187 (2012): 150–159. doi: 10.1016/j.cej.2012.01.113.

[51] Maatar, Waafa, Sabrine Alila, and Sami Boufi. "Cellulose based organogel as an adsorbent for dissolved organic compounds." *Industrial Crops and Products* 49 (2013): 33–42. doi: 10.1016/j.indcrop.2013.04.022.

[52] Yuan, Guohao, Mayakrishnan Prabakaran, Sun Qilong, Jung Soon Lee, Ill-Min Chung, Mayakrishnan Gopiraman, and Kyung-Hun Song, et al. "Cyclodextrin functionalized cellulose nanofiber composites for the faster adsorption of toluene from aqueous solution." *Journal of the Taiwan Institute of Chemical Engineering* 70 (2017): 352–358. doi: 10.1016/j.jtice.2016.10.028.

[53] Yang, Ruming, He Tan, Fanglin Wei, and Shuangfei Wang "Peroxidase conjugate of cellulose nanocrystals for the removal of chlorinated phenolic compounds in aqueous solution." *Biotechnology* 7 (2008): 233–241. doi: 10.3923/biotech.2008.233.241.

[54] Zhou, Hua, Bin Gao, Yanmei Zhou, Han Qiao, Wenli Gao, Haonan Qu, and Shanhu Liu, et al. "Facile preparation of 3D GO/CNCs composite with adsorption performance towards [BMIM][Cl] from aqueous solution." *Journal of Hazardous Materials* 337 (2017): 27–33. doi: 10.1016/j.jhazmat.2017.05.002.

[55] Sehaqui, Houssine, Andreas Mautner, Uxua Perez de Larraya, Numa Pfenninger, Philippe, Tingaut, and Tanja Zimmermann. "Cationic cellulose nanofibers from waste pulp residues and their nitrate, fluoride, sulphate and phosphate adsorption properties." *Carbohydrate Polymers* 135 (2016): 334–340. doi: 10.1016/j.carbpol.2015.08.091.

[56] Nair Vaishakh, and R. Vinu. "Peroxide-assisted microwave activation of pyrolysis char for adsorption of dyes from wastewater." *Bioresource Technology* 216 (2016): 511–519. doi: 10.1016/j.biortech.2016.05.070.

[57] Yao, Tong, Song Guo, Changfeng Zeng Chongqing Wang, and Lixiong Zhang. "Investigation on efficient adsorption of cationic dyes on porous magnetic polyacrylamide microspheres." *Journal of Hazardous Materials* 292 (2015): 90–97. doi: 10.1016/j.jhazmat.2015.03.014.

[58] Zhou, Chengjun, Qinglin Wu, Tingzhou Lei, and Ioan I. Negulescu. "Adsorption kinetic and equilibrium studies for methylene blue dye by partially hydrolyzed polyacrylamide/cellulose nanocrystal nanocomposite hydrogels." *Chemical Engineering Journal* 251 (2014): 17–24. doi: 10.1016/j.cej.2014.04.034.

[59] Yang, Xue, Hui Liu, Fuyi Han, Shuai Jiang, Lifang Liu, and Zhaopeng Xia. "Fabrication of cellulose nanocrystal from carexmeyeriana kunth and its application in the adsorption of methylene blue." *Carbohydrate Polymers* 175 (2017): 464–472. doi: 10.1016/j.carbpol.2017.08.007.

[60] Benmassaoud, Yassine, María J Villasenor, Rachid Salghi, Shehdeh Jodeh, Manuel Algarra, Mohammed Zougagh, and Angel Rios. "Magnetic/non-magnetic argan press cake nanocellulose for the selective extraction of Sudan dyes in food samples prior to the determination by capillary liquid chromatography." *Talanta* 166 (2017): 63–69. doi: 10.1016/j.talanta.2017.01.041.

[61] Tang, Juntao, Yang Song, Feiping Zhao, Stewart Spinney, Julianada Silva Bernardes, and Kam Chiu Tam. "Compressible cellulose nanofibril (CNF) based aerogels produced via a bio-inspired strategy for heavy metal ion and dye removal." *Carbohydrate Polymers* 208 (2019): 404–412. doi: 1016/j.carbpol.2018.12.079.

[62] Guan, Ying, Hou-Yong Yu, Somia Yassin Hussain Abdalkarim, Chuang Wang, Feng Tang, Jaromir Marek, and Wei-Lai Chen, et al. "Green one-step synthesis of ZnO/cellulose nanocrystal hybrids with modulated morphologies and superfast absorption of cationic dyes." *International Journal of Biological Macromolecules* 132 (2019): 51–62. doi: 10.1016/j.ijbiomac.2019.03.104.

[63] Kaili, Song, Helan Xu, Lan Xu, Kongliang Xie, and Yiqi Yang. "Cellulose nanocrystal-reinforced keratin bioadsorbent for effective removal of dyes from aqueous solution." *Bioresource Technology* 232 (2017): 254–262. doi: 10.1016/j.biortech.2017.01.070.

[64] Jin, Liqiang, Weigong Li, Qinghua Xu, and Qiucun Sun. "Amino-functionalized nanocrystalline cellulose as an adsorbent for anionic dyes." *Cellulose* 22 (2015): 2443–2456. doi: 10.1007/s10570-015-0649-4.

[65] Shanmugarajah, Bawaanii, Irene MeiLeng Chew, Nabisab Mujawar Mubarak, Thomas Shean Yaw Choong, Chang Kyoo Yoo, and Khang Wei Tan. "Valorization of palm oil agro-waste into cellulose biosorbents for highly effective textile effluent remediation." *Journal of Cleaner Production* 210 (2019): 697–709. doi: 10.1016/j.jclepro.2018.10.342.

[66] Karim, Zoheb, Aji P. Mathew, Mattias Grahn, Johanne Mouzon, and Kristiina Oksman. "Nanoporous membranes with cellulose nanocrystals as functional entity in chitosan: removal of dyes from water." *Carbohydrate Polymers* 112 (2014): 668–676. doi: 10.1016/j.carbpol.2014.06.048.

[67] Zhu, Wenjing, Lin Liu, Qian Liao, Xuan Chen, Zhouqi Qian, Junyan Shen, and Junlong Liang, et al. "Functionalization of cellulose with hyperbranched polyethylenimine for selective dye adsorption and separation." *Cellulose* 23 (2016): 3785–3797. doi: 10.1007/s10570-016-1045-4.

[68] Qiao, Han, Yanmei Zhou, Fang Yu, Enze Wang, Yinghao Min, Qi Huang, and Lanfang Pang, et al. "Effective removal of cationic dyes using carboxylate-functionalized cellulose nanocrystals." *Chemosphere* 141 (2015): 297–303. doi: 10.1016/j.chemosphere.2015.07.078.

[69] Beyki, Mostafa Hossein, Mehrnoosh Bayat, and Farzaneh Shemirani. "Fabrication of core-shell structured magnetic nanocellulose base polymeric ionic liquid for effective biosorption of Congo red dye." *Bioresource Technology* 218 (2016): 326–334. doi: 10.1016/j.biortech.2016.06.069.

[70] Maatar, Waafa, and Sami Boufi. "Microporous cationic nanofibrillar cellulose aerogel as promising adsorbent of acid dyes." *Cellulose* 24 (2016): 1001–1015. doi: 10.1007/s10570-016-1162-0.

[71] Luan, Jiabin, Jian Wu, Yudong Zheng, Wenhui Song, Guojie Wang, Jia Guo, and Xun Ding. "Impregnation of silver sulfadiazine into bacterial cellulose for antimicrobial and biocompatible wound dressing." *Biomedical Materials* 7 (2012): 065006. doi: 10.1088/1748-6041/7/6/065006.

[72] Andresen, Martin, Per Stenstad, Trond Moretro, Solveig Langsrud, Kristin Syverud, Leena-Sisko Johansson, and Per Stenius. "Nonleaching antimicrobial films prepared from surface-modified microfibrillated cellulose." *Biomacromolecules* 8 (2007): 2149–55. doi: 10.1021/bm070304e.

[73] Rai, Mahendra, Alka Yadav, and Aniket Gade. "Silver nanoparticles as a new generation of antimicrobials." *Biotechnology Advances* 27 (2009): 76–83. doi: 10.1016/j.biotechadv.2008.09.002.

[74] Xiong, Rui, Canhui Lu, Wei Zhang, Zehang Zhou, and Xinxing Zhang. "Facile synthesis of tunable silver nanostructures for antibacterial application using cellulose nanocrystals." *Carbohydrate Polymers* 95 (2013): 214–219. doi: 10.1016/j.carbpol.2013.02.077.

[75] Fortunati E., I. Armentano, Q. Zhou, A. Iannoni, E. Saino, L. Visai, L. A. Berglund, and J. M. Kenny. "Multifunctional bio-nanocomposite films of poly(lactic acid), cellulose nanocrystals and silver nanoparticles." *Carbohydrate Polymers* 87 (2012): 1596–1605. doi: 10.1016/j.carbpol.2011.09.066.

[76] Yu, Hou-Yong, Zong-Yi Qin, Bin Sun, Chen Feng Yan, and Ju-Ming Yao. "One-pot green fabrication and antibacterial activity of thermally stable corn-like CNC/Ag nanocomposites." *Journal of Nanoparticle Research* 16 (2014): 2202. doi: 10.1007/s11051-013-2202-4.

[77] Liu, He, Jie Song, Shibin Shang, Zhanqian Song, and Dan Wang. "Cellulose nanocrystal/silver nanoparticle composites as bifunctional nanofillers within waterborne polyurethane." *ACS Applied Materials & Interfaces* 4 (2012): 2413–2419. doi: 10.1021/am3000209.

[78] Martins, Natercia C. T., Carmen S. R. Freire, Ricardo J. B. Pinto, Susana C. M. Fernandes, Carlos Pascoal Neto, Armando J. D. Silvestre, and Jessica Causio, et al. "Electrostatic assembly of Ag nanoparticles onto nanofibrillated cellulose for antibacterial paper products." *Cellulose* 19 (2012): 1425–1436. doi: 10.1007/s10570-012-9713-5.

[79] Díez, Isabel, Paula Eronen, Monika Osterberg, Markus B. Linder, Olli Ikkala, and Robin H. A. Ras. "Functionalization of nanofibrillated cellulose with silver nanoclusters: fluorescence and antibacterial activity." *Macromolecular Bioscience* 11 (2011): 1185–1191. doi: 10.1002/mabi.201100099.

[80] Berndt, Sabrina, Falko Wesarg, Cornelia Wiegand, Dana Kralisch, and Frank A. Müller. "Antimicrobial porous hybrids consisting of bacterial nanocellulose and silver nanoparticles." *Cellulose* 20 (2013): 771–783. doi: 10.1007/s10570-013-9870-1.

[81] Rabie, Erika, June Cheptoo Serem, Hester Magdalena Oberholzer, Anabella Regina Marques Gaspar, and Megan Jean Bester. "How methylglyoxal kills bacteria: an ultrastructural study." *Ultrastructural Pathology* 40 (2016): 107–111. doi: 10.3109/01913123.2016.1154914.

[82] El-Wakil, Nahla A., Enas A. Hassan, Ragab E. Abou-Zeid, and Alain Dufresne. "Development of wheat gluten/nanocellulose/titanium dioxide nanocomposites for active food packaging." *Carbohydrate Polymers* 124 (2015): 337–346. doi: 10.1016/j.carbpol.2015.01.076.

[83] Galkina, O. L., K. Onneby, P. Huang, V. K. Ivanov, A. V. Agafonov, G. A. Seisenbaeva, and V. G. Kessler, "Antibacterial and photochemical properties of cellulose nanofiber-titania nanocomposites loaded with two different types of antibiotic medicines." *Journal of Materials Chemistry B* 3 (2015): 7125–7134. doi: 10.1039/C5TB01382H.

[84] Lefatshe, Kebadiretse, Cosmas M. Muiva, and Lemme P. Kebaabetswe. "Extraction of nanocellulose and *in-situ* casting of ZnO/cellulose nanocomposite with enhanced photocatalytic and antibacterial activity." *Carbohydrate Polymers* 164 (2017): 301–308. doi: 10.1016/j.carbpol.2017.02.020.

[85] Nath, B. K., C. Chaliha, E. Kalita, and M. C. Kalita. "Synthesis and characterization of ZnO:CeO$_2$: nanocellulose: PANI bionanocomposite. A bimodal agent for arsenic adsorption and antibacterial action." *Carbohydrate Polymers* 148 (2016): 397–405. doi: 10.1016/j.carbpol.2016.03.091.

[86] Anitha, S., B. Brabu, D. J. Thiruvadigal, C. Gopalakrishnan, and T. S. Natarajan. "Optical, bactericidal and water repellent properties of electrospun nano-composite membranes of cellulose acetate and ZnO." *Carbohydrate Polymers* 97 (2013): 856–863. doi: 10.1016/j.carbpol.2013.05.003.

[87] Abdalkarim, Somia Yassin Hussain, Hou-Yong Yu, Duanchao Wang, and Juming Yao. "Electrospun poly(3-hydroxybutyrate-*co*-3-hydroxy-valerate)/cellulose reinforced nanofibrous membranes with ZnO nanocrystals for antibacterial wound dressings." *Cellulose* 24 (2017): 2925–2938. doi: 10.1007/s10570-017-1303-0.

[88] Pal, Nidhi, Poornima Dubey, P. Gopinath, and Kaushik Pal. "Combined effect of cellulose nanocrystal and reduced graphene oxide into poly-lactic acid matrix nanocomposite as a scaffold and its antibacterial activity." *International Journal of Biological Macromolecules* 95 (2017): 94–105. doi: 0.1016/j.ijbiomac.2016.11.041.

[89] Xiao-Ning, Yang, Dong-Dong Xue, Jia-Ying Li, Miao Liu, Shi-Ru Jia, Li-Qiang Chu, and Fazli Wahid, et al. "Improvement of antimicrobial activity of graphene oxide/bacterial cellulose nanocomposites through the electrostatic modification." *Carbohydrate Polymers* 136 (2016): 1152–1160. doi: 10.1016/j.carbpol.2015.10.020.

16

State of the Art and Future Perspective of Nanocellulose in Water Treatment

Kritika S. Sharma
Central University of Gujarat, Gandhinagar, India

Rekha Sharma
Banasthali Vidyapith, Banasthali, India

Dinesh Kumar
Central University of Gujarat, Gandhinagar, India

CONTENTS

16.1 Introduction

Water is essential for the survival of life on Earth and also for domestic-industrial activities [1–3]. But potable clean water is one of the biggest challenges in the 21st century [4]. To address this issue, the use of water treatment has achieved an important consideration [5, 6]. Thus, drinking water, household wastewater, and industrial wastewater to eliminate contaminants are important for the use/reuse of water. Much research has been done for biomass use, as it is a renewable and sustainable resource. Cellulose is a potential source of biomass, and the most bountiful polymer has been used in many applications for hundreds of years [7, 8]. Cellulose structure shows a hierarchical assembly—elementary fibrils, macromolecules, nanofibrils, and microfibrils constitute cell walls of green plants along with hemicelluloses and lignin, as shown in Fig. 16.1 [9, 10].

The study on this subject began since it was discovered by mechanical and chemical treatment [11], various naturally existing sources of cellulose could be created by characteristics of fibrous matter one or two dimensions (1D or 2D) [12–14]. The words nanocellulose (NC) or cellulose nanomaterials (CN) are used to describe a range of materials obtained from cellulose, placing on a minimum of one dimension on a nanometer (nm) scale [15].

Various reports have been introduced on NC-enabled membrane technology based on nanoscale cellulose fibers that can be obtained from biomass [16]. The NC-based material and membranes can be created from low-cost, widely available, and sustainable materials. This can decrease membrane separation cost, as it [17] eliminates a range of contaminants in one step by size segregation and adsorption [18–20].

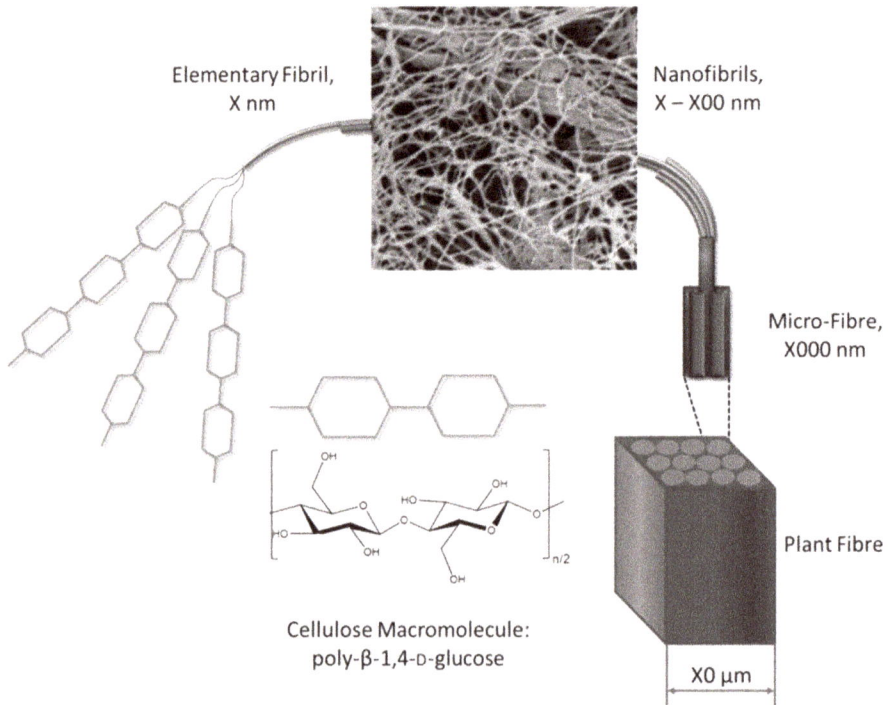

FIGURE 16.1   Hierarchical structure of plant fibers [9].

Membrane separation processes remain extremely energy efficient to eliminate water impurities [21] (particles of μm to Å size hydrated ions) [22–24]. But the present membrane technology is costly, thus is unreachable by the underprivileged [16]. This chronicle provides an approach to the fact that NC can be an important, safe, and low-cost nanomaterial for membrane applications. Exceptional capabilities and output were seen when simple types of nanocellulose were formed and used [25]. Fig. 16.2 shows a summary of categories of nanocellulose.

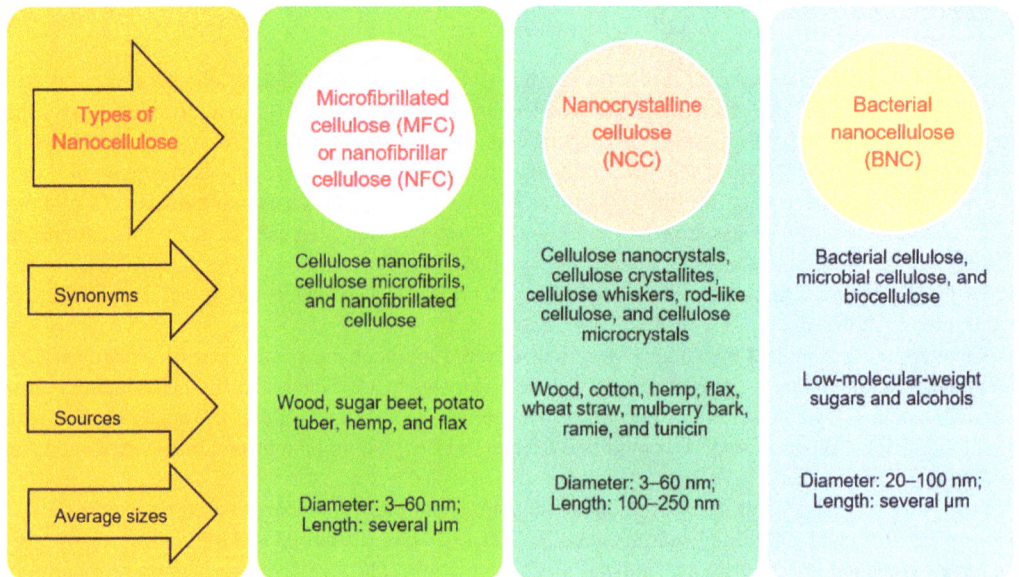

FIGURE 16.2   Summary of nanocellulose categories.

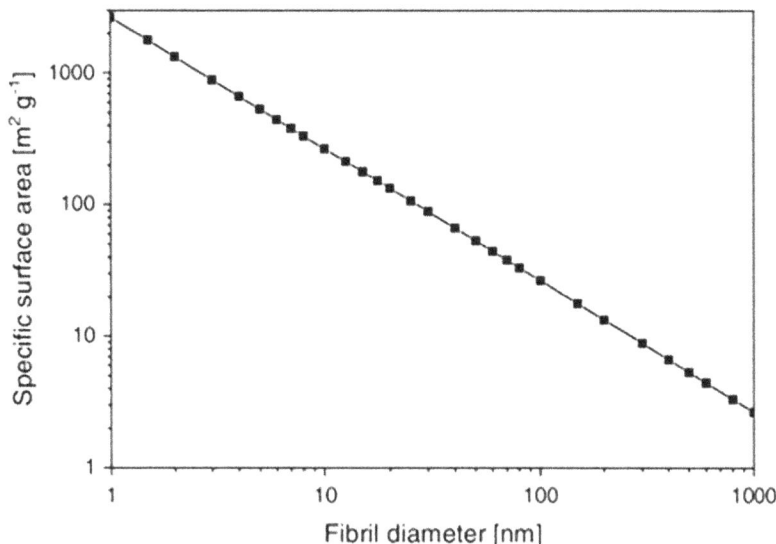

FIGURE 16.3   Relationship between fibril diameter and specific surface area [9].

The feasibility of NCs as base matter depends upon their excessive specific surface area (SSA) and wide availability of OH groups [9] that display some inclination toward contaminants (e.g., ionic structures/dyes) [26–28]. The NC is made up of elementary fibrils, SSA increases exponentially with diameter decrement. Fig. 16.3 shows the relationship between SSA and elementary fibril diameter [29]. The NCs in their several forms (bacterial cellulose, cellulose nanofibrils, and cellulose nanocrystals) are investigated and used in water treatment with filters and membranes, nanoparticles (NPs) filtration [30–32].

The NC has notable mechanical aspects and has high surface functional moieties that can be altered or crosslinked with various substances [33, 34]. There has been a curiosity for NC-based materials due to their extensive applications [35]. It is extensively used as a green strengthening agent in polymers. But when NCs are used in aqueous environments, hydrophilic characteristic reduces their mechanical strength. Thus, several studies were carried out on crosslinking techniques for the preparation of NC-based substances that are dimensionally strong in water and possess better physical-chemical characteristics [36–38].

16.2   State-of-the-Art Nanocellulose in Water Treatment

16.2.1   NC-Enabled Membranes, Composite Membranes, and Filters

A study introduced a new cheaper platform using nanocellulose-enabled membrane technology and pressure-guided filtration methods such as micro-, ultra-, nanofiltration techniques and reverse osmosis [39]. Some important factors regulating the filtration performance of nanocellulose-capable membranes were expressed in detail. Table 16.1 gives brief information on various membranes prepared from nanocellulose and their respective porosity/pore size. Such formats involve self-standing membranes, thin-film nanofibrous composite (TFNC) membranes, and nanocomposite barrier layers on differing scaffoldings.

Various articles discussed using NC in aqueous treatment emphasize membranes and filtration techniques produced entirely from NC and its composites, as shown in Fig. 16.4 [9]. Having wide usage and functions in domains such as composites [67, 42], functional additives, medical, and H_2O treatment [68–70], NCs are among the evolving products of the present century [71].

A complex membrane of polycaprolactone (PCL) and cellulose nanofibers (CNF) [1] in diverse proportions (PCL:CNF (v/v) ratios: (a) 80:20, (b) 60:40, and (c) 50:50) was formed, as shown in Fig. 16.5, to obtain organic membranes possessing superior mechanical and other characteristics to eliminate pollutants from H_2O [72, 73].

TABLE 16.1

Membranes Prepared from Nanocellulose.

Membrane		Reference	Porosity (%)/pore size (nm)
Self-standing membranes prepared from nanocellulose	*Cladophora* nanofibers	[40]	35%/19 nm
	Cationic CNF in H_2O	[41]	37%
	Cationic CNF solvent exchanged by C_2H_5OH	[40]	46%
	Cationic CNF freeze-dried	[40]	79%
	Cationic CNF supercritical CO_2 drying	[40]	73%
	CNC incorporated with chitosan, crosslinked with glutaraldehyde	[42]	13–10 nm
	Cladophora nanofibers	[43]	14–24 nm
	Homogenized-CNF solvent exchanged with C_2H_5OH	[44]	10 nm
	Incorporation of calcium particles in homogenized-CNF solvent exchanged with C_2H_5OH	[40]	10 nm
	TEMPO CNF and cellulose acetate membrane / lysine grafted	[45]	13.8 nm
	Cellulose in LiCl/DAM /graphene oxide	[46]	83%
	Nanocellulose fibers /NMMO /alumina	[47]	2.5 ± 12 nm
	Nanocellulose fibers /NMMO /interfacial polymerization PEI, TMC	[48]	0.45 nm
	Bacterial nanofibers	[49]	2.4 nm
	CNC	[40]	2.4 nm
	TEMPO-CNF	[40]	19 nm
	CNC /electrospun PVDF-HFP	[50]	0.2–0.45 µm
Nanocellulose used as a barrier layer in composite membranes	TEMPO-CNF	[51]	54.6 nm
	TEMPO-chitin	[40]	–
	TEMPO-NCC	[40]	–
	TEMPO-cellulose nanowhiskers /PAN	[52]	0.22 µm
	Microcrystalline CNFs /PAN	[53]	>100 nm
	Chemically modified CNFs /PAN	[40]	>100 nm
	CNF /PVDF membrane	[54]	–
	Meldrum acid-modified CNF /PVDF	[40]	–
	PVA and TEMPO-CNF	[55]	6.1 nm
	Cellulose solution in ionic liquid /PAN layer	[56]	50 nm
	TEMPO CNF /interfacial polymerization PIP, TEA	[57]	0.3 nm
	2,3-Dicarboxy CNFs crosslinked /$CaCl_2$	[58]	10–55 nm
	2,3-Dicarboxy CNFs crosslinked/$Na_3P_3O_9$	[40]	55 nm
	CNF	[59]	25 nm
	Chitin nanofiber	[40]	27 nm
	Blend of nanocellulose-nanofiber and chitin-nanofiber	[40]	14 nm
	Cellulose solution in urea/NaOH /cellulose acetate membrane	[60]	–
	2,3-Dicarboxy CNFs	[61]	–
	TEMPO CNF spray-coated	[62]	–
	TEMPO-CNF crosslinked /citric acid	[63]	23 µm
	NCC	[64]	0.78–0.22 µm
	TEMPO-CNF /polyvinyl amine grafted	[65]	0.73 µm
	CNF phase inversion method	[66]	70.9 nm

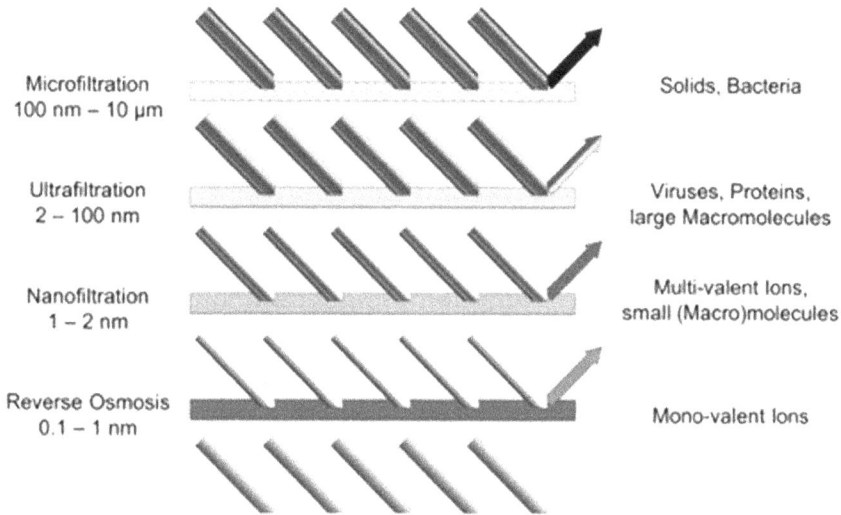

FIGURE 16.4 Membrane and filtration regimes and the size (right side of figure) and type of pollutants rejected (left side of figure) [9].

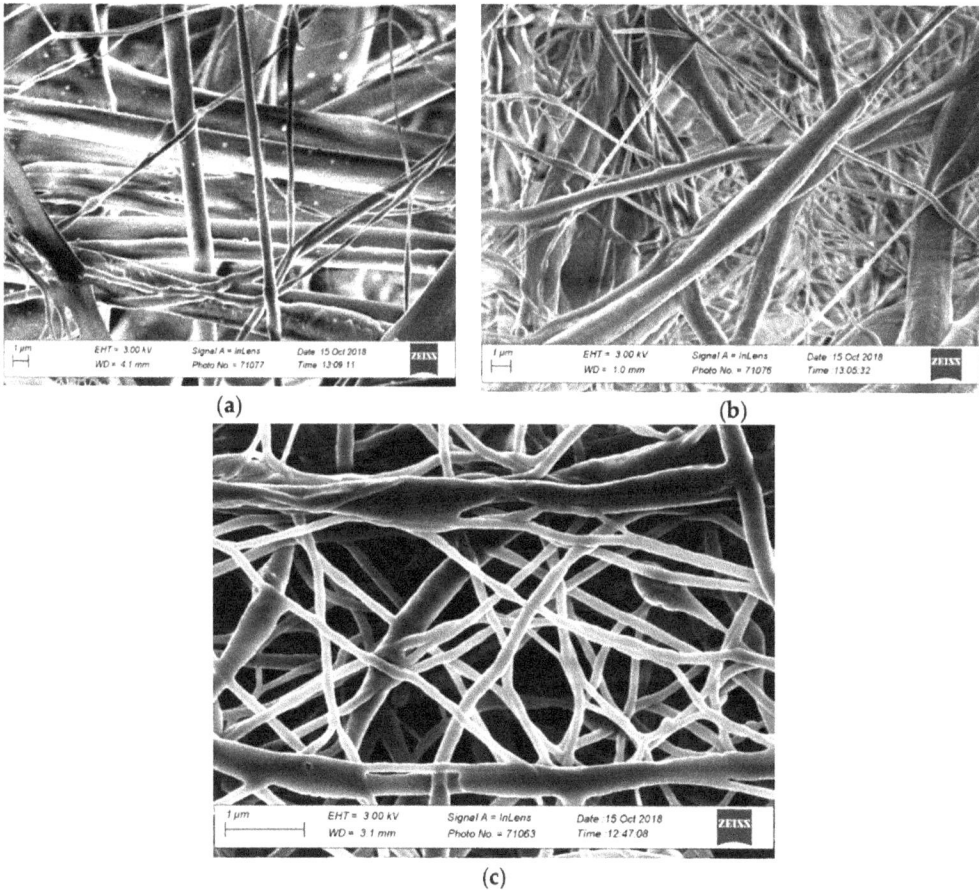

FIGURE 16.5 SEM images of electrospun nanofibrous membranes fabricated from polycaprolactone/cellulose nanofiber (PCL/CNF) copolymer blends at (v/v) ratios: (a) 80:20, (b) 60:40, and (c) 50:50 [1].

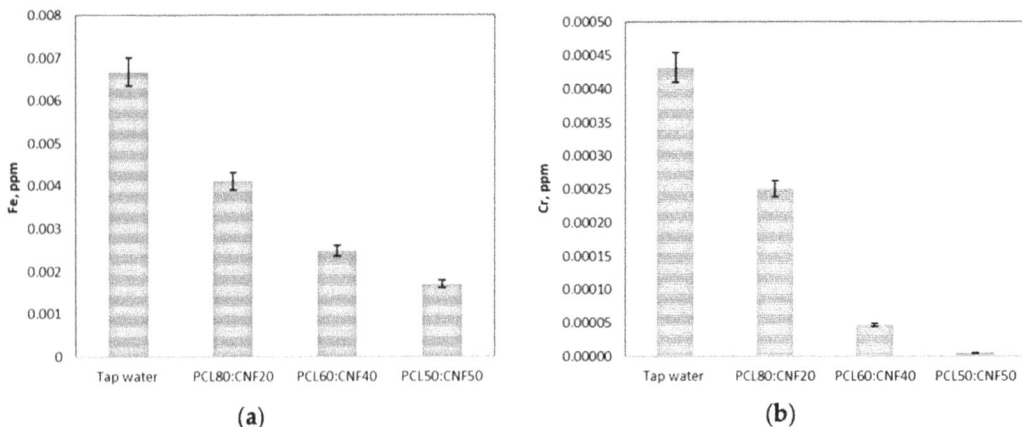

FIGURE 16.6 Removal of heavy metals in tap water using PCL/CNF (v/v) ratio (80:20, 60:40, and 50:50) composite electrospun membranes: (a) iron (Fe) and (b) chromium (Cr) [1].

Electrospinning technique was used to produce nanocellulose and PCL composite membranes. For tap water filtration, cellulose acquired from Agave bagasse and polycaprolactone nanofibers were utilized to produce membranes. Membranes formed depicted a porosity and structure on a nanometer range [74]. The electrospun PCL:CNF membrane accomplished the removal of heavy metals in tap water. PCL50:CNF50 showed excellent results among three membranes: 75% iron, 99% chromium, as shown in Figs. 16.6 and 16.7, and 100% turbidity and conductivity. CNF constitutes biowaste obtained from tequila manufacturing [75–78]. An environment-friendly and green filtration for H_2O treatment can be achieved by electrospun CNF and PCL membranes [79]. Thus, CNF is a suitable biowaste offshoot derived from tequila manufacturing and has added value in the manufacture of composite PCL/CNF electrospun membranes, which can be used in eco-friendly filtration systems for H_2O treatment [80, 81].

16.2.2 Crosslinked NCs-Based Material and Aerogels

The rapidly sprouting domain of NCs-based materials is crosslinked materials that have advantages such as upgraded mechanical, structural, aqueous, and thermic sturdiness. Also, possible uses of NC-based crosslinked substances in heavy metal adsorption cannot be denied [32]. Table 16.2 discusses various crosslinked NCs-based materials and aerogels and their respective target molecule or application.

A green method to prepare aerogel membranes based on CNFs with poly(ethylene imine) (PEI) crosslink was then decorated with silver nanoparticles (AgNPs), as formulated in a study. Magnificent constant catalytic discoloration of aqueous cationic and anionic dye solutions in batch and flow filtration was demonstrated by prepared composite aerogel membrane. Another alluring aspect noted while testing for long periods, aerogel membrane displayed magnificent recuperation of shape or shape-retaining aspect in an aqueous environment. Thus, high efficiency of catalytic discoloration efficiency and membrane stability when H_2O permeance or flux was also significant as much as 5×104 $L/m^2/h$. Hence, formed aerogel membranes displayed excellent capability in wastewater treatment and catalytic use [82].

16.2.3 NC-Based Adsorbent

NC materials have documentations (Table 16.3) of being feasible adsorbent. To put recent developments in a nutshell, methods under present usage offers an organized outline of technologies that endorse firm research in the prospect of NC-dependent adsorbents.

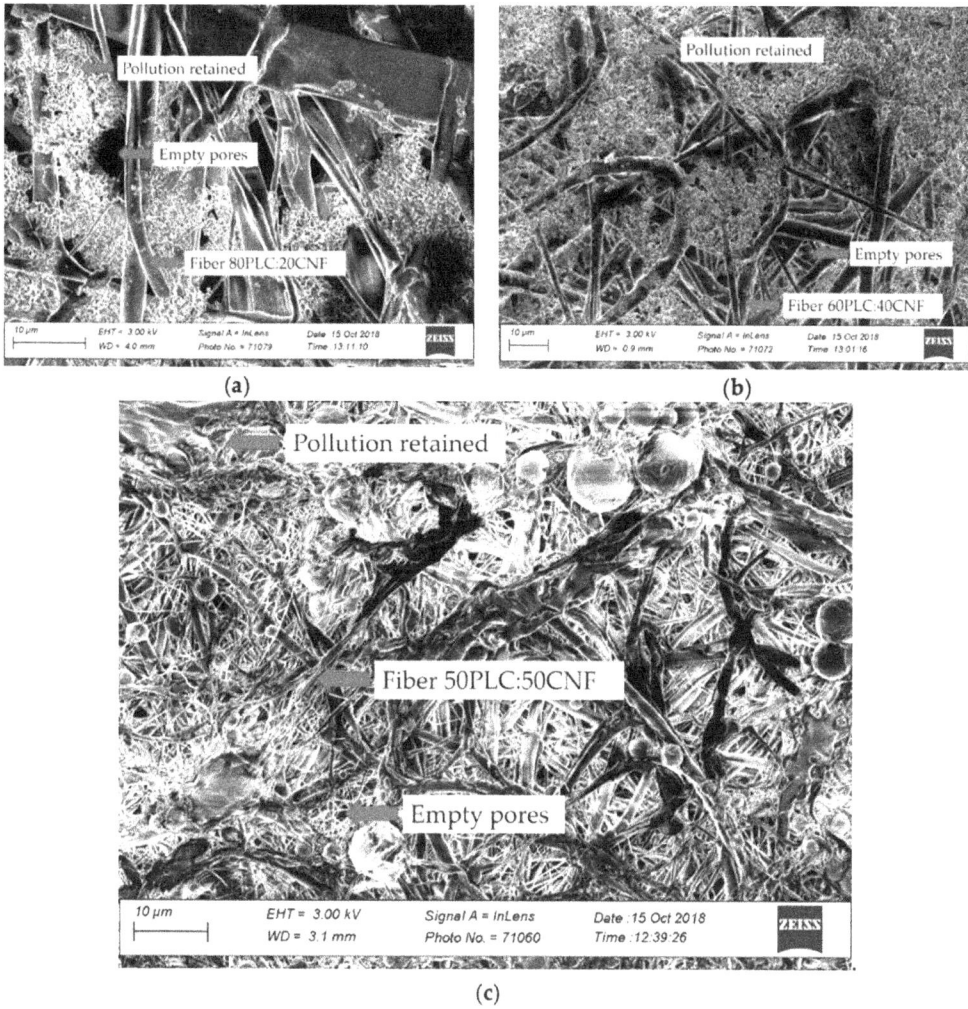

FIGURE 16.7 SEM images after filtration of tap water using electrospun membranes of PCL: CNF (v/v) ratios: (a) 80:20, (b) 60:40, and (c) 50:50 [1].

TABLE 16.2

Potential Applications of Crosslinked NCs-Based Material and Aerogels.

Fabrication	Target Molecule or Application	Reference
(i) Physically Crosslinked NC-Based Material		
CNFs/CTS film	–	[83]
MOF aerogels	Conventional filtration	[84]
CNCs/sodium alginate/Ca(II) hydrogel beads	Drug delivery	[85]
CNCs or CNFs/PVA/borax foams	Controllable scaffolds or templates for functional composite materials	[86]
Ca(II) crosslinked CNFs hydrogels	Biomaterials applications	[87]
		[88]
		[89]
CNCs/PVA/borax composite films		[90]
PAM/CNFs-Fe(III) hydrogels	Various applications	[91]
Collagen/CNCs films	Casing	[92]

(Continued)

TABLE 16.2 *(Continued)*

Potential Applications of Crosslinked NCs-Based Material and Aerogels.

Fabrication		Target Molecule or Application	Reference
(ii) Chemically Crosslinked NCs-Based Material			
(a) Amide-linkages	CNFs/MDI aerogels	Filters for hydrophobic liquids and water separation	[93]
	Gentamicin-modified CNFs/CAA/ APTES sponges	Biological applications	[94]
	TOCNFs/alginate hydrogels	Biomedical devices, drug delivery, sensors	[95]
	Chitosan/PVP/oxidized CNCs hydrogels	Biomedical applications	[96]
	P(AAm-*co*-AAc)/CNFs/Fe(III) hydrogels	Biomedical and tissue engineering	[97]
	CTS-TOCNFs film	–	[98]
	CNFs/PAE filaments	Replacement for naturally and industrially produced fibers	[99]
(b) Ester-linkages	CNFs/BTCA/MDPA spongelike aerogels	Thermal protective equipment	[100]
	CNFs/PVA/MTM-glutaraldehyde aerogels	Inorganic/organic hybrid foams	[101]
	TOCNFs/PAE hybrid films	Applied under high humid conditions	[102]
	Bifunctional reactive CNCs/PVA films	Reinforcing agent	[103]

TABLE 16.3

Nanocellulose-Based Adsorbent and Their Target Molecule or Applications.

Adsorbent	Target Molecule or Application	Reference
Amino-functionalized /nanocrystalline cellulose (CNC) grafted	Dye removal	[104]
Cellulose nanofibril (CNF)-based aerogels	Heavy metal and dye removal	[105]
CNC extract	Methylene blue	[106]
Interfacial polymerized CNC nanoporous membrane	Methyl orange	[107]
Magnetic NC membrane 3 A	Rhodamine B	[108]
TEMPO-oxidized CNF	Cu(II)	[109]
Glycidylmethacrylate (GMA) grafted magnetic nanocellulose	Ni(II)	[110]
S-Ligand tethered CNF	Pb(II)	[111]
CNC composite magnetic NPs		[112]
Hybrid crosslinked cellulose aerogels	Cr(VI)	[113]
NOCNFs	Cd(II)	[114]
SNC-3-MPA	Hg(II)	[115]
Silylated nanocellulose sponges	Oil removal	[116]
Surface imprinted magnetic carboxylated CNC	Adsorption of fluoroquinolones	[117]
Amino-functionalized CNC /chitosan grafted composite	Diclofenac adsorption	[118]
γ-Cyclodextrin CNF nanocomposite	Toluene	[119]
Surface-modified acetate CNF	DNA adsorption	[120]

16.3 Future Perspective of Nanocellulose in Water Treatment

Ligands are chosen as chemicals with the ability by which strong interactions can be established to obtain high adsorption capacity [111]. To our despair, low adsorption performance is shown by ligands from which target molecules can be easily eluted. All this can result in loss of performance, such as (i) failure to remove target molecule when weak eluent agent is used in regeneration processes, (ii) deformation of adsorbent when it is strong, and (iii) disturbing structure of isolated target molecule. It is generally through secondary interactions such as electrostatic [121], hydrophobic, and hydrogen bonding [122] that adsorption occurs [123]. Hence, eluent must have ionic strength able to break these interactions. This should be at the best level for the reasons mentioned above. Although NPs tend to agglomerate to reduce surface energy, they are more useful than compact structures such as hydrogels [112].

In some cases, in terms of internal diffusion and quantum size, nanosize structures that provide high surface area can be problematic. For structures with surface OH groups, such as nanocellulose, which permits a larger number of hydrogen bonds, this is a huge hurdle. Agglomeration decreases surface area, which causes reduced adsorption potential. Even though this can be defeated with hierarchical structures, e.g., cryogels and aerogels [124], collapse or contraction of network structure while removing the solvent by capillary forces should be circumvented [125]. Supercritical drying can resolve this problem, but the drawback is exorbitant [126].

During the improvement and execution phases, many hurdles, however, arise. Hence, preferably unaltered CN can be utilized efficiently for adsorption, surface improvement and alteration will potentially be essential to permit sorbent specificity, resistance to interference, large capacity, and enormous efficiency [104]. But, maximum improvement techniques are expensive, harmful secondary products, require elevated energy use, and acquired yield are lower than expected/desired in other techniques. The fact of less capacity of CN manufacturing and comparatively high cost curbing wide-scale utilization in inexpensive applications, e.g., drainage [127], industrial wastewater [128], and environmental remediation, are other hurdles in its materialization [129]. But for expensive applications (e.g., pharmaceutical wastewaters treatment) in which profit margins are high and volumes are low, the above hurdle is not noticed. With this in mind, more studies are needed on surface changes or improvements, cost reduction, adsorbent specification, adsorbent selection, adsorbing ability, etc. [130]. Further investigation on CN-based adsorbents might be done by concentrating on pharmaceutical and biomolecular use-specific adsorbent targets [131] because of the likely increment of production cost [132].

16.4 Conclusion

Effective segregation/elimination is vital, and its use ranges from the food industry to water remediation techniques [133]. The need for price and time efficiency must be met by a properly designed adsorption technique [134]. The desired raw material used in water/wastewater treatment to create adsorbents should be environment-friendly so that consequent ecological issues can be evaded [135]. Produced from the most widely available organic biopolymer, NC is an adsorbent that fulfills many high safe conditions and shows great effectiveness.

The fact is that the current commercial production of nanocellulosic materials is majorly from wood. However, new advancements associated with inexpensive nanocellulose extraction from non-wood plants are still in the primary phase [136].

Multiple characteristics and techniques used to customize nanocellulose-based materials have been analyzed to exhibit cost-effectiveness, desired results, time-efficient, bio-renewability, and sorbent maintenance [137]. As concluded from studies, CN possesses a greater affinity for positively charged atoms/molecules, also during the integration of CN and polymeric structures. They show enormous adsorption performance compared to divalent metal cations, e.g., lead, zinc, and copper [138]. Biomolecules and other sizable positively charged molecules lead to steric hindrance, decreasing adsorption potential [139]. Chemical modifications that improve performance do not always show expected adsorption performance due to an increase of non-specific entities, among other reasons, even though they are less favorable with respect to safety and security. Hence, there is an influence on desorption-adsorption efficiency. However, CN-based materials altogether are a potential substitute for conventional water treatment methods.

Acknowledgment

The authors are grateful to DST, New Delhi, for providing financial assistance for the work (Project Acceptance Order F. No. DST/TM/WTI/WIC/2K17/124(C).

REFERENCES

[1] Palacios Hinestroza, Hasbleidy, Hilary Urena-Saborio, Florentina Zurita, Aida Alejandra Guerrero de Leon, Gunasekaran Sundaram, and Belkis Sulbarán-Rangel. "Nanocellulose caprolactone nanospun composite membranes and their potential for the removal of pollutants from water." *Molecules* 25 (2020): 683. doi: 10.3390/molecules25030683.

[2] Garcia-Alonso, José Alfredo, Florentina Zurita-Martinez, Carlos Alberto Guzmán-González, Jorge Del Real-Olvera, and Belkis Coromoto Sulbarán-Rangel. "Nanostructured diatomite and its potential for the removal of an antibiotic from water." *Bioinspired, Biomimetic and Nanobiomaterials* 7 (2018): 167–173. doi: 10.1680/jbibn.18.00020.

[3] Tejeda, Allan, Angeles X. Torres-Bojorges, and Florentina Zurita. "Carbamazepine removal in three pilot-scale hybrid wetlands planted with ornamental species." *Ecological Engineering* 98 (2017): 410–417. doi: 10.1016/j.ecoleng.2016.04.012.

[4] United Nations Development Programme. Available at: https://www.undp.org/content/undp/en/home/sustainable-development-goals.html (accessed June 10, 2020).

[5] Hoslett, John, Theoni Maria Massara, Simos Malamis, Darem Ahmad, Inge van den Boogaert, Evina Katsou, and Balsam Ahmad, et al. "Surface water filtration using granular media and membranes: a review." *Science of the Total Environment* 639 (2018): 1268–1282. doi: 10.1016/j.scitotenv.2018.05.247.

[6] Molodkina, L. M., D. D. Kolosova, E. I. Leonova, M. F. Kudoyarov, M. Ya Patrova, and Yu V. Vedmetskii. "Track membranes in post-treatment of domestic wastewater." *Petroleum Chemistry* 52 (2012): 487–493. doi: 10.1134/S0965544112070092.

[7] Sponsler, O. L. "Orientation of cellulose space lattice in the cell wall: additional X-ray data from Valonia cell-wall." *Protoplasma* 12 (1931): 241–254. doi: 10.1007/BF01618718.

[8] Frey-Wyssling, A. "On the crystal structure of cellulose I." *Biochimica et Biophysica Acta* 18 (1955): 166–168. dbeso.es6-3002(55)90039-1.

[9] Mautner, Andreas. "Nanocellulose water treatment membranes and filters: a review." *Polymer International* 69 (2020): 741–751. doi: 10.1002/pi.5993.

[10] Nishiyama, Yoshiharu. "Structure and properties of the cellulose microfibril." *Journal of Wood Science* 55 (2009): 241–249. doi: 10.1007/s10086-009-1029-1.

[11] Dufresne, Alain. "Cellulose nanomaterial reinforced polymer nanocomposites." *Current Opinion in Colloid & Interface Science* 29 (2017): 1–8. doi: 10.1016/j.cocis.2017.01.004.

[12] Herrick, Franklin W., Ronald L. Casebier, J. Kelvin Hamilton, and Karen R. Sandberg. "Microfibrillated cellulose: morphology and accessibility." *Journal of Applied Polymer Science: Applied Polymer Symposia* 37 (1983). Available at: https://www.osti.gov/biblio/5039044.

[13] Nickerson, R. F., and J. A. Habrle. "Cellulose intercrystalline structure." *Industrial & Engineering Chemistry* 39 (1947): 1507–1512. doi: 10.1021/ie50455a024.

[14] Turbak, A. F., Snyder, F. W., and Sandberg, K. R. "Microfibrillated cellulose, a new cellulose product: properties, uses, and commercial potential." *Journal of Applied Polymer Science: Applied Polymer Symposia* 37 (1983). Available at: https/www.osti.gov/biblio/5062478.

[15] Dufresne, Alain. "Nanocellulose: a new ageless bionanomaterial." *Materials Today* 16 (2013): 220–227. doi: 10.1016/j.mattod.2013.06.004.

[16] Sharma, Priyanka R., Sunil K. Sharma, Tom Lindstrom, and Benjamin S. Hsiao. "Nanocellulose-enabled membranes for water purification: perspectives." *Advanced Sustainable Systems* 4 (2020): 1900114. doi: 10.1002/adsu.201900114.

[17] Ferraz, Natalia, Anastasiya Leschinskaya, Farshad Toomadj, Bengt Fellstrom, Maria Strømme, and Albert Mihagn. "Membrane characterization and solute diffusion in porous composite nanocellulose membranes for hemodialysis." *Cellulose* 20 (2013): 2959–2970. doi: 10.1007/s10570-013-0045-x.

[18] Cruz-Tato, Perla, Edwin O. Ortiz-Quiles, Karlene Vega-Figueroa, Liz Santiago-Martoral, Michael Flynn, Liz M. Diaz-Vazquez, and Eduardo Nicolau. "Metalized nanocellulose composites as a feasible material for membrane supports: design and applications for water treatment." *Environmental Science & Technology* 51 (2017): 4585–4595. doi: 10.1021/acs.est.6b05955.

[19] Leitch, Megan E., Chenkai Li, Olli Ikkala, Meagan S. Mauter, and Gregory V. Lowry. "Bacterial nanocellulose aerogel membranes: novel high-porosity materials for membrane distillation." *Environmental Science & Technology Letters* 3 (2016): 85–91. doi: 10.1021/acs.estlett.6b00030.

[20] Ferraz, Natalia, Daniel O. Carlsh IpHong, Rolf Larsson, Bengt Fellström, Leif Nyholm, Maria Strømme, and Albert Mihranyan. "Haemocompatibility and ion exchange capability of nanocellulose polypyrrole membranes intended for blood purification." *Journal of the Royal Society Interface* 9 (2012): 1943–1955. doi: 10.1098/rsif.2012.0019.

[21] Warsinger, David M., Sudip Chakraborty, Emily W. Tow, Megan H. Plumlee, Christopher Bellona, Savvina Loutatidou, and Leila Karimi, et al. "A review of polymeric membranes and processes for potable water reuse." *Progress in Polymer Science* 81 (2018): 209–237. doi: 10.1016/j.progpolymsci.2018.01.004.

[22] del Pino, Manuel P., and Bruce Durham. "Wastewater reuse through dual-membrane processes: opportunities for sustainable water resources." *Desalination* 124 (1999): 271–277. doi: 10.1016/s0011-9164(99)00112-5.

[23] Van der Bruggen, Bart, Carlo Vandecasteele, Tim Van Gestel, Wim Doyen, and Roger Leysen. "A review of pressure-driven membrane processes in wastewater treatment and drinking water production." *Environmental Progress* 22 (2003): 46–56. doi: 10.1002/ep.670220116.

[24] Geise, Geoffrey M., Hac-Seung Lee, Daniel J. Miller, Benny D. Freeman, James E. McGrath, and Donald R. Paul. "Water purification by membranes: the role of polymer science." *Journal of Polymer Science Part B: Polymer Physics* 48 (2010): 1685–1718. doi: 10.1002/polb.22037.

[25] Moon, Robert J., Ashlie Martini, John Nairn, John Simonsen, and Jeff Youngblood. "Cellulose nanomaterials review: structure, properties and nanocomposites." *Chemical Society Reviews* 40 (2011): 3941–3994. doi: 10.1039/C0CS00108B.

[26] Yao, Tong, Song Guo, Changfeng Zeng, Chongqing Wang, and Lixiong Zhang. "Investigation on efficient adsorption of cationic dyes on porous magnetic polyacrylamide microspheres." *Journal of Hazardous Materials* 292 (2015): 90–97. doi: 10.1016/j.jhazmat.2015.03.014.

[27] Zhang, Shouwei, Huihui Gao, Jiaxing Li, Yongshun Huang, Ahmed Alsaedi, Tasawar Hayat, Xijin Xu, and Xiangke Wang. "Rice husks as a sustainable silica source for hierarchical flower-like metal silicate architectures assembled into ultrathin nanosheets for adsorption and catalysis." *Journal of Hazardous Materials* 321 (2017): 92–102. doi: 10.1016/j.jhazmat.2016.09.004.

[28] Liu, Peng, Houssine Sehaqui, Philippe Tingaut, Adrian Wichser, Kristiina Oksman, and Aji P. Mathew. "Cellulose and chitin nanomaterials for capturing silver ions (Ag^+) from water via surface adsorption." *Cellulose* 21 (2014): 449–461. doi: 10.1007/s10570-013-0139-5.

[29] Page, D. H., and F. El-Hosseiny. "Mechanical properties of single wood pulp fibres. VI. Fibril angle and the shape of the stress-strain curve." *Pulp & Paper Canada* (1983). ISSN: 0316-4004.

[30] Klemm, Dieter, Brigitte Heublein, Hans-Peter Fink, and Andreas Bohn. "Cellulose: fascinating biopolymer and sustainable raw material." *Angewandte Chemie: International Edition* 44 (2005): 3358–3393. doi: 10.1002/anie.200460587.

[31] Klemm, Dieter, Friederike Kramer, Sebastian Moritz, Tom Lindström, Mikael Ankerfors, Derek Gray, and Annie Dorris. "Nanocelluloses: a new family of nature-based materials." *Angewandte Chemie: International Edition* 50 (2011): 5438–5466. doi: 10.1002/anie.201001273.

[32] Mautner, Andreas, K.-Y. Lee, P. Lahtinen, M. Hakalahti, T. Tammelin, K. Li, and A. Bismarck. "Nanopapers for organic solvent nanofiltration." *Chemical Communications* 50 (2014): 5778–5781. doi: 10.1039/C4CC00467A.

[33] Liang, Luna, Chen Huang, Naijia Hao, and Arthur J. Ragauskas. "Crosslinked poly(methyl vinyl ether-*co*-maleic acid)/poly(ethylene glycol)/nanocellulosics foams via directional freezing." *Carbohydrate Polymers* 213 (2019): 346–351. doi: 10.1016/j.carbpol.2019.02.073.

[34] H. Tayeb, Ali, and Mehdi Tajvidi. "Sustainable barrier system via self-assembly of colloidal montmorillonite and crosslinking resins on nanocellulose interfaces." *ACS Applied Materials & Interfaces* 11 (2018): 1604–1615. doi: 10.1021/acsami.8b16659.

[35] Yang, Xuan, and Emily D. Cranston. "Chemically crosslinked cellulose nanocrystal aerogels with shape recovery and superabsorbent properties." *Chemistry of Materials* 26 (2014): 6016–6025. doi: 10.1021/cm502873c.

[36] Liu, Yingying, Yulong Sui, Chao Liu, Changqing Liu, Meiyan Wu, Bin Li, and Youming Li. "A physically crosslinked polydopamine/nanocellulose hydrogel as potential versatile vehicles for drug delivery and wound healing." *Carbohydrate Polymers* 188 (2018): 27–36. doi: 10.1016/j.carbpol.2018.01.093.

[37] Herrera, Martha A., Aji P. Mathew, and Kristiina Oksman. "Barrier and mechanical properties of plasticized and crosslinked nanocellulose coatings for paper packaging applications." *Cellulose* 24 (2017): 3969–3980. doi: 10.1007/s10570-017-1405-8.

[38] Song, Meili, Houyong Yu, Jiping Gu, Shounuan Ye, and Yuwei Zhou. "Chemical crosslinked polyvinyl alcohol/cellulose nanocrystal composite films with high structural stability by spraying Fenton reagent as initiator." *International Journal of Biological Macromolecules* 113 (2018): 171–178. doi: 10.1016/j.ijbiomac.2018.02.117.

[39] Quellmalz, Arne, and Albert Mihranyan. "Citric acid crosslinked nanocellulose-based paper for size-exclusion nanofiltration." *ACS Biomaterials Science & Engineering* 1 (2015): 271–276. doi: 10.1021/ab500161x.

[40] Metreveli, Giorgi, Linus Wagberg, Eva Emmoth, Sándor Belák, Maria Strømme, and Albert Mihranyan. "A size-exclusion nanocellulose filter paper for virus removal." *Advanced Healthcare Materials* 3 (2014): 1546–1550. doi: 10.1002/adhm.201300641.

[41] Sehaqui, Houssine, Benjamin Michen, Eric Marty, Luca Schaufelberger, and Tanja Zimmermann. "Functional cellulose nanofiber filters with enhanced flux for the removal of humic acid by adsorption." *ACS Sustainable Chemistry & Engineering* 4 (2016): 4582–4590. doi: 10.1021/acssuschemeng.6b00698.

[42] Karim, Zoheb, Aji P. Mathew, Mattias Grahn, Johanne Mouzon, and Kristiina Oksman. "Nanoporous membranes with cellulose nanocrystals as functional entity in chitosan: removal of dyes from water." *Carbohydrate Polymers* 112 (2014): 668–676. doi: 10.1016/j.carbpol.2014.06.048.

[43] Gustafsson, Simon, and Albert Mihranyan. "Strategies for tailoring the pore-size distribution of virus retention filter papers." *ACS Applied Materials & Interfaces* 8 (2016): 13759–13767. doi: 10.1021/acsami.6b03093.

[44] Orsolini, Paola, Tommaso Marchesi D'Alvise, Cristiana Boi, Thomas Geiger, Walter R. Caseri, and Tanja Zimmermann. "Nanofibrillated cellulose templated membranes with high permeance." *ACS Applied Materials & Interfaces* 8 (2016): 33943–33954. doi: 10.1021/acsami.6b12107.

[45] Qin, Dujian, Dalun Zhang, Ziqiang Shao, Jianquan Wang, Keguang Mu, and Libin Zhao. "Short-chain amino acids functionalized cellulose nanofibers composite ultrafiltration membrane with enhanced properties." *RSC Advances* 6 (2016): 76336–76343. doi: 10.1039/C6RA14696A.

[46] Ao, Chenghong, Wei Yuan, Jiangqi Zhao, Xu He, Xiaofang Zhang, Qingye Li, Tian Xia, Wei Zhang, and Canhui Lu. "Superhydrophilic graphene oxide@ electrospun cellulose nanofiber hybrid membrane for high-efficiency oil/water separation." *Carbohydrate Polymers* 175 (2017): 216–222. doi: 10.1016/j.carbpol.2017.07.085.

[47] Zhang, Qiu Gen, Chao Deng, Faizal Soyekwo, Qing Lin Liu, and Ai Mei Zhu. "Sub-10 nm wide cellulose nanofibers for ultrathin nanoporous membranes with high organic permeation." *Advanced Functional Materials* 26 (2016): 792–800. doi: 10.1002/adfm.201503858.

[48] Soyekwo, Faizal, Qiugen Zhang, Runsheng Gao, Yan Qu, Chenxiao Lin, Xiaoling Huang, Aimei Zhu, and Qinglin Liu. "Cellulose nanofiber intermediary to fabricate highly-permeable ultrathin nanofiltration membranes for fast water purification." *Journal of Membrane Science* 524 (2017): 174–185. doi: 10.1016/j.memsci.2016.11.019.

[49] Dufresne, Alain. *Nanocellulose: from nature to high performance tailored materials.* Berlin: Walter de Gruyter GmbH & Co KG. 2017.

[50] Lalia, Boor Singh, Elena Guillen, Hassan A. Arafat, and Raed Hashaikeh. "Nanocrystalline cellulose reinforced PVDF-HFP membranes for membrane distillation application." *Desalination* 332 (2014): 134–141. doi: 10.1016/j.desal.2013.10.030.

[51] Ma, Hongyang, Christian Burger, Benjamin S. Hsiao, and Benjamin Chu. "Ultrafine polysaccharide nanofibrous membranes for water purification." *Biomacromolecules* 12 (2011): 970–976. doi: 10.1021/bm1013316.

[52] Ma, Hongyang, Christian Burger, Benjamin S. Hsiao, and Benjamin Chu. "Nanofibrous microfiltration membrane based on cellulose nanowhiskers." *Biomacromolecules* 13 (2012): 180–186. doi: 10.1021/bm201421g.

[53] Sato, Anna, Ran Wang, Hongyang Ma, Benjamin S. Hsiao, and Benjamin Chu. "Novel nanofibrous scaffolds for water filtration with bacteria and virus removal capability." *Journal of Electron Microscopy* 60 (2011): 201–209. doi: 10.1093/jmicro/dfr039.

[54] Gopakumar, Deepu A., Daniel Pasquini, Mariana Alves Henrique, Luis Carlos de Morais, Yves Grohens, and Sabu Thomas. "Meldrum's acid modified cellulose nanofiber-based polyvinylidene fluoride microfiltration membrane for dye water treatment and nanoparticle removal." *ACS Sustainable Chemistry & Engineering* 5 (2017): 2026–2033. doi: 10.1021/acssuschemeng.6b02952.

[55] Ma, Hongyang, Kyunghwan Yoon, Lixia Rong, Mina Shokralla, Andrey Kopot, Xiao Wang, Dufei Fang, Benjamin S. Hsiao, and Benjamin Chu. "Thin-film nanofibrous composite ultrafiltration membranes based on polyvinyl alcohol barrier layer containing directional water channels." *Industrial & Engineering Chemistry Research* 49 (2010): 11978–11984. doi: 10.1021/ie100545k.

[56] Ma, Hongyang, Kyunghwan Yoon, Lixia Rong, Yimin Mao, Zhirui Mo, Dufei Fang, Zachary Hollander, Joseph Gaiteri, Benjamin S. Hsiao, and Benjamin Chu. "High-flux thin-film nanofibrous composite ultrafiltration membranes containing cellulose barrier layer." *Journal of Materials Chemistry* 20 (2010): 4692–4704. doi: 10.1039/B922536F.

[57] Wang, Xiao, Dufei Fang, Benjamin S. Hsiao, and Benjamin Chu. "Nanofiltration membranes based on thin-film nanofibrous composites." *Journal of Membrane Science* 469 (2014): 188–197. doi: 10.1016/j.memsci.2014.06.049.

[58] Visanko, Miikka, Henrikki Liimatainen, Juho Antti Sirviö, and Osmo Hormi. "A crosslinked 2,3-dicarboxylic acid cellulose nanofibril network: a nanoporous thin-film layer with tailored pore size for composite membranes." *Separation and Purification Technology* 154 (2015): 44–50. doi: 10.1016/j.seppur.2015.09.026.

[59] Ma, Hongyang, Benjamin S. Hsiao, and Benjamin Chu. "Thin-film nanofibrous composite membranes containing cellulose or chitin barrier layers fabricated by ionic liquids." *Polymer* 52 (2011): 2594–2599. doi: 10.1016/j.polymer.2011.03.051.

[60] Huang, Weijuan, Yixiang Wang, Chao Chen, John Lok Man Law, Michael Houghton, and Lingyun Chen. "Fabrication of flexible self-standing all-cellulose nanofibrous composite membranes for virus removal." *Carbohydrate Polymers* 143 (2016): 9–17. doi: 10.1016/j.carbpol.2016.02.011.

[61] Visanko, Miikka, Henrikki Liimatainen, Juho Antti Sirviö, Antti Haapala, Rafal Sliz, Jouko Niinimäki, and Osmo Hormi. "Porous thin film barrier layers from 2,3-dicarboxylic acid cellulose nanofibrils for membrane structures." *Carbohydrate Polymers* 102 (2014): 584–589. doi: 10.1016/j.carbpol.2013.12.006.

[62] Wang, Xiao, Hongyang Ma, Benjamin Chu, and Benjamin S. Hsiao. "Thin-film nanofibrous composite reverse osmosis membranes for desalination." *Desalination* 420 (2017): 91–98. doi: 10.1016/j.desal.2017.06.029.

[63] Rohrbach, Kathleen, Yuanyuan Li, Hongli Zhu, Zhen Liu, Jiaqi Dai, Julia Andreasen, and Liangbing Hu. "A cellulose based hydrophilic, oleophobic hydrated filter for water/oil separation." *Chemical Communications* 50 (2014): 13296–13299. doi: 10.1039/C4CC04817B.

[64] Xu, Dandan, Xiaoting Zheng, and Ru Xiao. "Hydrophilic nanofibrous composite membrane prepared by melt-blending extrusion for effective separation of oil/water emulsion." *RSC Advances* 7 (2017): 7108–7115. doi: 10.1039/C6RA26068C.

[65] Wang, Ran, Sihui Guan, Anna Sato, Xiao Wang, Zhe Wang, Rui Yang, Benjamin S. Hsiao, and Benjamin Chu. "Nanofibrous microfiltration membranes capable of removing bacteria, viruses and heavy metal ions." *Journal of Membrane Science* 446 (2013): 376–382. doi: 10.1016/j.memsci.2013.06.020.

[66] Qu, Ping, Huanwei Tang, Yuan Gao, Liping Zhang, and Siqun Wang. "Polyethersulfone composite membrane blended with cellulose fibrils." *BioResources* 5 (2010): 2323–2336. doi: 10.15376/biores.5.4.2323-2336.

[67] Li, Kang. *Ceramic membranes for separation and reaction.* Hoboken, NJ: John Wiley & Sons, Inc. 2007.

[68] Saito, Tsuguyuki, and Akira Isogai. "TEMPO-mediated oxidation of native cellulose: the effect of oxidation conditions on chemical and crystal structures of the water-insoluble fractions." *Biomacromolecules* 5 (2004): 1983–1989. doi: 10.1021/bm0497769.

[69] Božič, Mojca, Peng Liu, Aji P. Mathew, and Vanja Kokol. "Enzymatic phosphorylation of cellulose nanofibers to new highly-ions adsorbing, flame-retardant and hydroxyapatite-growth induced natural nanoparticles." *Cellulose* 21 (2014): 2713–2726. doi: 10.1007/s10570-014-0281-8.

[70] Tibolla, Heloisa, Franciele M. Pelissari, Maria I. Rodrigues, and Florencia C. Menegalli. "Cellulose nanofibers produced from banana peel by enzymatic treatment: study of process conditions." *Industrial Crops and Products* 95 (2017): 664–674. doi: 10.1016/j.indcrop.2016.11.035.

[71] Klein, Elias. "Affinity membranes: a 10-year review." *Journal of Membrane Science* 179, 1-2 (2000): 1–27. doi: 10.1016/S0376-7388(00)00514-7.

[72] Yalcinkaya, Fatma. "Preparation of various nanofiber layers using wire electrospinning system." *Arabian Journal of Chemistry* 12 (2019): 5162–5172. doi: 10.1016/j.arabjc.2016.12.012.

[73] Technical Association of the Pulp Paper (TAPPI). Industry TAPPI Test Methods; TAPPI: Atlanta, GA, 1988.

[74] Astuti, N. H., N. A. Wibowo, and M. R. S. S. N. Ayub. "The porosity calculation of various types of paper using image analysis." *Jurnal Pendidikan Fisika Indonesia* 14 (2018): 46–51. doi: 10.15294/jpfi. v14i1.9878.

[75] Palacios Hinestroza, Hasbleidy, Javier A. Hernández Diaz, Marianelly Esquivel Alfaro, Guillermo Toriz, Orlando J. Rojas, and Belkis C. Sulbaran-Rangel. "Isolation and characterization of nanofibrillar cellulose from Agave tequilana Weber bagasse." *Advances in Materials Science and Engineering* 2019 (2019). doi: 10.1155/2019/1342547.

[76] Hernández, Javier Abraham, Víctor Hugo Romero, Alfredo Escalante, Guillermo Toriz, Orlando J. Rojas, and Belkis Coromoto Sulbaran. "Agave tequilana bagasse as source of cellulose nanocrystals via organosolv treatment." *BioResources* 13 (2018): 3603–3614. http://ncsu.edu/bioresources ISSN: 1930-2126.

[77] Robles, Eduardo, Javier Fernández-Rodríguez, Ananda M. Barbosa, Oihana Gordobil, Neftali LV Carreño, and Jalel Labidi. "Production of cellulose nanoparticles from blue Agave waste treated with environmentally friendly processes." *Carbohydrate Polymers* 183 (2018): 294–302. doi: 10.1016/j. carbpol.2018.01.015.

[78] Aleman-Nava, Gibrán S., Ilaria Alessandra Gatti, Roberto Parra-Saldivar, Jean-Francois Dallemand, Bruce E. Rittmann, and Hafiz MN Iqbal. "Biotechnological revalorization of tequila waste and by-product streams for cleaner production: a review from bio-refinery perspective." *Journal of Cleaner Production* 172 (2018): 3713–3720. doi: 10.1016/j.jclepro.2017.07.134.

[79] Carpenter, Alexis Wells, Charles-Francois de Lannoy, and Mark R. Wiesner. "Cellulose nanomaterials in water treatment technologies." *Environmental Science & Technology* 49 (2015): 5277–5287. doi: 10.1021/es506351r

[80] Lomelí-Ramírez, María Guadalupe, Edgar Mario Valdez-Fausto, Maite Rentería-Urquiza, Rosa María Jiménez-Amezcua, José Anzaldo Hernández, Jose Guillermo Torres-Rendon, and Salvador García Enriquez. "Study of green nanocomposites based on corn starch and cellulose nanofibrils from Agave tequilana Weber." *Carbohydrate Polymers* 201 (2018): 9–19. doi: 10.1016/j.carbpol.2018.08.045.

[81] Pech-Cohuo, Soledad-Cecilia, Gonzalo Canche-Escamilla, Alex Valadez-Gonzalez, Victor Vladimir Amilcar Fernández-Escamilla, and Jorge Uribe-Calderon. "Production and modification of cellulose nanocrystals from Agave tequilana Weber waste and its effect on the melt rheology of PLA." *International Journal of Polymer Science* 2018 (2018). doi: 10.1155/2018/3567901.

[82] Zhang, Weihua, Xiaoju Wang, Yongchao Zhang, Bas van Bochove, Ermei Mäkilä, Jukka Seppälä, Wenyang Xu, Stefan Willfor, and Chunlin Xu. "Robust shape-retaining nanocellulose-based aerogels decorated with silver nanoparticles for fast continuous catalytic discoloration of organic dyes." *Separation and Purification Technology* 242 (2020): 116523. doi: 10.1016/j.seppur.2020.116523.

[83] Toivonen, Matti S., Sauli Kurki-Suonio, Felix H. Schacher, Sami Hietala, Orlando J. Rojas, and Olli Ikkala. "Water-resistant, transparent hybrid nanopaper by physical crosslinking with chitosan." *Biomacromolecules* 16 (2015): 1062–1071. doi: 10.1021/acs.biomac.5b00145.

[84] Zhu, Luting, Lu Zong, Xiaochen Wu, Mingjie Li, Haisong Wang, Jun You, and Chaoxu Li. "Shapeable fibrous aerogels of metal-organic-frameworks templated with nanocellulose for rapid and large-capacity adsorption." *ACS Nano* 12 (2018): 4462–4468. doi: 10.1021/acsnano.8b00566.

[85] Supramaniam, Jagadeesen, Rohana Adnan, Noor Haida Mohd Kaus, and Rani Bushra. "Magnetic nano-cellulose alginate hydrogel beads as potential drug delivery system." *International Journal of Biological Macromolecules* 118 (2018): 640–648. doi: 10.1016/j.ijbiomac.2018.06.043.

[86] Han, Jingquan, Yiying Yue, Qinglin Wu, Chaobo Huang, Hui Pan, Xianxu Zhan, Changtong Mei, and Xinwu Xu. "Effects of nanocellulose on the structure and properties of poly(vinyl alcohol)-borax hybrid foams." *Cellulose* 24 (2017): 4433–4448. doi: 10.1007/s10570-017-1409-4.

[87] Basu, Alex, Karen Heitz, Maria Strømme, Ken Welch, and Natalia Ferraz. "Ion-crosslinked wood-derived nanocellulose hydrogels with tunable antibacterial properties: candidate materials for advanced wound care applications." *Carbohydrate Polymers* 181 (2018): 345–350. doi: 10.1016/j.carbpol.2017.10.085.

[88] Basu, Alex, Jonas Lindh, Eva Ålander, Maria Strømme, and Natalia Ferraz. "On the use of ion-crosslinked nanocellulose hydrogels for wound healing solutions: physicochemical properties and application-oriented biocompatibility studies." *Carbohydrate Polymers* 174 (2017): 299–308. doi: 10.1016/j.carbpol.2017.06.073.

[89] Basu, Alex, Maria Strømme, and Natalia Ferraz. "Towards tunable protein-carrier wound dressings based on nanocellulose hydrogels crosslinked with calcium ions." *Nanomaterials* 8 (2018): 550. doi: 10.3390/nano8070550.

[90] Tanpichai, Supachok, and Kristiina Oksman. "Crosslinked poly(vinyl alcohol) composite films with cellulose nanocrystals: mechanical and thermal properties." *Journal of Applied Polymer Science* 135 (2018): 45710. doi: 10.1002/app.45710.

[91] Niu, Jiabao, Jianquan Wang, Xiaofu Dai, Ziqiang Shao, and Xiaonan Huang. "Dual physically cross-linked healable polyacrylamide/cellulose nanofibers nanocomposite hydrogels with excellent mechani-cal properties." *Carbohydrate Polymers* 193 (2018): 73–81. doi: 10.1016/j.carbpol.2018.03.086.

[92] Long, Keying, Ruitao Cha, Yapei Zhang, Juanjuan Li, Fangping Ren, and Xingyu Jiang. "Cellulose nanocrystals as reinforcements for collagen-based casings with low gas transmission." *Cellulose* 25 (2018): 463–471. doi: 10.1007/s10570-017-1569-2.

[93] Jiang, Feng, and You-Lo Hsieh. "Cellulose nanofibril aerogels: synergistic improvement of hydropho-bicity, strength, and thermal stability via crosslinking with diisocyanate." *ACS Applied Materials & Interfaces* 9 (2017): 2825–2834. doi: 10.1021/acsami.6b13577.

[94] Xiao, Yongmei, Liduo Rong, Bijia Wang, Zhiping Mao, Hong Xu, Yi Zhong, Linping Zhang, and Xiaofeng Sui. "A light-weight and high-efficacy antibacterial nanocellulose-based sponge via cova-lent immobilization of gentamicin." *Carbohydrate Polymers* 200 (2018): 595–601. doi: 10.1016/j.carbpol.2018.07.091.

[95] Leppiniemi, Jenni, Panu Lahtinen, Antti Paajanen, Riitta Mahlberg, Sini Metsaa-Kortelainen, Tatu Pinomaa, Heikki Pajari, Inger Vikholm-Lundin, Pekka Pursula, and Vesa P. Hytoonen. "3D-printable bioactivated nanocellulose-alginate hydrogels." *ACS Applied Materials & Interfaces* 9 (2017): 21959–21970. doi: 10.1021/acsami.7b02756.

[96] Sheng, Chao, Yiming Zhou, Jiannan Lu, Xinyu Zhang, and Guoxin Xue. "Preparation and characteriza-tion of chitosan based hydrogels chemical crosslinked by oxidized cellulose nanowhiskers." *Polymer Composites* 40 (2019): 2432–2440. doi: 10.1002/pc.25109.

[97] Huang, Shu, Zhan Zhao, Chuang Feng, Edwin Mayes, and Jie Yang. "Nanocellulose reinforced P (AAm-co-AAc) hydrogels with improved mechanical properties and biocompatibility." *Composites Part A: Applied Science and Manufacturing* 112 (2018): 395–404. doi: 10.1016/j.compositesa.2018.06.028.

[98] Tang, Ruilin, Zhiming Yu, Scott Renneckar, and Yang Zhang. "Coupling chitosan and TEMPO-oxidized nanofibrillated cellulose by electrostatic attraction and chemical reaction." *Carbohydrate Polymers* 202 (2018): 84–90. doi: 10.1016/j.carbpol.2018.08.097.

[99] Geng, Lihong, Binyi Chen, Xiangfang Peng, and Tairong Kuang. "Strength and modulus improvement of wet-spun cellulose I filaments by sequential physical and chemical crosslinking." *Materials & Design* 136 (2017): 45–53. doi: 10.1016/j.matdes.2017.09.054.

[100] Guo, Limin, Zhilin Chen, Shaoyi Lyu, Feng Fu, and Siqun Wang. "Highly flexible crosslinked cellulose nanofibril sponge-like aerogels with improved mechanical property and enhanced flame retardancy." *Carbohydrate Polymers* 179 (2018): 333–340. doi: 10.1016/j.carbpol.2017.09.084.

[101] Liu, Andong, Lilian Medina, and Lars A. Berglund. "High-strength nanocomposite aerogels of ternary composition: poly(vinyl alcohol), clay, and cellulose nanofibrils." *ACS Applied Materials & Interfaces* 9 (2017): 6453–6461. doi: 10.1021/acsami.6b15561.

[102] Yang, Weisheng, Huiyang Bian, Liang Jiao, Weibing Wu, Yulin Deng, and Hongqi Dai. "High wet-strength, thermally stable and transparent TEMPO-oxidized cellulose nanofibril film via crosslinking with poly-amide epichlorohydrin resin." *RSC Advances* 7 (2017): 31567–31573. doi: 10.1039/C7RA05009G.

[103] Sirvio, Juho Antti, Samuli Honkaniemi, Miikka Visanko, and Henrikki Liimatainen. "Composite films of poly(vinyl alcohol) and bifunctional crosslinking cellulose nanocrystals." *ACS Applied Materials & Interfaces* 7 (2015): 19691–19699. doi: 10.1021/acsami.5b04879.

[104] Jin, Liqiang, Weigong Li, Qinghua Xu, and Qiucun Sun. "Amino-functionalized nanocrystalline cellulose as an adsorbent for anionic dyes." *Cellulose* 22 (2015): 2443–2456. doi: 10.1007/s10570-015-0649-4.

[105] Tang, Juntao, Yang Song, Feiping Zhao, Stewart Spinney, Juliana da Silva Bernardes, and Kam Chiu Tam. "Compressible cellulose nanofibril (CNF) based aerogels produced via a bio-inspired strategy for heavy metal ion and dye removal." *Carbohydrate Polymers* 208 (2019): 404–412. doi: 10.1016/j.carbpol.2018.12.079.

[106] Yang, Xue, Hui Liu, Fuyi Han, Shuai Jiang, Lifang Liu, and Zhaopeng Xia. "Fabrication of cellulose nanocrystal from *Carex meyeriana* Kunth and its application in the adsorption of methylene blue." *Carbohydrate Polymers* 175 (2017): 464–472. doi: 10.1016/j.carbpol.2017.08.007.

[107] Bai, Langming, Yatao Liu, An Ding, Nanqi Ren, Guibai Li, and Heng Liang. "Fabrication and characterization of thin-film composite (TFC) nanofiltration membranes incorporated with cellulose nanocrystals (CNCs) for enhanced desalination performance and dye removal." *Chemical Engineering Journal* 358 (2019): 1519–1528. doi: 10.1016/j.cej.2018.10.147.

[108] Amiralian, Nasim, Mislav Mustapic, Md Shahriar A. Hossain, Chaohai Wang, Muxina Konarova, Jing Tang, Jongbeom Na, Aslam Khan, and Alan Rowan. "Magnetic nanocellulose: a potential material for removal of dye from water." *Journal of Hazardous Materials* (2020): 122571. doi: 10.1016/j.jhazmat.2020.122571.

[109] Maaloul, Najeh, Paula Oulego, Manuel Rendueles, Achraf Ghorbal, and Mario Diaz. "Novel biosorbents from almond shells: characterization and adsorption properties modeling for Cu(II) ions from aqueous solutions." *Journal of Environmental Chemical Engineering* 5 (2017): 2944–2954. doi: 10.1016/j.jece.2017.05.037.

[110] Donia, A. M., A. A. Atia, and F. I. Abouzayed. "Preparation and characterization of nano-magnetic cellulose with fast kinetic properties towards the adsorption of some metal ions." *Chemical Engineering Journal* 191 (2012): 22–30. doi: 10.1016/j.cej.2011.08.034.

[111] Abu-Danso, Emmanuel, Sirpa Peräniemi, Tiina Leiviskä, and Amit Bhatnagar. "Synthesis of S-ligand tethered cellulose nanofibers for efficient removal of Pb(II) and Cd(II) ions from synthetic and industrial wastewater." *Environmental Pollution* 242 (2018): 1988–1997. doi: 10.1016/j.envpol.2018.07.044.

[112] Lu, Jiao, Ru-Na Jin, Chao Liu, Yan-Fei Wang, and Xiao-kun Ouyang. "Magnetic carboxylated cellulose nanocrystals as adsorbent for the removal of Pb(II) from aqueous solution." *International Journal of Biological Macromolecules* 93 (2016): 547–556. doi: 10.1016/j.ijbiomac.2016.09.004.

[113] Bo, Shaoguo, Wenjing Ren, Chao Lei, Yuanbo Xie, Yurong Cai, Shunli Wang, Junkuo Gao, Qingqing Ni, and Juming Yao. "Flexible and porous cellulose aerogels/zeolitic imidazolate framework (ZIF-8) hybrids for adsorption removal of Cr(IV) from water." *Journal of Solid State Chemistry* 262 (2018): 135–141. doi: 10.1016/j.jssc.2018.02.022.

[114] Sharma, Priyanka R., Aurnov Chattopadhyay, Sunil K. Sharma, Lihong Geng, Nasim Amiralian, Darren Martin, and Benjamin S. Hsiao. "Nanocellulose from spinifex as an effective adsorbent to remove cadmium (II) from water." *ACS Sustainable Chemistry & Engineering* 6 (2018): 3279–3290. doi: 10.1021/acssuschemeng.7b03473.

[115] Ram, Bhagat, and Ghanshyam S. Chauhan. "New spherical nanocellulose and thiol-based adsorbent for rapid and selective removal of mercuric ions." *Chemical Engineering Journal* 331 (2018): 587–596. doi: 10.1016/j.cej.2017.08.128.

[116] Zhang, Zheng, Gilles Sèbe, Daniel Rentsch, Tanja Zimmermann, and Philippe Tingaut. "Ultralightweight and flexible silylated nanocellulose sponges for the selective removal of oil from water." *Chemistry of Materials* 26 (2014): 2659–2668. doi: 10.1021/cm5004164.

[117] Hu, Zhao-Hong, Yan-Fei Wang, Ahmed Mohamed Omer, and Xiao-Kun Ouyang. "Fabrication of ofloxacin imprinted polymer on the surface of magnetic carboxylated cellulose nanocrystals for highly selective adsorption of fluoroquinolones from water." *International Journal of Biological Macromolecules* 107 (2018): 453–462. doi: 10.1016/j.ijbiomac.2017.09.009.

[118] Hu, Dalin, Ran Jiang, Nan Wang, Hongping Xu, Yang-Guang Wang, and Xiao-kun Ouyang. "Adsorption of diclofenac sodium on bilayer amino-functionalized cellulose nanocrystals/chitosan composite." *Journal of Hazardous Materials* 369 (2019): 483–493. doi: 10.1016/j.jhazmat.2019.02.057.

[119] Yuan, Guohao, Mayakrishnan Prabakaran, Sun Qilong, Jung Soon Lee, Ill-Min Chung, Mayakrishnan Gopiraman, Kyung-Hun Song, and Ick Soo Kim. "Cyclodextrin functionalized cellulose nanofiber composites for the faster adsorption of toluene from aqueous solution." *Journal of the Taiwan Institute of Chemical Engineers* 70 (2017): 352–358. doi: 10.1016/j.jtice.2016.10.028.

[120] Demirci, Serkan, Asli Celebioglu, and Tamer Uyar. "Surface modification of electrospun cellulose acetate nanofibers via raft polymerization for DNA adsorption." *Carbohydrate Polymers* 113 (2014): 200–207. doi: 10.1016/j.carbpol.2014.06.086.

[121] Anirudhan, T. S., and S. R. Rejeena. "Poly(acrylic acid)-modified poly(glycidylmethacrylate)-grafted nanocellulose as matrices for the adsorption of lysozyme from aqueous solutions." *Chemical Engineering Journal* 187 (2012): 150–159. doi: 10.1016/j.cej.2012.01.113.

[122] Jiang, Nan, Ran Shang, Sebastiaan G. J. Heijman, and Luuk C. Rietveld. "High-silica zeolites for adsorption of organic micro-pollutants in water treatment: a review." *Water Research* 144 (2018): 145–161. doi: 10.1016/j.watres.2018.07.017.

[123] Ruiz-Palomero, Celia, M. Laura Soriano, and Miguel Valcarcel. "β-Cyclodextrin decorated nanocellulose: a smart approach towards the selective fluorimetric determination of danofloxacin in milk samples." *Analyst* 140 (2015): 3431–3438. doi: 10.1039/C4AN01967A.

[124] De France, Kevin J., Todd Hoare, and Emily D. Cranston. "Review of hydrogels and aerogels containing nanocellulose." *Chemistry of Materials* 29 (2017): 4609–4631. doi: 10.1021/acs.chemmater.7b00531.

[125] Pourfadakari, Sudabeh, Sahand Jorfi, Mehdi Ahmadi, and Afshin Takdastan. "Experimental data on adsorption of Cr(VI) from aqueous solution using nanosized cellulose fibers obtained from rice husk." *Data in Brief* 15 (2017): 887–895. doi: 10.1016/j.dib.2017.10.043.

[126] Korhonen, Juuso T., Marjo Kettunen, Robin H. A. Ras, and Olli Ikkala. "Hydrophobic nanocellulose aerogels as floating, sustainable, reusable, and recyclable oil absorbents." *ACS Applied Materials & Interfaces* 3 (2011): 1813–1816. doi: 10.1021/am200475b.

[127] Erol, Kadir, Kazım Köse, Lokman Uzun, Rıdvan Say, and Adil Denizli. "Polyethyleneimine assisted-two-step polymerization to develop surface imprinted cryogels for lysozyme purification." *Colloids and Surfaces B: Biointerfaces* 146 (2016): 567–576. doi: 10.1016/j.colsurfb.2016.06.060.

[128] González, Joaquín A., María E. Villanueva, Lidia L. Piehl, and Guillermo J. Copello. "Development of a chitin/graphene oxide hybrid composite for the removal of pollutant dyes: adsorption and desorption study." *Chemical Engineering Journal* 280 (2015): 41–48. doi: 10.1016/j.cej.2015.05.112.

[129] Lu, Shuguang, and Stuart W. Gibb. "Copper removal from wastewater using spent-grain as biosorbent." *Bioresource Technology* 99 (2008): 1509–1517. doi: 10.1016/j.biortech.2007.04.024.

[130] Garg, Krishan Kishor, and Basheshwar Prasad. "Treatment of toxic pollutants of purified terephthalic acid waste water: a review." *Environmental Technology & Innovation* 8 (2017): 191–217. doi: 10.1016/j.eti.2017.07.001.

[131] Orona-Navar, Carolina, Raul García-Morales, Rodrigo Rubio-Govea, Jurgen Mahlknecht, Raul I. Hernandez-Aranda, Juan Gabriel Ramírez, K. D. P. Nigam, and Nancy Ornelas-Soto. "Adsorptive removal of emerging pollutants from groundwater by using modified titanate nanotubes." *Journal of Environmental Chemical Engineering* 6 (2018): 5332–5340. doi: 10.1016/j.jece.2018.08.010.

[132] Martínez-Huitle, Carlos Alberto, and Marco Panizza. "Electrochemical oxidation of organic pollutants for wastewater treatment." *Current Opinion in Electrochemistry* 11 (2018): 62–71. doi: 10.1016/j.coelec.2018.07.010.

[133] Ruiz-Palomero, Celia, M. Laura Soriano, and Miguel Valcárcel. "Nanocellulose as analyte and analytical tool: opportunities and challenges." *TrAC Trends in Analytical Chemistry* 87 (2017): 1–18. doi: 10.1016/j.trac.2016.11.007.

[134] Crini, Gregorio. "Non-conventional low-cost adsorbents for dye removal: a review." *Bioresource Technology* 97 (2006): 1061–1085. doi: 10.1016/j.biortech.2005.05.001.

[135] Chenampulli, Sangeetha, G. Unnikrishnan, A. Sujith, Sabu Thomas, and Tania Francis. "Cellulose nanoparticles from Pandanus: viscometric and crystallographic studies." *Cellulose* 20 (2013): 429–438. doi: 10.1007/s10570-012-9831-0.

[136] Khan, Ayub, Xiangke Wang, Kashif Gul, Fazli Khuda, Zaynab Aly, and A. M. Elseman. "Microwave-assisted spent black tea leaves as cost-effective and powerful green adsorbent for the efficient removal of

Eriochrome black T from aqueous solutions." *Egyptian Journal of Basic and Applied Sciences* 5 (2018): 171–182. doi: 10.1016/j.ejbas.2018.04.002.

[137] Zimnitsky, Dmitry S., Tatiana L. Yurkshtovich, and Pavel M. Bychkovsky. "Multilayer adsorption of amino acids on oxidized cellulose." *Journal of Colloid and Interface Science* 285 (2005): 502–508. doi: 10.1016/j.jcis.2004.12.021.

[138] Voisin, Hugo, Lennart Bergström, Peng Liu, and Aji P. Mathew. "Nanocellulose-based materials for water purification." *Nanomaterials* 7 (2017): 57. doi: 10.3390/nano7030057.

[139] Abou-Zeid, Ragab E., Sawsan Dacrory, Korany A. Ali, and Samir Kamel. "Novel method of preparation of tricarboxylic cellulose nanofiber for efficient removal of heavy metal ions from aqueous solution." *International Journal of Biological Macromolecules* 119 (2018): 207–214. doi: 10.1016/j.ijbiomac.2018.07.127.

Index

For Product Safety Concerns and Information please contact our EU
representative GPSR@taylorandfrancis.com
Taylor & Francis Verlag GmbH, Kaufingerstraße 24, 80331 München, Germany

www.ingramcontent.com/pod-product-compliance
Lightning Source LLC
Chambersburg PA
CBHW061413210326
41598CB00035B/6201